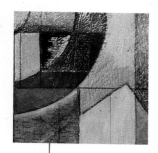

A GRAPHICAL APPROACH TO
Trigonometry

A GRAPHICAL APPROACH TO
Trigonometry

E. John Hornsby, Jr.
University of New Orleans

Margaret L. Lial
American River College

with the assistance of William A. Armstrong
Lakeland Community College

 **ADDISON-WESLEY**

An imprint of Addison Wesley Longman, Inc.

Reading, Massachusetts • Menlo Park, California • New York • Harlow, England
Don Mills, Ontario • Sydney • Mexico City • Madrid • Amsterdam

Sponsoring Editor: Anne Kelly
Developmental Editor: Lynn Mooney
Project Editor: Ann-Marie Buesing
Design Administrator: Jess Schaal
Text Design: Lesiak/Crampton Design: Cindy Crampton
Cover Design: Lesiak/Crampton Design: Cindy Crampton
Cover Art: Tom James
Production Administrator: Randee Wire
Compositor: Interactive Composition Corporation
Printer and Binder: R. R. Donnelley & Sons Company
Cover Printer: The Lehigh Press, Inc.

A Graphical Approach to Trigonometry

Library of Congress Cataloging-in-Publication Data

Hornsby, E. John.
    A graphical approach to trigonometry/E. John Hornsby, Jr.,
Margaret L. Lial; with the assistance of William A. Armstrong.
        p.   cm.
    Includes index.
    ISBN 0–673–99904–1
    1. Trigonometry.   I. Lial, Margaret L.   II. Armstrong, William A.
III. Title.
QA531.H6726   1996
516.24—dc20   96–18916
                                                           CIP

1 2 3 4 5 6 7 8 9 10—DOW—99 98 97 96

TO JIM, AND TO THE MEMORY OF ANNIE BOYLE

C O N T E N T S

P R E F A C E

This book, intended for a college trigonometry course, is a culmination of four years of teaching experience with the graphing calculator and more years of teaching mathematics than we both like to admit. In it, we treat the standard topics of trigonometry with complete integration of modern graphing calculators. We pursued this project with the firm commitment that it was not to be merely an adaptation of our traditional text to graphing calculator analysis. In so doing, we realized that a completely new approach would be necessary, based on the premise that all students would have graphing calculators on the first day of class and would use this technology while studying the mathematical concepts throughout the book.

Development of the Project

We began teaching algebra and trigonometry classes with a graphing calculator requirement during the 1991–1992 academic year. During the first year we attempted to adapt a traditional college algebra/trigonometry text to this new approach and soon realized that a book designed to implement the graphing calculator at all stages was necessary. The approach used for so many years in our classes was just no longer applicable in light of the power of the technology available. The new technology demands that graphing be covered as early as possible. Furthermore, we now feel that *interpretation of graphs* plays a more important role than *graph-sketching techniques*. The modern approach to mathematics spawned by the statements of NCTM and AMATYC needed to be addressed. We felt it natural that, as mathematics authors for so many years in the traditional market, we could write a series of books to accomplish our goals. During the summer of 1993, we began writing the material and continued throughout the academic year 1993–1994, teaching the material after writing it. The class-testing and review processes continued during 1994–1995, while initial production commenced in late 1994.

   The book you are reading is one of a series of texts that completely integrates graphing calculators in algebra, trigonometry, and precalculus. It is the culmination of many hours of work by the authors, reviewers, answer-checkers, editors, class-testers, and students.

The Role of Chapter 1

The college algebra text in this series was published in 1996. In the preface for that book, we included the following statement of philosophy:

   Throughout the first five chapters, we present the various classes of functions covered in a standard college algebra text. The first chapter introduces functions and relations, using the linear function as the basis for our presentation. In the chapter, we present the approach that follows throughout the succeeding chapters. After introducing a class of function and examining the nature of its graph, we discuss analytic methods of solution of equations based on the function and show how to provide graphical support using the graphing calculator. Once we establish methods of solving these equations, we move on to the analytic methods of solving the associated inequalities and how their solutions can be supported using graphs. We use two approaches to graphical analysis of equation and inequality solving: the x-intercept method and the intersection-of-graphs method. These two methods are constantly reviewed and reinforced. Finally, once the student has a feel for the particular class of function under consideration, we use our analytic and graphical methods to solve applications that lead to equations or inequalities involving that function. By consistently using this approach, students become aware that we are really doing the same thing, but just applying it to a new kind of function.

Because we realize that this trigonometry text will be the first opportunity for some users to fully integrate graphing calculators, the first chapter, titled *A Review of Basic Concepts: Analysis of Equations, Inequalities, and Graphs of Functions*, provides an overview of the necessary background. The concepts covered in this chapter were carefully chosen so that first-time users have the opportunity to experience how graphing calculators will be used later in the text when function-graphing and solving of equations and inequalities are presented.

Overview of the Content

As mentioned above, Chapter 1 is a summary of important algebra topics from a graphing calculator point of view. *It need not be covered if the class has experience with graphing calculator-required courses.* In Chapter 2 we introduce the geometric foundations of trigonometry. This text provides an early introduction to the unit circle approach, and immediately focuses on the graphs of the circular functions. In Chapter 3 we investigate identities, equations, and inequalities, using the graphing calculator to support our analytic results whenever possible. We also introduce the inverse circular functions and their graphs, and conclude with a section on applications (many involving modeling periodic data with circular function graphs). Chapter 4 presents the right triangle approach to trigonometry, and investigates the laws of sines and cosines. Applications of this approach are found in the final section. In Chapter 5 we introduce vectors and complex numbers, and the final chapter, Chapter 6, looks at graphs of polar curves and parametric equations and their applications.

The Approach to Technology

Critics of technology often state that they oppose the idea of "pushing a button to get the answer." Anyone who has made an effort to teach with the graphing calculator knows that this is not the way it works. We constantly emphasize that it is essential to understand the mathematical concepts and apply them hand-in-hand with the calculator. The calculator helps us to understand the concepts, and we wrote the book with this idea. We agree with Peg Crider of Tomball College: "Your brain is the most powerful tool in the whole process."

Due to the ever-changing world of technology, we felt strongly that this text should not attempt to teach the student how to use a particular model of calculator. We do not provide specific keystrokes for various models, because experience has taught us that by the time a text is published, the technology is often out of date. However, we did decide to use actual graphing calculator-generated art in addition to the traditional art found in standard textbooks. We chose to use art generated by the TI-82 graphing calculator, manufactured by Texas Instruments. The TI-Graph Link software produced the graphics art, and we are grateful for the cooperation of Texas Instruments in this effort.

Features of the Text

Approach

The graphing calculator is introduced in Chapter 1 and is used consistently throughout the book. Due to the nature of the material in the course, certain chapters are more dependent on graphing calculators than others. It is assumed *always,* however, that the student has access to one.

Pedagogy

Technological Notes supplement the exposition by discussing the power and the limitations of learning with technology. ***For Group Discussion*** activities allow instructors to use cooperative learning in their classes. ***Chapter summaries*** provide a quick review of the key ideas and terms in the chapter.

Exercises

There are many examples and exercises that draw students into the heart of mathematical concepts. True conceptual exercises are balanced by a suitable ratio of drill and applications. Many exercises involve real data and require brief essay answers. Some exercise sets include a group of exercises labeled ***Relating Concepts*** that make students aware of the connections between topics they studied earlier and the ones they are studying currently. ***Further Explorations*** are found in many sections, and offer further insight into the capabilities of the TABLE feature of the TI-82 and TI-83 graphing calculators.

Preview to the Text

The next four pages of the preface are a short guide through the text. These sample pages were chosen to show the features and pedagogical aids in the book that help reinforce the graphical approach to learning trigonometry.

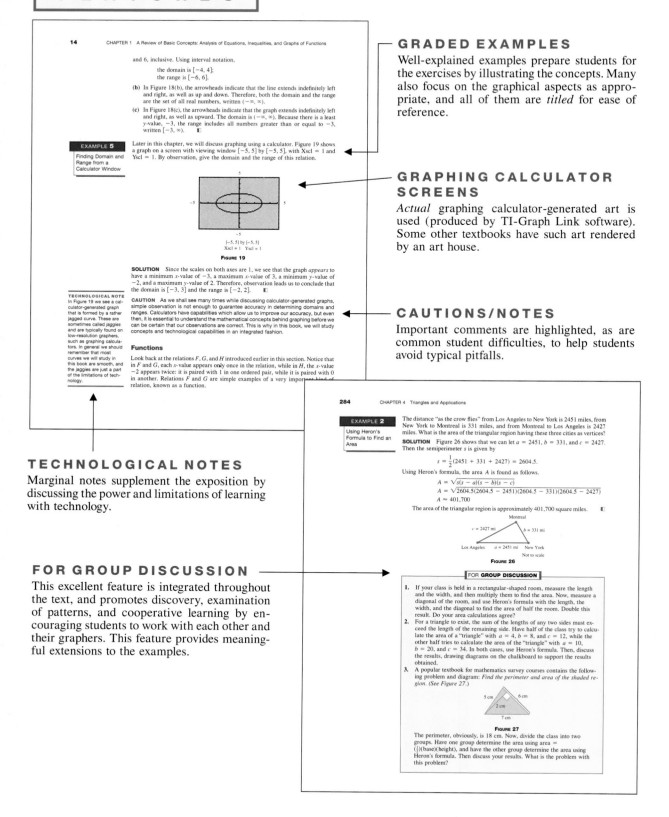

14 CHAPTER 1 A Review of Basic Concepts: Analysis of Equations, Inequalities, and Graphs of Functions

and 6, inclusive. Using interval notation,

the domain is $[-4, 4]$;
the range is $[-6, 6]$.

(b) In Figure 18(b), the arrowheads indicate that the line extends indefinitely left and right, as well as up and down. Therefore, both the domain and the range are the set of all real numbers, written $(-\infty, \infty)$.

(c) In Figure 18(c), the arrowheads indicate that the graph extends indefinitely left and right, as well as upward. The domain is $(-\infty, \infty)$. Because there is a least y-value, -3, the range includes all numbers greater than or equal to -3, written $[-3, \infty)$.

EXAMPLE 5
Finding Domain and Range from a Calculator Window

Later in this chapter, we will discuss graphing using a calculator. Figure 19 shows a graph on a screen with viewing window $[-5, 5]$ by $[-5, 5]$, with Xscl $= 1$ and Yscl $= 1$. By observation, give the domain and the range of this relation.

$[-5, 5]$ by $[-5, 5]$
Xscl $= 1$ Yscl $= 1$
FIGURE 19

SOLUTION Since the scales on both axes are 1, we see that the graph *appears* to have a minimum x-value of -3, a maximum x-value of 3, a minimum y-value of -2, and a maximum y-value of 2. Therefore, observation leads us to conclude that the domain is $[-3, 3]$ and the range is $[-2, 2]$.

TECHNOLOGICAL NOTE
In Figure 19 we see a calculator-generated graph that is formed by a rather jagged curve. These are sometimes called *jaggies* and are typically found on low-resolution graphers, such as graphing calculators. In general we should remember that most curves we will study in this book are smooth, and the jaggies are just a part of the limitations of technology.

CAUTION As we shall see many times while discussing calculator-generated graphs, simple observation is not enough to guarantee accuracy in determining domains and ranges. Calculators have capabilities which allow us to improve our accuracy, but even then, it is essential to understand the mathematical concepts behind graphing before we can be certain that our observations are correct. This is why in this book, we will study concepts and technological capabilities in an integrated fashion.

Functions

Look back at the relations F, G, and H introduced earlier in this section. Notice that in F and G, each x-value appears only once in the relation, while in H, the x-value -2 appears twice: it is paired with 1 in one ordered pair, while it is paired with 0 in another. Relations F and G are simple examples of a very important kind of relation, known as a function.

---

284 CHAPTER 4 Triangles and Applications

EXAMPLE 2
Using Heron's Formula to Find an Area

The distance "as the crow flies" from Los Angeles to New York is 2451 miles, from New York to Montreal is 331 miles, and from Montreal to Los Angeles is 2427 miles. What is the area of the triangular region having these three cities as vertices?

SOLUTION Figure 26 shows that we can let $a = 2451$, $b = 331$, and $c = 2427$. Then the semiperimeter s is given by

$$s = \frac{1}{2}(2451 + 331 + 2427) = 2604.5.$$

Using Heron's formula, the area A is found as follows.

$$A = \sqrt{s(s - a)(s - b)(s - c)}$$
$$A = \sqrt{2604.5(2604.5 - 2451)(2604.5 - 331)(2604.5 - 2427)}$$
$$A \approx 401{,}700$$

The area of the triangular region is approximately 401,700 square miles.

Montreal
$c = 2427$ mi $b = 331$ mi
Los Angeles $a = 2451$ mi   New York
Not to scale
FIGURE 26

FOR GROUP DISCUSSION

1. If your class is held in a rectangular-shaped room, measure the length and the width, and then multiply them to find the area. Now, measure a diagonal of the room, and use Heron's formula with the length, the width, and the diagonal to find the area of half the room. Double this result. Do your area calculations agree?
2. For a triangle to exist, the sum of the lengths of any two sides must exceed the length of the remaining side. Have half of the class try to calculate the area of a "triangle" with $a = 4$, $b = 8$, and $c = 12$, while the other half tries to calculate the area of the "triangle" with $a = 10$, $b = 20$, and $c = 34$. In both cases, use Heron's formula. Then, discuss the results, drawing diagrams on the chalkboard to support the results obtained.
3. A popular textbook for mathematics survey courses contains the following problem and diagram: *Find the perimeter and area of the shaded region. (See Figure 27.)*

5 cm   6 cm
2 cm
7 cm
FIGURE 27

The perimeter, obviously, is 18 cm. Now, divide the class into two groups. Have one group determine the area using area $= (\frac{1}{2})$(base)(height), and have the other group determine the area using Heron's formula. Then discuss your results. What is the problem with this problem?

GRADED EXAMPLES

Well-explained examples prepare students for the exercises by illustrating the concepts. Many also focus on the graphical aspects as appropriate, and all of them are *titled* for ease of reference.

GRAPHING CALCULATOR SCREENS

Actual graphing calculator-generated art is used (produced by TI-Graph Link software). Some other textbooks have such art rendered by an art house.

CAUTIONS/NOTES

Important comments are highlighted, as are common student difficulties, to help students avoid typical pitfalls.

TECHNOLOGICAL NOTES

Marginal notes supplement the exposition by discussing the power and limitations of learning with technology.

FOR GROUP DISCUSSION

This excellent feature is integrated throughout the text, and promotes discovery, examination of patterns, and cooperative learning by encouraging students to work with each other and their graphers. This feature provides meaningful extensions to the examples.

CALCULUS PREPARATORY DISCUSSIONS

This example illustrates a concept students will encounter in the study of calculus.

EXAMPLE 5

Writing a Function Value in Terms of u

Write $\sin(\tan^{-1} u)$ as an algebraic expression in u, $u > 0$.

SOLUTION　Let $\theta = \tan^{-1} u$. Then $\tan \theta = u$, and $0 < \theta < \frac{\pi}{2}$.

$$1 + \tan^2 \theta = \sec^2 \theta$$
$$1 + u^2 = \sec^2 \theta$$
$$\sec \theta = \sqrt{1 + u^2} \qquad \begin{array}{l}\sec \theta > 0 \\ \text{since } 0 < \theta < \frac{\pi}{2}.\end{array}$$
$$\cos \theta = \frac{1}{\sqrt{1 + u^2}} \qquad \cos \theta = \frac{1}{\sec \theta}$$

Now use $\sin^2 \theta = 1 - \cos^2 \theta$ to find $\sin \theta$.

$$\sin^2 \theta = 1 - \cos^2 \theta$$
$$= 1 - \left(\frac{1}{\sqrt{1 + u^2}}\right)^2$$
$$= 1 - \frac{1}{1 + u^2}$$
$$= \frac{1 + u^2}{1 + u^2} - \frac{1}{1 + u^2} \qquad \text{Get a common denominator.}$$
$$= \frac{u^2}{1 + u^2}$$
$$\sin \theta = \frac{u}{\sqrt{1 + u^2}} \qquad \text{Find the square root.}$$
$$= \frac{u\sqrt{1 + u^2}}{1 + u^2} \qquad \text{Rationalize the denominator.}$$

Therefore, $\sin(\tan^{-1} u) = \frac{u\sqrt{1 + u^2}}{1 + u^2}$ for $u > 0$.　■

3.5 EXERCISES

Use the graph of the appropriate inverse circular function, found in Figure 25(a), 29(a), or 33(a), to find the exact value of each of the following.

1. $\arcsin\left(-\frac{1}{2}\right)$　　2. $\arccos \frac{\sqrt{3}}{2}$　　3. $\tan^{-1} 1$　　4. $\sin^{-1} 0$

5. $\cos^{-1}(-1)$　　6. $\cos^{-1} \frac{1}{2}$　　7. $\sin^{-1}\left(-\frac{\sqrt{3}}{2}\right)$　　8. $\cos^{-1} 0$

9. $\arctan(-1)$　　10. $\arccos\left(-\frac{1}{2}\right)$　　11. $\arcsin \frac{\sqrt{2}}{2}$　　12. $\arcsin\left(-\frac{\sqrt{2}}{2}\right)$

13. $\arccos\left(-\frac{\sqrt{3}}{2}\right)$　　14. $\arcsin\left(-\frac{\sqrt{3}}{2}\right)$　　15. $\tan^{-1} 0$　　16. $\tan^{-1}(-\sqrt{3})$

---

*C*hapter 4　　SUMMARY

In this chapter we see how the trigonometric functions can be defined in terms of ratios of lengths of the sides of a right triangle. For an acute angle θ in a right triangle, the sides can be identified as *opposite* θ, *adjacent to* θ, and the *hypotenuse*. Ratios of these sides form the six trigonometric functions of θ, as seen in the first section.

　　Solving a right triangle means finding the measures of all its angles and sides. We do this by using the trigonometric function ratios. The degree of accuracy depends on the accuracy of the given information.

　　Practical problems often require using or finding the angle of elevation or depression. These angles are always measured from the horizontal. Some applications involve bearing, which is measured in two ways. One way is to measure in a clockwise direction from due north. With another method we measure the angle from due north to the west or east, or from due south to the west or east.

　　The law of sines and the law of cosines are used to solve oblique triangles. By the law of sines the ratio of any side of a triangle to the sine of the opposite angle equals the corresponding ratio of another side to the sine of its opposite angle. If we are given two sides and the angle opposite one of them, the information may lead to no triangle, one triangle, or two triangles. This situation is called the ambiguous case of the law of sines. It is useful to remember that the largest angle must be opposite the largest side, and the smallest angle must be opposite the smallest side.

　　The law of cosines says that the square of a side of a triangle is equal to the sum of the squares of the other two sides, minus twice the product of the two sides and the cosine of the included angle. Note the similarity to the Pythagorean theorem. When using the law of cosines, remember that the sum of the lengths of any two sides of a triangle must be greater than the length of the third side.

　　The trigonometric functions often appear in formulas from mechanics, engineering, and other fields, leading to meaningful applications. Using the laws of sines and cosines, we can derive two formulas for the area of a triangle, one of which is known as Heron's formula. Heron's formula is used when three side lengths of a triangle are known. The other area formula is used when two sides and the included angle are known.

　　The linear velocity of a point moving along a circle tells how fast the point is moving. As the point moves along the circle, the positive angle in standard position also changes. The rate at which the angle changes is the angular velocity. Linear and angular velocity have many practical applications.

Key Terms

SECTION 4.1	SECTION 4.2
hypotenuse	congruent triangles
side opposite an angle	oblique triangle
side adjacent to an angle	
significant digits	**SECTION 4.4**
exact number	
angle of elevation	linear velocity
angle of depression	angular velocity
bearing	

CHAPTER SUMMARY & KEY TERMS

At the end of each chapter, a review of concepts gives students an overview of the chapter (a built-in study guide). The friendly narrative (as opposed to listing only formulas) is focused on the connection of concepts.

different angle of elevation at each location. On April 29, 1976, at 11:35 A.M., the lunar angles of elevation during a partial solar eclipse at Bochum in upper Germany and at Donaueschingen in lower Germany were measured as 52.6997° and 52.7430°, respectively. The two cities are 398 kilometers apart.* Calculate the distance to the moon from Bochum on this day. Disregard the curvature of the earth in this calculation.

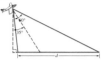

ground distance d in feet that will be shown in this photograph. † See the figure. (The 60° angle is bisected.)

31. (See Exercise 30.) A camera lens with a 6-inch focal length has an angular coverage of 86°. Suppose an aerial photograph is taken vertically with no tilt at an altitude of 3500 feet over the horizon line with an increasing slope of 5° as shown in the figure. Calculate the ground distance CB that would appear in the resulting photograph.

30. The distance covered by an aerial photograph is determined by both the focal length of the camera and the tilt of the camera from the perpendicular to the ground. Although the tilt is usually small, both archaeological and Canadian photographs often use larger tilts. A camera lens with a 12-inch focal length will have an angular coverage of 60°. If an aerial photograph is taken with this camera tilted $\theta = 35°$ at an altitude of 5000 feet, calculate the

32. Repeat Exercise 31 if the camera lens has an 8.25-inch focal length with an angular coverage of 72°.

In each set of data for triangle ABC, two sides and an angle opposite one of them are given. As explained in Examples 4–6, such a set of data may lead to 0, 1, or 2 triangles. Solve for all possible triangles.

33. $A = 42.5°$, $a = 15.6$ ft, $b = 8.14$ ft

34. $C = 52.3°$, $a = 32.5$ yd, $c = 59.8$ yd

35. $B = 72.2°$, $b = 78.3$ m, $c = 145$ m

36. $C = 68.5°$, $c = 258$ cm, $b = 386$ cm

37. $A = 38°\ 40'$, $a = 9.72$ km, $b = 11.8$ km

38. $C = 29°\ 50'$, $a = 8.61$ m, $c = 5.21$ m

39. $B = 32°\ 50'$, $a = 7540$ cm, $b = 5180$ cm

40. $C = 22°\ 50'$, $b = 159$ mm, $c = 132$ mm

41. $A = 96.80°$, $b = 3.589$ ft, $a = 5.818$ ft

42. $C = 88.70°$, $b = 56.87$ yd, $c = 112.4$ yd

43. Apply the law of sines to the following: $a = \sqrt{5}$, $c = 2\sqrt{5}$, $A = 30°$. What is the value of $\sin C$? What is the measure of C? Based on its angle measures, what kind of triangle is triangle ABC?

44. In your own words, explain the condition that must exist to determine that there is no triangle satisfying the given values of a, b, and B, once the value of $\sin B$ is found.

* Schlosser, W., T. Schmidt-Kaler, and E. Milone, *Challenges of Astronomy* (New York: Springer-Verlag, 1991).
† Brooks, R. and Dieter Johannes, *Phytoarchaeology* (Portland, OR: Dioscorides Press, 1990). Moffitt, F.,*Photogrammetry* (Scranton: International Textbook Company, 1967).

APPLICATIONS WITH REAL DATA AND REFERENCES

Well thought-out, up-to-date applications help answer the students' question: "What is this good for?" Some also sharpen writing skills by requiring brief essay responses. Many also have multiple steps that lead students to a conclusion gradually.

---

184 CHAPTER 3 Identities, Equations, Inequalities, and Applications

3.2 EXERCISES

The graph of $y = -2 \cos x + 1$ is shown in both screens below. It is graphed in the window $[0, 2\pi]$ by $[-4, 4]$. Use the screens, along with the facts that

$$\frac{\pi}{3} \approx 1.0471976 \quad and \quad \frac{5\pi}{3} \approx 5.2359878$$

to respond to Exercises 1–6.

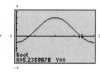

First View   Second View

1. Find the solution set of
$$-2 \cos x + 1 = 0$$
over the interval $[0, 2\pi]$. Give exact values.

2. Find the solution set of
$$-2 \cos x + 1 > 0$$
over the interval $[0, 2\pi]$. Give exact values for endpoints.

3. Find the solution set of
$$-2 \cos x + 1 \leq 0$$
over the interval $[0, 2\pi]$. Give exact values for endpoints.

4. Over the interval $[0, 2\pi]$, how many solutions does the equation
$$-2 \cos x + 1 = 2$$
have? (*Hint:* Imagine that the graph of $y = 2$ also appears on the screen.)

5. Over the interval $[0, 2\pi]$, how many solutions does the equation
$$-2 \cos x + 1 = -2$$
have? (*Hint:* Imagine that the graph of $y = -2$ also appears on the screen.)

6. Find the solution set of $-2 \cos x + 1 = 0$ over all real numbers.

For the function f in Exercises 7–16.
(a) *Solve $f(x) = 0$ analytically over the interval $[0, 2\pi)$, and give solutions in exact forms.*
(b) *Graph $y = f(x)$ in the window $[0, 2\pi]$ by $[-4, 4]$.*
(c) *Use the results of (a) and the graph in (b) to give the exact solution set of $f(x) > 0$ over $[0, 2\pi)$.*
(d) *Use the results of (a) and the graph in (b) to give the exact solution set of $f(x) < 0$ over $[0, 2\pi)$.*

7. $f(x) = 2 \cos x + 1$

8. $f(x) = 2 \sin x + 1$

9. $f(x) = \tan^2 x - 3$

10. $f(x) = \sec^2 x - 1$

BALANCED EXERCISES

There are many exercises that draw students into the heart of mathematical concepts. True conceptual exercises are balanced by a suitable ratio of drill, application, and writing exercises.

Use the negative number identities to write each of the following as a circular function of a positive number. (For example, sin(−3.4) = −sin 3.4.)

37. cos(−4.38)   **38.** cos(−5.46)   **39.** sin(−.5)

40. sin(−2.5)   **41.** $\tan\left(-\dfrac{\pi}{7}\right)$ **42.** $\tan\left(-\dfrac{4\pi}{7}\right)$

43. sec(−8)   **44.** sec(−.055)   **45.** $\csc\left(-\dfrac{1}{4}\right)$

46. $\csc\left(-\dfrac{5}{9}\right)$ **47.** cot(−10⁵)   **48.** cot(−5⁸)

*F*urther Explorations

While radian mode is very important for connecting angle measure and arc length, the graphing calculator cannot display values like π, $\frac{\pi}{2}$, and 2π in their exact forms. Instead the graphing calculator can only display and evaluate decimal approximations. In other words, it is unable to manipulate irrational numbers; it must first convert to a rational approximation. For this reason, it is sometimes preferable to use DEGREE mode when evaluating trigonometric functions. If your graphing calculator has a TABLE feature, it can be used to find patterns in trigonometric functions. The figure shows a TABLE where $Y_1 = \cos x$ and $Y_2 = \sin x$. The calculator is in DEGREE MODE and ΔTbl = 15°.

Note some patterns:

cos 45° = sin 45°

cos x decreases from 1 to 0 as x goes from 0° to 90°.

sin x increases from 0 to 1 as x goes from 0° to 90°.

1. Create the TABLE from the figure in your calculator. Scroll through the TABLE at least as far as $x = 360°$ in order to answer the following.
   (a) Find all values of x where cos x = sin x.
   (b) Describe all intervals where cos x is decreasing and all intervals where cos x is increasing.
   (c) Describe all intervals where sin x is decreasing and all intervals where sin x is increasing.
   (d) What are the maximum and minimum values for cos x and sin x?
   (e) Find a ΔTbl and Tblmin where cos x always will equal 1. For what values of x does cos x = 1?
   (f) Find a ΔTbl and Tblmin where cos x always will equal 0. For what values of x does cos x = 0?
   (g) Find a ΔTbl and Tblmin where sin x always will equal 1. For what values of x does sin x = 1?
   (h) Find a ΔTbl and Tblmin where sin x always will equal 0. For what values of x does sin x = 0?

---

*R*elating Concepts

From the law of cosines, in triangle ABC, $\cos A = \frac{b^2 + c^2 - a^2}{2bc}$. Use this equation to show that each of the following is true, and from these exercises, prove Heron's formula. Work Exercises 27–32 in order.

27. $1 + \cos A = \dfrac{(b + c + a)(b + c - a)}{2bc}$ **28.** $1 - \cos A = \dfrac{(a - b + c)(a + b - c)}{2bc}$

29. $\cos\dfrac{A}{2} = \sqrt{\dfrac{s(s-a)}{bc}}$ $\left(Hint: \cos\dfrac{A}{2} = \sqrt{\dfrac{1 + \cos A}{2}}\right)$

30. $\sin\dfrac{A}{2} = \sqrt{\dfrac{(s-b)(s-c)}{bc}}$ $\left(Hint: \sin\dfrac{A}{2} = \sqrt{\dfrac{1 - \cos A}{2}}\right)$

31. The area of a triangle having sides b and c and angle A is given by $(\frac{1}{2})bc \sin A$. Show that this result can be written as
$$\sqrt{\frac{1}{2}bc(1 + \cos A) \cdot \frac{1}{2}bc(1 - \cos A)}.$$

32. Use the results of Exercises 27–31 to prove Heron's area formula.

Use the figures below to derive the area formula $\mathcal{A} = \frac{1}{2}bc \sin A$. Work Exercises 33–36 in order.

33. Using triangle ABD, find an expression for sin A in terms of h and c.

34. Solve for h.

35. The familiar area formula for a triangle, area $= \frac{1}{2}$(base)(height), can now be applied. Since the height to base b is h, use the expression for h found in Exercise 34 and write the formula in terms of b and that expression.

36. Explain why, if $A = 90°$, the formula $\mathcal{A} = \frac{1}{2}bc \sin A$ becomes the familiar area formula.

Solve the following problems involving linear and angular velocity.

37. A tire is rotating 600 times per minute. Through how many degrees does a point on the edge of the tire move in $\frac{1}{2}$ second?

38. An airplane propeller rotates 1000 times per minute. Find the number of degrees that a point on the edge of the propeller will rotate in 1 second.

39. A pulley rotates through 75° in one minute. How many rotations does the pulley make in an hour?

40. **(a)** How many inches will the weight in the figure rise if the pulley is rotated through an angle of 71° 50′?

(b) Through what angle, to the nearest minute, must the pulley be rotated to raise the weight 6 in.?

9.27 in.

The Supplement Package for *A Graphical Approach to Trigonometry*

The text is supported with an extensive package of supplements for both the instructor and the student.

For the Instructor

Instructor's Annotated Exercises   With this volume, instructors have immediate access to the answers to every exercise in the text. Each answer is printed in bold type next to or below the pertinent exercise.

The *Instructor's Testing Manual*   offers four to six test items per text section for the instructor to use.

The *Instructor's Solution Manual*   provides worked solutions to all the even-numbered exercises in the text.

The Test Generator/Editor for Mathematics with QuizMaster   Available in IBM (both DOS and Windows applications) and Macintosh versions, the *Test Generator* is fully networkable. The *Test Generator* enables instructors to select questions by objective, section, or chapter, or to use a ready-made test for each chapter. The *Editor* enables instructors to edit any preexisting data or to create their own questions easily. The software is algorithm-driven, allowing the instructor to regenerate constants while maintaining problem type, providing a very large number of test or quiz items in multiple-choice and/or open-response formats for one or more test forms. The system features printed graphics and accurate mathematical symbols. *QuizMaster* enables instructors to create tests and quizzes using the *Test Generator/ Editor* and save them to disk so students can take the test or quiz on a stand-alone computer or network. *QuizMaster* then grades the test or quiz and allows the instructor to create reports on individual students or entire classes.

Graph Explorer   software, available for IBM and Macintosh hardware, allows students to learn through exploration. With this tool-oriented approach to algebra, students can graph rectangular, conic, polar, and parametric equations, zoom, transform functions, and experiment with families of equations. Students have the option to experiment with different solutions, display multiple representations, and print all work.

Videotapes   features lessions on using appropriate operations using keystrokes on the TI-82® and TI-85® graphing calculators.

For the Student

The Student's Solution Manual   written by Norma James of New Mexico State University, contains worked solutions to all of the odd-numbered exercises in the text.

A *Graphing Calculator Keystroke Guide Expanded Edition*   by Stuart Moskowitz of Humboldt State University, provides keystroke operations for the following calculator models: TI-82®, TI-83®, TI-85®, HP38G®, and Casio CFX-9800G®. Examples are taken from the text.

Interactive Mathematics Tutorial Software with Management System   This innovative package is available in DOS, Windows, and Macintosh versions and is fully networkable. As with the *Test Generator/Editor,* this software is algorithm-driven, which automatically regenerates constants so that the numbers rarely repeat in a problem type when students revisit any particular section. The tutorial is objective-based, self-paced, and provides unlimited opportunities to review lessons and to practice problem solving. If students give a wrong answer, they can ask to see the problem worked out and get a textbook page reference.

Many problems include hints for first incorrect responses. Tools such as an online glossary and Quick Reviews provide definitions and examples, and an online calculator aids students in computation. The program is menu-driven for ease of use, and on-screen help can be obtained at any time with a single keystroke. Students' scores are calculated at the end of each lesson and can be printed for a permanent record. The optional *Management System* lets instructors record student scores on disk and print diagnostic reports for individual students or classes. This software may also be purchased by students for home use. Student versions include record keeping and practice tests.

Acknowledgments

A project of this magnitude cannot be completed without the help of countless other individuals who offer support, suggestions, and criticisms. We would like to thank the following individuals who reviewed and checked the accuracy of the text.

John J. Avioli, Christopher Newport College ▌ Janis M. Cimperman, St. Cloud State University ▌ Julane B. Crabtree, Johnson County Community College ▌ James R. Fryxell, College of Lake County ▌ Richard Leedy, Polk Community College ▌ Judy S. McInerney, Sandhills Community College ▌ Gabrielle McIntosh, New Mexico State University ▌ Willard J. Raiffeisen, Texas State Technical College ▌ J. Doug Richey, Northeast Texas Community College ▌ Robyn Serven, University of Central Arkansas ▌ Ron Smith, Edison Community College ▌ James Trefzger, Parkland College ▌ Paula G. Young, Salem College ▌

To the hundreds of students who studied during the various stages of the project, we offer our thanks. Susan Danielson, an instructor at the University of New Orleans, offered great support and many excellent suggestions, and her support during the past few years has been most welcome. She is a true innovator, and our profession could use many more like her. William Armstrong of Lakeland Community College wrote portions of Chapter 1 and Chapter 6, and assisted in the production process during the Spring of 1996. His work was a great help in allowing us to finish the project on time. The ***Further Explorations*** exercises were written by Stuart Moskowitz of Butte College, who did an outstanding job. To Brent Simon, we extend our thanks just for being there for about twenty years now. Gwen Hornsby provided great assistance in the preparation of the graphing art. As always, our families gave up husband/father and wife/mother to many hours of work, and our thanks and love go out to them.

We feel that we worked with the finest editorial, marketing, and production staff ever assembled. The support of the Glenview Group was unwavering from the first day we proposed this project. Anne Kelly signed the book. Ed Moura believed in it. Lisa Kamins, Emily Barman, and Judy Rudnick were fantastic in helping to provide manuscript copies for class testing. Lynn Mooney did a wonderful job in the development of the text. Ann-Marie Buesing did her usual excellent job in production. To these individuals and all the others who worked up on the second floor, we offer our sincerest gratitude, and wish them the best in their future endeavors.

A Final Word

We hope that this book and its companion texts begin to make a difference in the manner in which precalculus mathematics is presented and learned as we move into the twenty-first century. We ask that both instructors and students pursue the contents with an open mind, ready to teach and to learn in a manner that only now, after so many thousands of years, is possible. We, like Newton, can do so only because we "have stood on the shoulders of giants."

We welcome your comments, suggestions, and criticisms. Please write to us at Addison-Wesley, One Jacob Way, Reading, MA 01867, and let us know what you think.

E. John Hornsby, Jr.
Margaret L. Lial

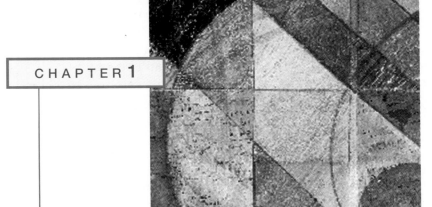

CHAPTER 1

A Review of Basic Concepts: Analysis of Equations, Inequalities, and Graphs of Functions

1.1 REAL NUMBERS AND COORDINATE SYSTEMS

Sets of Real Numbers ▌ Roots ▌ Coordinate Systems ▌ Viewing Windows

Sets of Real Numbers

The idea of counting goes back to the beginning of our civilization. When people first counted they used only the **natural numbers,** written in set notation as

$$\{1, 2, 3, 4, 5, \ldots\}.$$

Much more recent is the idea of counting *no* object—that is, the idea of the number 0. Including 0 with the set of natural numbers gives the set of **whole numbers.**

$$\{0, 1, 2, 3, 4, 5, \ldots\}$$

(These and other sets of numbers are summarized later in this section.)

About 500 years ago, people came up with the idea of counting backwards, from 4 to 3 to 2 to 1 to 0. There seemed no reason not to continue this process, calling the new numbers $-1, -2, -3$, and so on. Including these numbers with the set of whole numbers gives the very useful set of **integers,**

$$\{\ldots, -4, -3, -2, -1, 0, 1, 2, 3, \ldots\}.$$

1

Integers can be shown pictorially with a **number line.** (A number line is similar to a thermometer on its side.) As an example, the elements of the set $\{-3, -1, 0, 1, 3, 5\}$ are located on the number line in Figure 1.

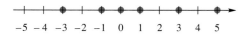

FIGURE 1

The result of dividing two integers, with a nonzero divisor, is called a *rational number.* By definition, the **rational numbers** are the elements of the set

$$\left\{ \frac{p}{q} \,\middle|\, p, q \text{ are integers and } q \neq 0 \right\}.$$

This definition, which is given in *set-builder notation,* is read "the set of all elements p/q such that p and q are integers and $q \neq 0$." Examples of rational numbers include $\frac{3}{4}, -\frac{5}{8}, \frac{7}{2}$, and $-\frac{14}{9}$. All integers are rational numbers, since any integer can be written as the quotient of itself and 1.

Rational numbers can be located on a number line by a process of subdivision. For example, $\frac{5}{8}$ can be located by dividing the interval from 0 to 1 into 8 equal parts, then labeling the fifth part $\frac{5}{8}$. Several rational numbers are located on the number line in Figure 2.

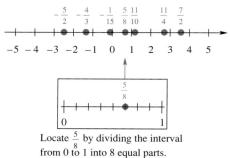

Locate $\dfrac{5}{8}$ by dividing the interval from 0 to 1 into 8 equal parts.

FIGURE 2

The set of all numbers that correspond to points on a number line is called the set of **real numbers.** The set of real numbers is shown in Figure 3.

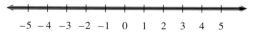

FIGURE 3

A real number that is not rational is called an **irrational number.** The set of irrational numbers includes $\sqrt{3}$ and $\sqrt{5}$ but not $\sqrt{1}, \sqrt{4}, \sqrt{9}, \ldots$, which equal $1, 2, 3, \ldots$, and hence are rational numbers. Another irrational number is π, which is approximately equal to 3.14159. The numbers in the set $\{-\frac{2}{3}, 0, \sqrt{2}, \sqrt{5}, \pi, 4\}$ can be located on a number line as shown in Figure 4. (Only $\sqrt{2}, \sqrt{5}$, and π are irrational here. The others are rational.)

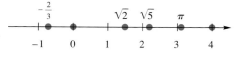

FIGURE 4

Real numbers can also be defined in another way, in terms of decimals. Using repeated subdivisions, any real number can be located (at least in theory) as a point on a number line. By this process, the set of real numbers can be defined as the set of all decimals. Every rational number has a decimal representation that either terminates (comes to an end) or repeats in a fixed "block" of digits. Here are some examples.

Rational Numbers Whose Decimals Terminate	**Rational Numbers Whose Decimals Repeat**
$\frac{1}{4} = .25$	$\frac{1}{3} = .3333\ldots$
$\frac{3}{8} = .375$	$\frac{5}{6} = .8333\ldots$
$\frac{7}{4} = 1.75$	$\frac{3}{7} = .428571428571\ldots$

The three dots at the end of the repeating decimals indicate that the pattern of digits established continues indefinitely. Another way of indicating the repeating digits is the use of a bar over the part that repeats. Thus, we would write the following.

$$\frac{1}{3} = .\overline{3} \qquad \frac{5}{6} = .8\overline{3} \qquad \text{and} \qquad \frac{3}{7} = .\overline{428571}$$

If at any time we use an approximation for a rational number, we use the $\approx$ symbol to indicate "is approximately equal to." Thus it would technically be incorrect to write $\frac{2}{3} = .67$; we should write $\frac{2}{3} \approx .67$ if an approximation is warranted. We call .67 an *approximation of $\frac{2}{3}$ to the nearest hundredth*, while $.\overline{6}$ is the *exact decimal representation for $\frac{2}{3}$*.

NOTE In this text we will often make distinctions about whether an approximation or an exact value is required. As we progress in our work, more will be said about this.

The decimal representation of an irrational number will neither terminate nor repeat. The locations of $\sqrt{2}$, $\sqrt{5}$, and π on the number line in Figure 4 were determined by observing these calculator approximations:

$$\sqrt{2} \approx 1.414213562 \qquad \sqrt{5} \approx 2.236067977 \qquad \pi \approx 3.141592654.$$

Let set $A = \{-8, -6, -.75, 0, .\overline{09}, \sqrt{2}, \sqrt{5}, 6, \frac{107}{4}\}$. List the elements from set A that belong to each of the following sets: (a) real numbers, (b) integers, (c) rational numbers, (d) irrational numbers, (e) whole numbers, and (f) natural numbers.

SOLUTION

(a) Because every element of A can be represented by a point on a number line, all elements are real numbers.

(b) The integers are -8, -6, 0, and 6.

(c) The rational numbers are -8, -6, $-.75$, 0, $.\overline{09}$, 6, and $\frac{107}{4}$.

(d) The irrational numbers are $\sqrt{2}$ and $\sqrt{5}$.

(e) The whole numbers are 0 and 6.

(f) The only natural number in the set is 6. ▌

The relationships among the various subsets of the real numbers, along with examples in the sets, are shown in Figure 5.

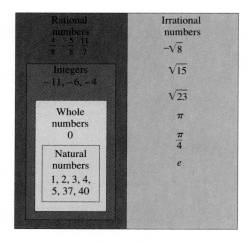

FIGURE 5
The Real Numbers

TECHNOLOGICAL NOTE
Various models of calculators use different keystrokes for finding decimal values for roots. Be aware of how to find roots using your particular model.

Roots

Some common irrational numbers that we will encounter in this course are roots—**square roots, cube roots,** and so forth. You should be able to use your calculator to find roots. Graphing calculators usually have dedicated keys for square and cube roots, and have functions that allow for other roots. Consult your owner's manual to see how to find roots on your particular model.

EXAMPLE 2

Finding Roots on a Calculator

Use your calculator to verify the following decimal approximations for roots that are irrational numbers. (Depending on the model you have and how it is set for the number of digits displayed, there may be a slight discrepancy in the final digit.)

SOLUTION
(a) $\sqrt{23} \approx 4.795831523$
(b) $\sqrt[3]{87} \approx 4.431047622$
(c) $\sqrt[4]{12} \approx 1.861209718$ ▫

It is often convenient to use the fact that $\sqrt[n]{a} = a^{1/n}$ for appropriate values of n and a to find roots.

FOR **GROUP DISCUSSION**

Experiment with your calculator to show that the following pairs of expressions give the same approximations. You may need to refer to your owner's manual to see how to raise a number to a power.

1. $\sqrt{98.4}, 98.4^{1/2}$ **2.** $\sqrt[3]{12.9}, 12.9^{1/3}$ **3.** $\sqrt[4]{86}, 86^{.25}$

Coordinate Systems

Figure 6 shows a number line with the points corresponding to several different numbers marked on the line. A number that corresponds to a particular point on a line is called the **coordinate** of the point. For example, the leftmost marked point in Figure 6 has coordinate -4. The correspondence between points on a line and the real numbers is called a **coordinate system** for the line. (The phrase "the point on a number line with coordinate a" will be abbreviated as "the point with coordinate a," or simply "the point a.")

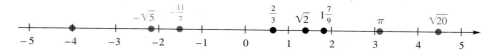

FIGURE 6

If the real number a is to the left of the real number b on a number line, then ***a* is less than *b*,** written $a < b$. If a is to the right of b, then ***a* is greater than *b*,** written $a > b$. For example, in Figure 6, $-\sqrt{5}$ is to the left of $-\frac{11}{7}$ on the number line, so $-\sqrt{5} < -\frac{11}{7}$, while $\sqrt{20}$ is to the right of π, indicating $\sqrt{20} > \pi$.

NOTE   Remember that the "point" of the inequality symbol goes toward the smaller number.

As an alternative to this geometric definition of "is less than" or "is greater than," there is an algebraic definition: if a and b are two real numbers and if the difference $a - b$ is positive, then $a > b$. If $a - b$ is negative, then $a < b$. The geometric and algebraic statements of order are summarized as follows.

Statement	Geometric Form	Algebraic Form
$a > b$	a is to the right of b	$a - b$ is positive
$a < b$	a is to the left of b	$a - b$ is negative

The symbols $<$ and $>$ can be combined with the symbol for equality. The statement $a \leq b$ means "a is less than or equal to b" and is true if either $a < b$ or $a = b$. Similarly, $a \geq b$ means "a is greater than or equal to b" and is true if either $a > b$ or $a = b$. Statements involving $<$ or $>$ are called **strict** inequalities, while those involving $\leq$ or $\geq$ are called **nonstrict** inequalities. We can negate any of these symbols by using a slash bar ($/$).

EXAMPLE 3
Showing why Inequality Statements are True

The list below shows several statements and the reason why each is true.

Statement	Reason
$8 \leq 10$	$8 < 10$
$8 \leq 8$	$8 = 8$
$-9 \geq -14$	$-9 > -14$
$-8 \not> -2$	$-8 < -2$
$4 \not< 2$	$4 > 2$

∎

The inequality $a < b < c$ says that b is *between* a and c, since

$$a < b < c$$

means $\quad a < b \quad$ and $\quad b < c.$

In the same way, $\quad a \leq b \leq c$

means $\quad a \leq b \quad$ and $\quad b \leq c.$

CAUTION When writing these "between" statements, make sure that both inequality symbols point in the same direction, toward the smallest number. For example,

both $2 < 7 < 11$ and $5 > 4 > -1$

are true statements, but $3 < 5 > 2$ is meaningless. Generally, it is best to rewrite statements such as $5 > 4 > -1$ as $-1 < 4 < 5$, which is the order of these numbers on the number line.

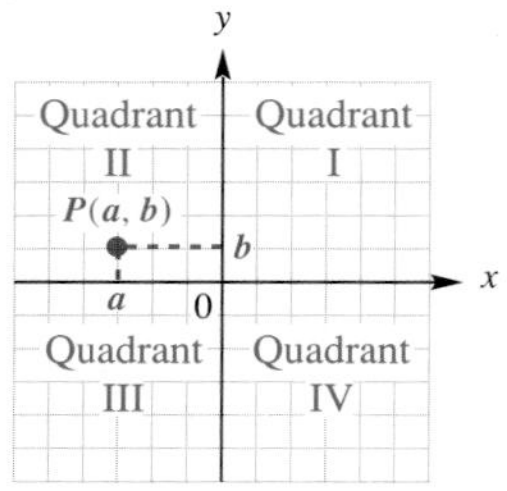

FIGURE 7

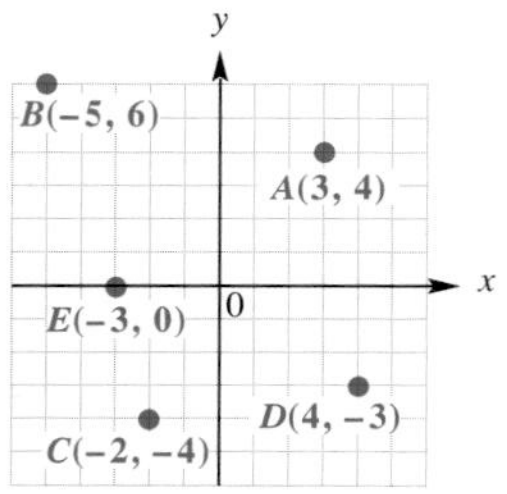

FIGURE 8

TECHNOLOGICAL NOTE
You should read your owner's manual to see how to alter the viewing window on your screen. Remember that different settings will result in different views of graphs. When you look out of your window at home, you may see something completely different than when your neighbor looks out of his or her window; the same goes for graphs generated by calculators.

A number line is an example of a one-dimensional coordinate system, and it is sufficient to graph real numbers. If we place two number lines at right angles, intersecting at their origins, we obtain a two-dimensional **rectangular coordinate system.** It is customary to have one of these lines vertical and the other horizontal. They intersect at the **origin** of the system, designated 0. The horizontal line is called the ***x*-axis,** and the vertical line is called the ***y*-axis.** On the x-axis, positive numbers are located to the right of the origin, while negative numbers are located to the left. On the y-axis, positive numbers are located above the origin, negative numbers below.

This rectangular coordinate system is also called the **cartesian coordinate system,** named after Rene Descartes (1596–1650). The plane into which the coordinate system is introduced is the **coordinate plane,** or ***xy*-plane.** The x-axis and y-axis divide the plane into four regions, or **quadrants,** labeled as shown in Figure 7. The points on the x-axis and y-axis belong to no quadrant.

Each point P in the xy-plane corresponds to a unique ordered pair (a, b) of real numbers. The numbers a and b are the **coordinates** of point P. We call a the x-coordinate and b the y-coordinate. To locate on the xy-plane the point corresponding to the ordered pair $(3, 4)$, for example, draw a vertical line through 3 on the x-axis and a horizontal line through 4 on the y-axis. These two lines intersect at point A in Figure 8. Point A corresponds to the ordered pair $(3, 4)$. Also in Figure 8, B corresponds to the ordered pair $(-5, 6)$, C to $(-2, -4)$, D to $(4, -3)$, and E to $(-3, 0)$. The point P corresponding to the ordered pair (a, b) often is written as $P(a, b)$ as in Figure 7 and referred to as "the point (a, b)."

Viewing Windows

The characteristic that distinguishes this text from traditional trigonometry texts is that it features full integration of modern-day graphing calculators. A graphing calculator differs from a typical scientific calculator in many ways, the most obvious of which is that it allows the user to plot points and a variety of graphs at the touch of keys.

NOTE In this text, all references to calculators are made with the understanding that the reader has access to a modern graphing calculator. Therefore, the term "calculator" is understood to mean "graphing calculator."

The rectangular (cartesian) coordinate system theoretically extends indefinitely in all directions. We are limited to illustrating only a portion of such a system in a text figure. Similar limitations are found in portraying coordinate systems on calculator screens. For this reason, the student should become familiar with the key on the calculator that sets the limits for x- and y-coordinates. Some calculators use the word "range" for this function, while others use a more appropriate designation, "window." (There also may be other designations.) Figure 9 shows a calculator screen that has been set to have a minimum x-value of -10, a maximum x-value of 10, a minimum y-value of -10, and a maximum y-value of 10. Additionally, the tick marks on the axes have been set to be 1 unit apart. Throughout this book, this window will be called the *standard viewing window*.

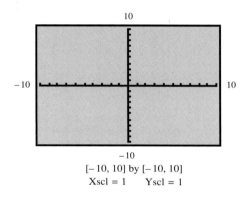

$[-10, 10]$ by $[-10, 10]$
Xscl = 1   Yscl = 1

FIGURE 9

In order to convey important information about a viewing window, in this text we will use the following abbreviations:

Xmin: minimum value of x Ymin: minimum value of y

Xmax: maximum value of x Ymax: maximum value of y

Xscl: scale (distance between   Yscl: scale (distance between
    tick marks) on the x-axis   tick marks) on the y-axis

To further condense this information, we will often use the following symbolism:

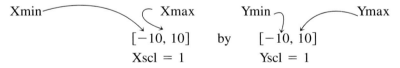

Xmin   Xmax   Ymin   Ymax
$[-10, 10]$ by $[-10, 10]$
Xscl = 1   Yscl = 1

The symbols above indicate the viewing window information for the window in Figure 9.

All calculators have a standard viewing window. Viewing windows may be changed by manually entering the information, or by using the zoom feature of the calculator. The graphing screen is made up of pixels, which are small areas that, when illuminated, will represent points in the plane. The coordinates of the pixels may be found by using the trace feature of the calculator.

Figure 10 on the following page shows several other viewing windows, with the important information. Notice that (b) and (c) look exactly alike, and unless we are told what the settings are, we have no way of distinguishing between them. Paying careful attention to window settings will be an important part of our work in this text.

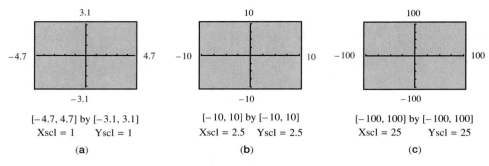

$$[-4.7, 4.7] \text{ by } [-3.1, 3.1]$$
Xscl = 1   Yscl = 1
(a)

$$[-10, 10] \text{ by } [-10, 10]$$
Xscl = 2.5   Yscl = 2.5
(b)

$$[-100, 100] \text{ by } [-100, 100]$$
Xscl = 25   Yscl = 25
(c)

FIGURE 10

1.1 EXERCISES

For each of the following sets, list all elements that belong to (a) natural numbers, (b) whole numbers, (c) integers, (d) rational numbers, (e) irrational numbers, and (f) real numbers.

1. $\left\{-6, -\dfrac{12}{4}, -\dfrac{5}{8}, -\sqrt{3}, 0, .31, .\overline{3}, 2\pi, 10, \sqrt{17}\right\}$ **2.** $\left\{-8, -\dfrac{14}{7}, -.245, 0, \dfrac{6}{2}, 8, \sqrt{81}, \sqrt{12}\right\}$

3. $\left\{-\sqrt{100}, -\dfrac{13}{6}, -1, 5.23, 9.\overline{14}, 3.14, \dfrac{22}{7}\right\}$ **4.** $\{-\sqrt{49}, -.405, -.\overline{3}, .1, 3, 18, 6\pi, 56\}$

Graph each set of numbers on a number line.

5. $\{2, 3, 4, 5\}$ **6.** $\{0, 2, 4, 6, 8\}$ **7.** $\{-4, -3, -2, -1, 0, 1\}$

8. $\{-6, -5, -4, -3, -2\}$ **9.** $\left\{-.5, .75, \dfrac{5}{3}, 3.5\right\}$ **10.** $\left\{-.6, \dfrac{9}{8}, 2.5, \dfrac{13}{4}\right\}$

Each rational number written in common fraction form in Exercises 11–20 has its decimal equivalent appearing in the column on the right. Without using a calculator, if possible, match the fraction with its decimal equivalent.

11. $\dfrac{1}{5}$ **12.** $\dfrac{2}{3}$ **A.** .5   **B.** .67

13. $\dfrac{67}{100}$ **14.** $\dfrac{12}{10}$ **C.** .75   **D.** .2

15. $\dfrac{3}{4}$ **16.** $\dfrac{10}{3}$ **E.** $.\overline{6}$ **F.** .125

17. $\dfrac{50}{100}$ **18.** $\dfrac{3}{11}$ **G.** $.\overline{27}$ **H.** $3.\overline{3}$

19. $\dfrac{7}{5}$ **20.** $\dfrac{1}{8}$ **I.** 1.2   **J.** 1.4

Each rational number in Exercises 21–28 has a decimal equivalent that repeats. Use the bar symbolism to write the decimal. Use a calculator.

21. $\dfrac{5}{6}$ **22.** $\dfrac{1}{9}$ **23.** $-\dfrac{13}{3}$ **24.** $-\dfrac{9}{11}$

25. $\dfrac{6}{27}$ **26.** $\dfrac{5}{33}$ **27.** $\dfrac{9}{110}$ **28.** $\dfrac{77}{990}$

29. Explain the difference between the rational numbers .87 and .$\overline{87}$.

30. A student, using her powerful new calculator, found the decimal 1.414213562 when she evaluated $\sqrt{2}$. Is this decimal the exact value of $\sqrt{2}$, or just an approximation? Should she write $\sqrt{2} = 1.414213562$ or $\sqrt{2} \approx 1.414213562$?

Use a calculator to find a decimal approximation of each root or power. Give as many decimal places as your calculator shows.

31. $\sqrt{58}$ **32.** $\sqrt{97}$ **33.** $\sqrt[3]{33}$ **34.** $\sqrt[3]{91}$

35. $\sqrt[4]{86}$ **36.** $\sqrt[4]{123}$ **37.** $5^{1/2}$ **38.** $19^{1/2}$

39. $18^{1/3}$ **40.** $29^{1/3}$ **41.** $76^{.3}$ **42.** $98^{.275}$

Decide which of the following symbols may be placed in the blank to make a true statement: $<, \leq, >, \geq$. There may be more than one correct answer.

43. -5 _____ -4 **44.** -1.3 _____ $-.6$ **45.** 8 _____ 4

46. 9 _____ 8.9 **47.** -6 _____ -6 **48.** 2 _____ 2

Locate the following points on a rectangular coordinate system. Identify the quadrant, if any, in which the point lies.

49. $(2, 3)$ **50.** $(-1, 2)$

51. $(-3, -2)$ **52.** $(1, -4)$

53. $(0, 5)$ **54.** $(-2, -4)$

55. $(-2, 4)$ **56.** $(3, 0)$

57. $(-2, 0)$ **58.** $(3, -3)$

Recall from elementary algebra that the product of two numbers with the same signs is positive and the product of two numbers with different signs is negative. A similar rule holds for quotients. Name the possible quadrants in which the point (x, y) can lie if the given condition is true.

59. $xy > 0$ **60.** $xy < 0$ **61.** $\dfrac{x}{y} < 0$ **62.** $\dfrac{x}{y} > 0$

63. If the x-coordinate of a point is 0, the point must lie on which axis?

64. If the y-coordinate of a point is 0, the point must lie on which axis?

It is important to become familiar with your graphing calculator so that you can operate it efficiently. Answer the following questions about your particular calculator, using a complete sentence or sentences.

65. How do you set the screen in order to obtain the standard viewing window?

66. What are the minimum and maximum values of x and y in your standard viewing window?

Using the notation described in the text, set the viewing window of your calculator to the following specifications.

67. $[-10, 10]$ by $[-10, 10]$
Xscl = 1   Yscl = 1

68. $[-40, 40]$ by $[-30, 30]$
Xscl = 5   Yscl = 5

69. $[-5, 10]$ by $[-5, 10]$
Xscl = 3   Yscl = 3

70. $[-3.5, 3.5]$ by $[-4, 10]$
Xscl = 1   Yscl = 1

71. $[-100, 100]$ by $[-50, 50]$
Xscl = 20   Yscl = 25

72. $[-4.7, 4.7]$ by $[-3.1, 3.1]$
Xscl = .5   Yscl = .5

73. Set your viewing window to $[-10, 10]$ by $[-10, 10]$ and then set Xscl to 0 and Yscl to 0. Do you notice any tick marks on the axes? Make a conjecture as to how to set a screen with no tick marks on the axes.

74. Set your viewing window to $[-50, 50]$ by $[-50, 50]$ and then set Xscl to 1 and Yscl to 1. Observe this screen and describe the appearance of the axes as compared to those seen in the standard window. Why do you think they appear this way? How can you change your scale settings so that this "problem" is alleviated?

1.2 INTRODUCTION TO RELATIONS AND FUNCTIONS

Set-builder and Interval Notation ▮ Relations, Domain, and Range ▮ Functions ▮ Function Notation

In this section we introduce some of the most important concepts in the study of mathematics: relation, function, domain, and range. In order to make our work simpler, various types of set notation are useful. We begin by discussing two types: set-builder and interval notation.

Set-builder and Interval Notation

Inequalities and variables can be used to specify sets of real numbers. Suppose we wish to symbolize the set of real numbers greater than -2. One way to symbolize this is $\{x \mid x > -2\}$, read "the set of all x such that x is greater than -2." This is called set-builder notation, since the variable x is used to "build" the set. On a number line, we show the elements of this set (the set of all real numbers to the right of -2) by drawing a line from -2 to the right. We use a parenthesis at -2 since -2 is not an element of the given set. The result, shown in Figure 11, is called the **graph** of the set $\{x \mid x > -2\}$.

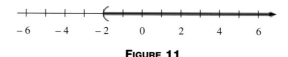

FIGURE 11

The set of numbers greater than -2 is an example of an **interval** on the number line. A simplified notation, called **interval notation,** is used for writing intervals. For example, using this notation, the interval of all numbers greater than -2 is written as $(-2, \infty)$. The **infinity symbol** ∞ does not indicate a number; it is used to show that the interval includes all real numbers greater than -2. The left parenthesis indicates that -2 is not included. A parenthesis is always used next to the infinity symbol in interval notation. The set of all real numbers is written in interval notation as $(-\infty, \infty)$.

EXAMPLE 1

Graphing an Inequality and Using Interval Notation

Write $\{x \mid x < 4\}$ in interval notation and graph the interval.

SOLUTION   The interval is written as $(-\infty, 4)$. The graph is shown in Figure 12. Since the elements of the set are all the real numbers *less* than 4, the graph extends to the left.   ▮

FIGURE 12

The set $\{x \mid x \leq -6\}$ contains all the real numbers less than or equal to -6. To show that -6 itself is part of the set, a *square bracket* is used at -6, as shown in Figure 13. In interval notation, this set is written as $(-\infty, -6]$.

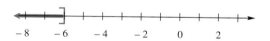

FIGURE 13

EXAMPLE **2**

Graphing an
Inequality and Using
Interval Notation

Write $\{x \mid x \geq -4\}$ in interval notation and graph the interval.

SOLUTION This set is written in interval notation as $[-4, \infty)$. The graph is shown in Figure 14. A square bracket is used at -4 since -4 is part of the set. ▯

FIGURE 14

It is common to graph sets of numbers that are *between* two given numbers. For example, the set $\{x \mid -2 < x < 4\}$ is made up of all those real numbers between -2 and 4, but not the numbers -2 and 4 themselves. This set is written in interval notation as $(-2, 4)$. The graph has a heavy line between -2 and 4 with parentheses at -2 and 4. See Figure 15. The inequality $-2 < x < 4$ is read "-2 is less than x and x is less than 4," or "x is between -2 and 4." It is an example of a *compound inequality*.

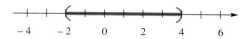

FIGURE 15

EXAMPLE **3**

Graphing a
Compound
Inequality and Using
Interval Notation

Write in interval notation and graph $\{x \mid 3 < x \leq 10\}$.

SOLUTION Use a parenthesis at 3 and a square bracket at 10 to get $(3, 10]$ in interval notation. The graph is shown in Figure 16. Read the inequality $3 < x \leq 10$ as "3 is less than x and x is less than or equal to 10," or "x is between 3 and 10, excluding 3 and including 10." ▯

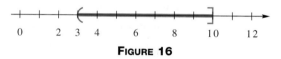

FIGURE 16

A chart summarizing the names of various types of intervals follows. Whenever two real numbers a and b are used to write an interval in the chart, it is assumed that $a < b$.

Type of Interval	Set-Builder Notation	Interval Notation	Graph
Open interval	$\{x \mid a < x < b\}$	(a, b)	
Closed interval	$\{x \mid a \le x \le b\}$	$[a, b]$	
Half-open (or half-closed) interval	$\{x \mid a < x \le b\}$	$(a, b]$	
	$\{x \mid a \le x < b\}$	$[a, b)$	
Unbounded interval	$\{x \mid x > a\}$	(a, ∞)	
	$\{x \mid x \ge a\}$	$[a, \infty)$	
	$\{x \mid x < a\}$	$(-\infty, a)$	
	$\{x \mid x \le a\}$	$(-\infty, a]$	
All real numbers	$\{x \mid x \text{ is real}\}$	$(-\infty, \infty)$	

CAUTION   Notice how the interval notation for the open interval (*a*, *b*) looks exactly like the notation for the ordered pair (*a*, *b*). While this does not usually cause confusion, as the interpretation is determined by the context of the use, we will, when the need arises, distinguish between them by using "the interval (*a*, *b*)" or "the point (*a*, *b*)."

Relations, Domain, and Range

Suppose that you have made a study and found that in your new car, driving 55 miles per hour yields a mileage of 31 miles per gallon, while driving 65 miles per hour reduces your mileage to 28 miles per gallon. You have observed a relationship between speed and miles per gallon, and this information may be described using the ordered pairs (55, 31) and (65, 28). These are only two of infinitely many possible ordered pairs that you could have found. If we consider the set of all such ordered pairs, we have an example of a relation.

RELATION

A **relation** is a set of ordered pairs.

Here are three examples of relations.

$$F = \{(1, 2), (-2, 5), (3, -1)\}$$
$$G = \{(-2, 1), (-1, 0), (0, 1), (1, 2), (2, 2)\}$$
$$H = \{(-4, 1), (-2, 1), (-2, 0)\}$$

If we denote the ordered pairs of a relation by (x, y), the set of all x-values is called the **domain** of the relation and the set of all y-values is called the **range** of the relation. In the relations above,

Domain of $F = \{1, -2, 3\}$ Range of $F = \{2, 5, -1\}$

Domain of $G = \{-2, -1, 0, 1, 2\}$ Range of $G = \{1, 0, 2\}$

Domain of $H = \{-4, -2\}$ Range of $H = \{1, 0\}$.

Since a relation is a set of ordered pairs, it may be represented graphically in the rectangular coordinate plane. The graphs of F, G, and H are shown in Figure 17.

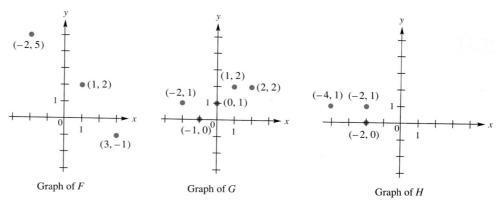

Graph of F Graph of G Graph of H

FIGURE 17

By observing the graph of a relation, we can determine its domain and range, as explained in Example 4.

EXAMPLE 4

Finding Domains and Ranges from Graphs

Three relations are graphed in Figure 18. Give the domain and range of each.

SOLUTION

(a) In Figure 18(a), the x-values of the points on the graph include all numbers between -4 and 4, inclusive. The y-values include all numbers between -6

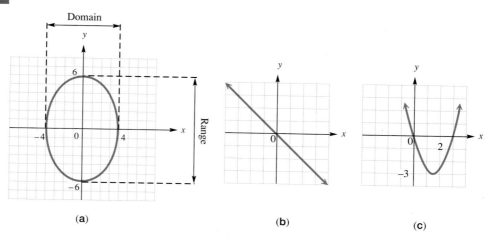

(a)   (b)   (c)

FIGURE 18

and 6, inclusive. Using interval notation,

the domain is $[-4, 4]$;
the range is $[-6, 6]$.

(b) In Figure 18(b), the arrowheads indicate that the line extends indefinitely left and right, as well as up and down. Therefore, both the domain and the range are the set of all real numbers, written $(-\infty, \infty)$.

(c) In Figure 18(c), the arrowheads indicate that the graph extends indefinitely left and right, as well as upward. The domain is $(-\infty, \infty)$. Because there is a least y-value, -3, the range includes all numbers greater than or equal to -3, written $[-3, \infty)$. ▯

EXAMPLE **5**
Finding Domain and Range from a Calculator Window

Later in this chapter, we will discuss graphing using a calculator. Figure 19 shows a graph on a screen with viewing window $[-5, 5]$ by $[-5, 5]$, with Xscl $= 1$ and Yscl $= 1$. By observation, give the domain and the range of this relation.

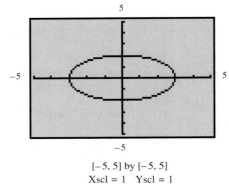

[−5, 5] by [−5, 5]
Xscl = 1   Yscl = 1

FIGURE 19

SOLUTION   Since the scales on both axes are 1, we see that the graph *appears* to have a minimum x-value of -3, a maximum x-value of 3, a minimum y-value of -2, and a maximum y-value of 2. Therefore, observation leads us to conclude that the domain is $[-3, 3]$ and the range is $[-2, 2]$. ▯

TECHNOLOGICAL NOTE
In Figure 19 we see a calculator-generated graph that is formed by a rather jagged curve. These are sometimes called *jaggies* and are typically found on low-resolution graphers, such as graphing calculators. In general we should remember that most curves we will study in this book are smooth, and the jaggies are just a part of the limitations of technology.

CAUTION   As we shall see many times while discussing calculator-generated graphs, simple observation is not enough to guarantee accuracy in determining domains and ranges. Calculators have capabilities which allow us to improve our accuracy, but even then, it is essential to understand the mathematical concepts behind graphing before we can be certain that our observations are correct. This is why in this book, we will study concepts and technological capabilities in an integrated fashion.

Functions

Look back at the relations F, G, and H introduced earlier in this section. Notice that in F and G, each x-value appears only once in the relation, while in H, the x-value -2 appears twice: it is paired with 1 in one ordered pair, while it is paired with 0 in another. Relations F and G are simple examples of a very important kind of relation, known as a function.

> **FUNCTION**
>
> A **function** is a relation in which each element in the domain corresponds to exactly one element in the range. *

Suppose a group of students get together each Monday evening to study algebra (and perhaps watch football). A number giving the student's weight to the nearest kilogram can be associated with each member of this set of students. Since each student has only one weight at a given time, the relationship between the students and their weights is a function. The domain is the set of all students in the group, while the range is the set of all the weights of the students.

If x represents any element in the domain, the set of students, x is called the **independent variable.** If y represents any element in the range, the weights, then y is called the **dependent variable,** because the value of y *depends on* the value of x. That is, each weight depends on the student associated with it.

In most mathematical applications of functions, the correspondence between the domain and range elements is defined with an equation, like $y = 5x - 11$. The equation is usually solved for y, as it is here, because y is the dependent variable. As we choose values from the domain for x, we can easily determine the corresponding y values of the ordered pairs of the function. (These equations need not use only x and y as variables; any appropriate letters may be used. In physics, for example, t is often used to represent the independent variable *time.*)

EXAMPLE 6

Deciding Whether a Relation is a Function

Decide whether the following sets are functions. Give the domain and range of each relation.

SOLUTION

(a) $\{(1, 2), (3, 4), (5, 6), (7, 8), (9, 10)\}$

The domain is the set $\{1, 3, 5, 7, 9\}$, and the range is $\{2, 4, 6, 8, 10\}$. Since each element in the domain corresponds to just one element in the range, this set is a function. The correspondence is shown below using D for the domain and R for the range.

$$D = \{1, 3, 5, 7, 9\}$$
$$\downarrow \ \downarrow \ \downarrow \ \downarrow \ \downarrow$$
$$R = \{2, 4, 6, 8, 10\}$$

(b) $\{(1, 1), (1, 2), (1, 3), (2, 4)\}$

The domain here is $\{1, 2\}$, and the range is $\{1, 2, 3, 4\}$. As shown in the correspondence below, one element in the domain, 1, has been assigned three different elements from the range, so this relation is not a function.

$$D = \{1, 2\}$$
$$R = \{1, 2, 3, 4\}$$

(c) $\{(-5, 2), (-4, 2), (-3, 2), (-2, 2), (-1, 2)\}$

Here, the domain is $\{-5, -4, -3, -2, -1\}$, and the range is $\{2\}$. Although every element in the domain corresponds to the same range element, this is a

---

*An alternate definition of function based on the idea of correspondence is given later in the section.

function because each element in the domain has exactly one range element as-signed to it.

(d) $\{(x, y) \mid y = x - 2, x \text{ any real number}\}$

Since y is always found by subtracting 2 from x, each x corresponds to just one y, so this relation is a function. Any number can be used for x, and each x will give a number 2 smaller for y; thus, both the domain and the range are the set of real numbers, or in interval notation, $(-\infty, \infty)$.   ◻

There is a quick way to tell whether a given graph is the graph of a function. Figure 20 shows two graphs. In the graph for part (a), each value of x leads to only one value of y, so that this is the graph of a function. On the other hand, the graph in part (b) is not the graph of a function. For example, if $x = x_1$, the vertical line through x_1 intersects the graph at two points, showing that there are two values of y that correspond to this x-value. This idea is known as the *vertical line test* for a function.

VERTICAL LINE TEST

If each vertical line intersects a graph in no more than one point, the graph is the graph of a function.

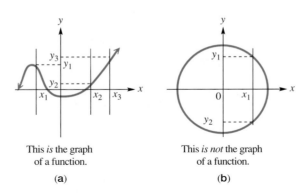

This *is* the graph   This *is not* the graph
of a function.   of a function.

(a)   **(b)**

FIGURE 20

EXAMPLE 7

Using the Vertical Line Test and Determining Domain and Range

(a)   Is the graph in Figure 21 the graph of a function? Specify the domain and the range using interval notation.

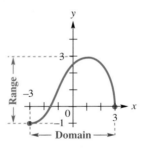

FIGURE 21

SOLUTION   The graph satisfies the vertical line test, and is therefore the graph of a function. As indicated by the annotations at the left and below the graph, the domain appears to be $[-3, 3]$ and the range appears to be $[-1, 3]$.

(b)   Assuming the graph in Figure 22 extends left and right indefinitely and upward indefinitely, does it appear to be the graph of a function? What are the domain and the range if Xscl = 1, Yscl = 1 (use observation)?

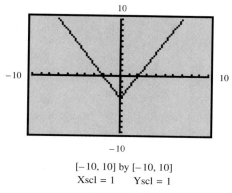

[-10, 10] by [-10, 10]
Xscl = 1   Yscl = 1

FIGURE 22

SOLUTION   It appears that no vertical line will intersect the graph more than once, so we may conclude that it is the graph of a function. Since we are told that it extends left and right indefinitely, the domain is $(-\infty, \infty)$. It appears that the lowest point on the graph has the ordered pair $(0, -4)$, and since we know that the graph extends upward indefinitely, the range appears to be the interval $[-4, \infty)$.

(Graphs that are generated by graphing calculators will not exhibit arrowheads, and thus we will need to be aware of the type of function we are observing in order to determine the domain and the range.)   ◻

TECHNOLOGICAL NOTE
Some later model calculators have the capability of using function notation (as seen in *Further Explorations* in this section). Check to see if your model has it.

Function Notation

While the concept of function is crucial to the study of mathematics, the definition of function may vary in wording from text to text. We now give an alternate definition of function that will be helpful in understanding the function notation that follows.

> **ALTERNATE DEFINITION OF FUNCTION**
>
> A function is a correspondence in which each element x from a set called the domain is paired with one and only one element y from a set called the range.

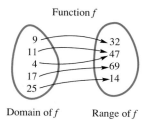

Function f

Domain of f Range of f

FIGURE 23

This idea of correspondence, or mapping, can be illustrated as shown in Figure 23, where the function $f = \{(9, 32), (11, 47), (4, 47), (17, 69), (25, 14)\}$ is depicted. In function f, the x-value 9 is paired with the y-value 32, the x-value 11 is paired with the y-value 47, and so on. Notice that each x-value is used only once in this correspondence. (There is a y-value, 47, that is used more than once, but this does not violate the definition of function.)

To say that y is a function of x means that for each value of x from the domain of the function, there is exactly one value of y. To emphasize that y *is a function of x*, or that y depends on x, it is common to write

$$y = f(x),$$

with $f(x)$ read "f of x." This notation is called **function notation.** For the function f illustrated in Figure 23, we have

$$f(9) = 32 \quad \text{because (9, 32) belongs to the correspondence,}$$
$$f(11) = 47 \quad \text{because (11, 47) belongs to the correspondence,}$$

and so on.

Function notation is used frequently when functions are defined by equations. For example, if a function is defined by the equation $y = 9x - 5$, we may name this function f and write

$$f(x) = 9x - 5.$$

Note that $f(x)$ is simply another name for y. In this function f, if $x = 2$, then we find y, or $f(2)$, by replacing x with 2.

$$f(2) = 9 \cdot 2 - 5$$
$$= 18 - 5$$
$$= 13$$

The statement "if $x = 2$, then $y = 13$" is abbreviated with function notation as

$$f(2) = 13.$$

Also, $f(0) = 9 \cdot 0 - 5 = -5$, and $f(-3) = -32$.

These ideas and the symbols used to represent them can be explained as follows.

Name of the function   Defining expression

$$y = \underbrace{f(x)}_{} = \overbrace{9x - 5}$$

Value of the function   Name of the independent variable

CAUTION The symbol $f(x)$ does *not* indicate "f times x," but represents the y-value for the indicated x-value. As shown above, $f(2)$ is the y-value that corresponds to the x-value 2.

EXAMPLE 8

Using Function Notation

In each of the following, find $f(3)$.

(a) $f(x) = 3x - 7$

SOLUTION Replace x with 3 to get

$$f(3) = 3(3) - 7$$
$$= 9 - 7$$
$$= 2.$$

This result means that the ordered pair (3, 2) belongs to the function, and this ordered pair lies on the graph of this function.

(b) function f depicted in Figure 24

SOLUTION In the correspondence shown, 3 in the domain is paired with 5 in the range, so $f(3) = 5$.

(c) the function f graphed in Figure 25

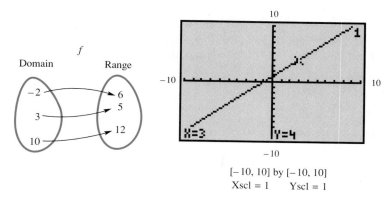

$[-10, 10]$ by $[-10, 10]$
Xscl = 1   Yscl = 1

FIGURE 24   **FIGURE 25**

SOLUTION From the information displayed at the bottom of the screen, when $x = 3$, $y = 4$, so $f(3) = 4$. ∎

EXAMPLE 9

Using Function Notation

If $f(x) = x^2 + 2x - 3$, find $f(a + 1)$ and simplify.

SOLUTION We replace x with $a + 1$ in the expression $x^2 + 2x - 3$.

$$f(x) = x^2 + 2x - 3$$
$$f(a + 1) = (a + 1)^2 + 2(a + 1) - 3$$
$$= a^2 + 2a + 1 + 2a + 2 - 3$$
$$= a^2 + 4a$$

Notice that in our simplification, we squared the expression $a + 1$ and used the distributive property to write $2(a + 1)$ as $2a + 2$. These operations are covered in algebra courses. You should be able to perform them in this course. ∎

1.2 EXERCISES

Write each of the following using interval notation, and then graph each set on the real number line.

1. $\{x \mid -1 < x < 4\}$ **2.** $\{x \mid x \geq -3\}$ **3.** $\{x \mid x < 0\}$ **4.** $\{x \mid 8 > x > 3\}$

5. $\{x \mid 1 \leq x < 2\}$ **6.** $\{x \mid -5 < x \leq -4\}$ **7.** $\{x \mid -9 > x\}$ **8.** $\{x \mid 6 \leq x\}$

Using the variable x, write each of the following using set-builder notation.

9. $(-4, 3)$ **10.** $[2, 7)$ **11.** $(-\infty, -1]$ **12.** $(3, \infty)$

13.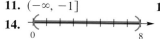

14.

15.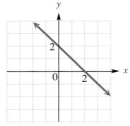

16.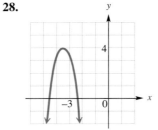

17. Explain how to determine whether a parenthesis or a square bracket is used when graphing an inequality on a number line.

18. The three-part inequality $a < x < b$ means "a is less than x and x is less than b." Which one of the following inequalities is not satisfied by some real number x?
   (a) $-3 < x < 5$ **(b)** $0 < x < 4$ **(c)** $-3 < x < -2$ **(d)** $-7 < x < -10$

Determine the domain and the range of each relation, and tell whether the relation is a function. If it is calculator-generated, assume that the graph extends indefinitely.

19. $\{(5, 1), (3, 2), (4, 9), (7, 6)\}$

20. $\{(8, 0), (5, 4), (9, 3), (3, 8)\}$

21. $\{(2, 4), (0, 2), (2, 5)\}$

22. $\{(9, -2), (-3, 5), (9, 2)\}$

23. $\{(-3, 1), (4, 1), (-2, 7)\}$

24. $\{(-12, 5), (-10, 3), (8, 3)\}$

25. $\{(1, 3), (4, 7), (0, 6), (7, 2)\}$

26. $\{(8, 5), (3, 9), (-2, 11), (5, 3)\}$

27.

28.

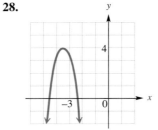

29.

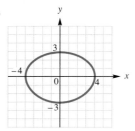

30.

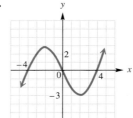

31.

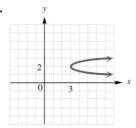

32.

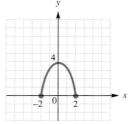

33.

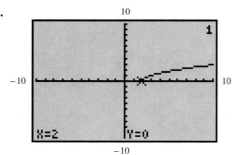

34.

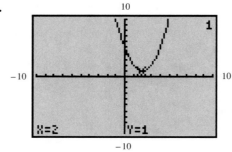

35.

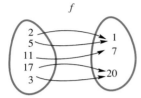

36.

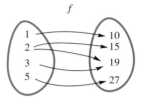

37.

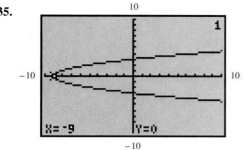

38.

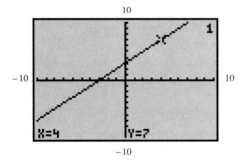

39.

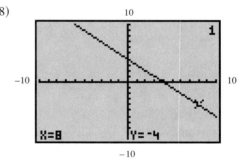

40. In your own words, explain what is meant by a function, using the terms *domain* and *range* in your explanation.

Find each of the following function values.

41. $f(3)$, if $f(x) = -2x + 9$

42. $f(6)$, if $f(x) = -2x + 8$

43. $f(0)$, if $f(x) = 4x + 8$

44. $f(0)$, if $f(x) = -8x + 1$

45. $f(11)$, for the function f in Exercise 37

46. $f(2)$, for the function f in Exercise 34

47. $f(2)$, for the function f in Exercise 33

48. $f(4)$

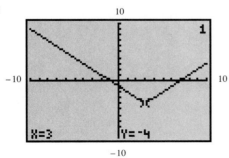

49. $f(8)$

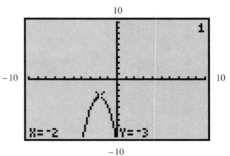

50. $f(3)$

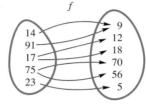

51. $f(-2)$

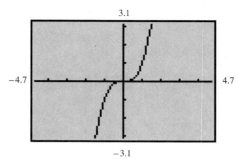

52. If f is the function graphed in Exercise 28 and $f(x) = 4$, what is x?

53. If f is the function graphed in Exercise 30, and $f(x) = 0$, what are the possible values of x?

54. Suppose that you were told the following: "For the function f graphed in Exercise 31, $f(0) = 3$ and $f(0) = -3$." What is wrong with this statement?

In Exercises 55–60, a function is given. Find the simplified form of the function value specified.

55. $f(x) = 3x - 7$; find $f(a + 2)$

56. $f(x) = -9x + 2$; find $f(t - 3)$

57. $g(x) = 2x^2 - 3x + 4$; find $g(r + 1)$

58. $g(x) = -3x^2 + 4x - 9$; find $g(s - 3)$

59. $F(x) = x$; find $F(7p + 2s - 1)$

60. $F(x) = -x$; find $F(\sqrt{2})$

Further Explorations

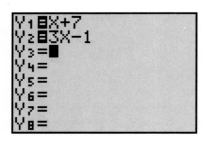

FIGURE 1

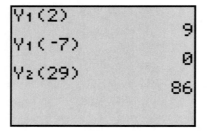

FIGURE 2

Some graphing calculators have the capability of evaluating functions using function notation. If your graphing calculator has such capabilities, enter the function in the $Y =$ menu (Figure 1), then use the home screen to evaluate the function for various values of X (Figure 2). Using the function notation feature of your graphing calculator, find each of the following function values.

1. $Y_1(4)$, if $Y_1 = -8x + 17$

2. $Y_1(-63)$, if $Y_1 = -8x + 17$

3. $Y_2(29)$, if $Y_2 = 27x - 82$

4. $Y_2(322)$, if $Y_2 = 27x - 82$

1.3 ANALYTIC AND GRAPHICAL SOLUTION OF EQUATIONS AND INEQUALITIES

Solving Linear Equations Analytically ▮ Graphical Support for Solutions of Linear Equations ▮ Solving Linear Inequalities Analytically ▮ Graphical Support for Solutions of Linear Inequalities ▮ Solving Quadratic Equations and Inequalities

Solving Linear Equations Analytically

In this text we will use two distinct approaches to equation solving. The **analytic approach** is the method you have probably seen in previous courses, where paper and pencil are used to transform complicated equations into simpler ones. In so doing, mathematical concepts are applied, and there is no reliance upon graphical representation. Most of the equations that we will encounter in this text are solvable by strictly analytic methods, and the student must realize that this approach *is not to be downplayed* just because the text is based on graphing calculator use.

The **graphical approach** is the method that distinguishes this text from most of the texts you have previously encountered. We will often *support* our analytic solutions by using graphical techniques. Occasionally we will encounter equations that are very difficult or even impossible to solve using analytic methods, and for those we may choose to present a solution that is strictly graphical in nature. It is important to realize that both analytic and graphical approaches will be used, and part of becoming a successful mathematics student is learning when to use and when not to use the two approaches.

An **equation** is a statement that two expressions are equal. Examples of equations are

$$3(2x - 4) = 7 - (x + 5) \qquad \text{and} \qquad .06x + .09(15 - x) = .07(15).$$

To **solve** an equation means to find all numbers that make the equation a true statement. Such numbers are called **solutions** or **roots** of the equation. A number that is a solution of an equation is said to **satisfy** the equation, and the solutions of an equation make up its **solution set.**

The simplest type of equation we encounter in algebra is the *linear equation.*

LINEAR EQUATION IN ONE VARIABLE

A **linear equation** in one variable is an equation that can be written in the form

$$ax + b = 0,$$

where $a \neq 0$.

One way to solve an equation is to rewrite it as a series of simpler equations, each of which has the same solution set as the original one. Such equations are said to be **equivalent equations.** These simpler equations are obtained by using the addition and multiplication properties of equality.

ADDITION AND MULTIPLICATION PROPERTIES OF EQUALITY

For real numbers a, b, and c:

$a = b$ and $a + c = b + c$ are **equivalent.** (*The same number may be added to both sides of an equation without changing the solution set.*)

If $c \neq 0$, then $a = b$ and $ac = bc$ are **equivalent.** (*Both sides of an equation may be multiplied by the same nonzero number without changing the solution set.*)

Extending the addition property of equality allows us to subtract the same number from both sides. Similarly, extending the multiplication property of equality allows us to divide both sides of an equation by the same nonzero number.

EXAMPLE 1

Solving a Linear
Equation Analytically

Solve $10 + 3(2x - 4) = 17 - (x + 5)$.

SOLUTION   Use the distributive property and then collect like terms to get the following series of simpler equivalent equations.

$$10 + 3(2x - 4) = 17 - (x + 5)$$

$$10 + 6x - 12 = 17 - x - 5 \qquad \text{Distributive property}$$

$$-2 + 7x = 12 \qquad \text{Add } x \text{ to each side; combine terms.}$$

$$7x = 14 \qquad \text{Add 2 to each side.}$$

$$x = 2 \qquad \text{Multiply both sides by } \tfrac{1}{7}.$$

An *analytic* check to determine whether 2 is indeed the solution of this equation requires that we substitute 2 for x in the original equation to see if a true statement is obtained.

$$10 + 3(2x - 4) = 17 - (x + 5) \qquad \text{Original equation}$$

$$10 + 3(2 \cdot 2 - 4) = 17 - (2 + 5) \qquad ? \quad \text{Let } x = 2.$$

$$10 + 3(4 - 4) = 17 - 7 \qquad ?$$

$$10 = 10 \qquad \text{True}$$

Since replacing x with 2 results in a true statement, 2 is the solution of the given equation. The solution set is therefore $\{2\}$. ∎

When fractions appear in an equation, our work can be made simpler by multiplying both sides by the least common denominator of all the fractions in the equation. Example 2 illustrates this type of equation.

EXAMPLE 2

Solving a Linear
Equation with
Fractional
Coefficients
Analytically

Solve $\dfrac{x + 7}{6} + \dfrac{2x - 8}{2} = -4$.

SOLUTION   Start by eliminating the fractions. Multiply both sides by 6.

$$6\left[\frac{x + 7}{6} + \frac{2x - 8}{2}\right] = 6 \cdot (-4)$$

$$6\left(\frac{x + 7}{6}\right) + 6\left(\frac{2x - 8}{2}\right) = 6(-4) \qquad \text{Distributive property}$$

$$x + 7 + 3(2x - 8) = -24$$

$$x + 7 + 6x - 24 = -24 \qquad \text{Distributive property}$$

$$7x - 17 = -24 \qquad \text{Combine terms.}$$

$$7x = -7 \qquad \text{Add 17.}$$

$$x = -1 \qquad \text{Divide by 7.}$$

Analytic check:

$$\frac{x+7}{6} + \frac{2x-8}{2} = -4 \qquad \text{Original equation}$$

$$\frac{(-1)+7}{6} + \frac{2(-1)-8}{2} = -4 \qquad ? \quad \text{Let } x = -1.$$

$$\frac{6}{6} + \frac{-10}{2} = -4 \qquad ?$$

$$1 + (-5) = -4 \qquad ?$$

$$-4 = -4 \qquad \text{True}$$

Our analytic check indicates that $\{-1\}$ is the solution set. ◧

The equations solved in Examples 1 and 2 each have a single solution. They are examples of **conditional equations.**

Graphical Support for Solutions of Linear Equations

Let us go back to the linear equation solved in Example 1. We found that the number 2 makes the equation $10 + 3(2x - 4) = 17 - (x + 5)$ a true statement. We can also look at this situation from a standpoint of functions. The statement

"Solve $10 + 3(2x - 4) = 17 - (x + 5)$"

can be re-worded as follows:

"Find the value(s) in the domain of the functions
$$f(x) = 10 + 3(2x - 4)$$
and $\quad g(x) = 17 - (x + 5)$
that give the same function value(s) (i.e., range values)."

From a graphing perspective, this can be interpreted as follows:

"Find the x-value(s) of the point(s) of intersection of the graphs of $f(x) = 10 + 3(2x - 4)$ and $g(x) = 17 - (x + 5)$."

Since the graphs of both functions are straight lines (because they are linear functions), they will intersect in a single point, no point, or infinitely many points. See Figure 26.

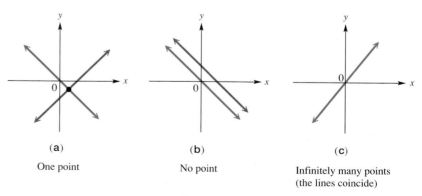

(a)
One point

(b)
No point

(c)
Infinitely many points
(the lines coincide)

FIGURE 26

Let us consider the first of these situations, as seen in Figure 26(a). To support graphically our analytic work in Example 1, we graph the lines $y_1 = 10 + 3(2x - 4)$ and $y_2 = 17 - (x + 5)$ in the same viewing window, and use the capabilities of our calculator to find the point of intersection. We see in Figure 27 that the point of intersection is $(2, 10)$. The x-coordinate here, 2, is the solution of the equation, while the y-coordinate, 10, is the value that we get when we substitute 2 for x in both of the original expressions. (See the analytic check in Example 1.)

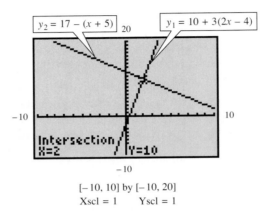

$[-10, 10]$ by $[-10, 20]$
Xscl = 1   Yscl = 1

FIGURE 27

EXAMPLE 3

Providing Visual
Support for an
Analytic Solution

Support the result of Example 2: that is,

$$\frac{x + 7}{6} + \frac{2x - 8}{2} = -4$$

has solution set $\{-1\}$, using the graphical method described above.

SOLUTION   Enter $y_1 = \frac{x + 7}{6} + \frac{2x - 8}{2}$ and $y_2 = -4$. See Figure 28. The two straight lines intersect at $(-1, -4)$, providing support that $\{-1\}$ is the solution set. Remember, if

$$(-1, -4)$$

is the point of intersection , -1 is the solution, and -4 is the value obtained when -1 is substituted into both expressions.   ■

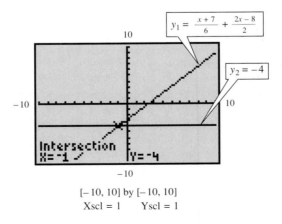

$[-10, 10]$ by $[-10, 10]$
Xscl = 1   Yscl = 1

FIGURE 28

We now state the first of two methods of equation solving by graphical methods that we will use in this book.

INTERSECTION-OF-GRAPHS METHOD OF GRAPHICAL SOLUTION

To solve the equation

$$f(x) = g(x)$$

graphically, graph $y_1 = f(x)$ and $y_2 = g(x)$. The x-coordinate of any point of intersection of the two graphs is a solution of the equation.

At this point we must make two important observations. First, since we are illustrating *linear* and *quadratic* functions in this introductory chapter, our examples and exercises will consist of functions whose graphs are straight lines and parabolas. The intersection-of-graphs method of solving equations will be applied to such functions now, but we will, in later chapters, apply it to the circular functions.

Second, we should realize that the most modern graphing calculators have the capability of determining the coordinates of the point of intersection of two graphs to a great degree of accuracy, and will in many cases give the exact decimal values. Of course, if a coordinate is an irrational number, the decimal shown will be only an approximation. Figure 29(a) shows the former case, while Figure 29(b) shows the latter.

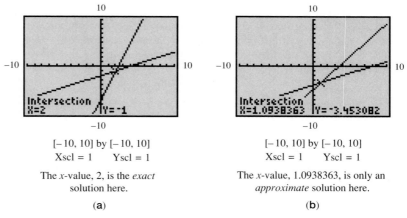

[−10, 10] by [−10, 10]
Xscl = 1   Yscl = 1

The x-value, 2, is the *exact* solution here.

(a)

[−10, 10] by [−10, 10]
Xscl = 1   Yscl = 1

The x-value, 1.0938363, is only an *approximate* solution here.

(b)

FIGURE 29

TECHNOLOGICAL NOTE
The words *root* and *solution* mean the same thing. If there is one thing that you should remember when studying methods of solving equations, it is this: *The real solutions (or roots) of an equation of the form f(x) = 0 correspond to the x-intercepts of the graph of y = f(x).* For this reason, modern graphing calculators are programmed so that *x*-intercepts (roots) can be located. (One new model uses the term *zero*, as in "zero of a function.")

There is a second method of graphical support for solving equations. Suppose that we once again wish to solve

$$f(x) = g(x).$$

By subtracting $g(x)$ from both sides, we obtain $f(x) - g(x) = 0$.

Notice that $f(x) - g(x)$ is simply a function itself. Let us call it $F(x)$. Then we only need to solve

$$F(x) = 0$$

to obtain the solution set of the original equation. From algebra, we know that any number that satisfies this equation is an x-intercept of the graph of $y = F(x)$. Using this idea, we now state another method of solving an equation graphically.

> ### x-INTERCEPT METHOD OF GRAPHICAL SOLUTION
>
> To solve the equation
>
> $$f(x) = g(x)$$
>
> graphically, graph $y = f(x) - g(x) = F(x)$. Any x-intercept of the graph of $y = F(x)$ is a solution of the equation.

The x-intercept method is used in the next example.

EXAMPLE 4

Providing Visual Support for an Analytic Solution

It can be shown analytically that the solution set of

$$6x - 4(3 - 2x) = 5(x - 4) - 10$$

is $\{-2\}$. Use the x-intercept method of graphical solution to support this result.

SOLUTION   Begin by letting

$$f(x) = 6x - 4(3 - 2x)$$

and

$$g(x) = 5(x - 4) - 10.$$

Then, find $f(x) - g(x)$ and enter it as y_1 in a calculator.

$$y_1 = f(x) - g(x)$$

$$= 6x - 4(3 - 2x) - \left(5(x - 4) - 10\right)$$

Graph this function to get the straight line shown in Figure 30. By choosing a standard viewing window, we see that the x-intercept is -2, supporting the information given in the statement of the problem.   ◼

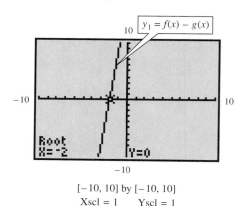

$[-10, 10]$ by $[-10, 10]$
Xscl = 1   Yscl = 1

The x-intercept, -2, is the **solution** (or **root**) of the equation $f(x) = g(x)$.

FIGURE 30

CAUTION   When using the x-intercept method of graphical solution, as in Example 4, it is a common error to forget to enter symbols of inclusion around the expression that is being subtracted. Notice how we used parentheses around $g(x) = 5(x - 4) - 10$ when we determined the expression for y_1. Graphing calculator technology provides us a new respect for the need to use symbols of inclusion correctly!

> ┌─ FOR **GROUP DISCUSSION** ─┐
>
> Repeat Example 4, but this time, rather than graphing
> $$y_1 = f(x) - g(x),$$
> graph
> $$y_2 = g(x) - f(x).$$
> Observe the graph of y_2. Does it have the same x-intercept as y_1? Can you make a conjecture concerning the order in which the two functions are subtracted when using the x-intercept method of solution?

As stated earlier, we can use graphical methods to find approximate solutions of equations that, for one reason or another, are difficult to solve algebraically. In this book, we will occasionally require approximate solutions. We now state a method of rounding decimal numbers to a particular place value.

TECHNOLOGICAL NOTE
Consult your owner's manual to see how to set the number of decimal places your calculator will display.

RULES FOR ROUNDING

To round a number to a place value to the right of the decimal point:

Step 1   Locate the **place** to which the number is being rounded.

Step 2   Look at the next **digit to the right** of the place to which the number is being rounded.

Step 3A   If this digit is **less than 5,** drop all digits to the right of the place to which the number is being rounded. Do *not change* the digit in the place to which the number is being rounded.

Step 3B   If this digit is **5 or greater,** drop all digits to the right of the place to which the number is being rounded. *Add 1* to the digit in the place to which the number is being rounded.

To round a number to a place value to the left of the decimal point:

Step 1   Same as above.

Step 2   Same as above.

Step 3A   If this digit is **less than 5,** do *not change* the digit, and replace the digits to the right with zeros.

Step 3B   If this digit is **5 or greater,** *add 1* to it and replace all digits to the right with zeros.

If either situation requires that 1 be added to 9, replace the 9 with a 0 and add 1 to the digit to the left of 9.

Solving Linear Inequalities Analytically

An equation says that two expressions are equal, while an **inequality** says that one expression is greater than, greater than or equal to, less than, or less than or equal to, another. As with equations, a value of the variable for which the inequality is true is a solution of the inequality, and the set of all such solutions is the solution set of the inequality. Two inequalities with the same solution set are **equivalent inequalities.**

Inequalities are solved with the following properties of inequality.

PROPERTIES OF INEQUALITY

For real numbers a, b, and c:

a. $a < b$ and $a + c < b + c$ are equivalent.
 (The same number may be added to both sides of an inequality without changing the solution set.)

b. If $c > 0$, then $a < b$ and $ac < bc$ are equivalent.
 (Both sides of an inequality may be multiplied by the same positive number without changing the solution set.)

c. If $c < 0$, then $a < b$ and $ac > bc$ are equivalent.
 (Both sides of an inequality may be multiplied by the same negative number without changing the solution set, as long as the direction of the inequality symbol is reversed.)

Replacing $<$ with $>$, $\leq$, or $\geq$ results in similar properties.

NOTE Because division is defined in terms of multiplication, the word "multiplied" may be replaced by "divided" in parts (b) and (c) of the properties of inequality. Similarly, in part (a) the words "added to" may be replaced with "subtracted from."

Part (c) of the properties of inequality requires that we remember the following important rule.

When multiplying or dividing both sides of an inequality by a negative number, we must reverse the direction of the inequality symbol to obtain an equivalent inequality.

A linear inequality in one variable is defined in a way similar to a linear equation in one variable.

LINEAR INEQUALITY IN ONE VARIABLE

A **linear inequality** in one variable is an inequality that can be written in one of the following forms, where $a \neq 0$:

$$ax + b > 0 \qquad ax + b < 0$$
$$ax + b \geq 0 \qquad ax + b \leq 0.$$

We solve a linear inequality analytically using the same steps as those used to solve a linear equation.

EXAMPLE 5

Solving a Linear
Inequality
Analytically

Solve the inequality $3x - 2(2x + 6) \leq 2(x + 3)$. Express the solution set using interval notation.

SOLUTION

$$3x - 2(2x + 6) \leq 2(x + 3)$$

$$3x - 4x - 12 \leq 2x + 6 \qquad \text{Distributive property}$$

$$-x - 12 \leq 2x + 6 \qquad \text{Combine like terms.}$$

$$-3x \leq 18 \qquad \begin{array}{l}\text{Subtract } 2x \text{ and add 12;} \\ \text{combine like terms.}\end{array}$$

$$x \geq -6 \qquad \begin{array}{l}\text{Divide by } -3 \text{ and reverse the} \\ \text{direction of the inequality} \\ \text{symbol.}\end{array}$$

The solution set is $[-6, \infty)$.   ▯

Graphical Support for Solutions of Linear Inequalities

Earlier in this section we learned two methods of graphical support for solutions of equations. We will now extend these methods to solutions of inequalities. Suppose that two linear functions f and g are graphed, as shown in Figure 31, and the equation $f(x) = g(x)$ is conditional. Then, according to the figure, the solution set is $\{x_1\}$, since by applying the intersection-of-graphs method, it is the x-coordinate of the point of intersection of the two lines.

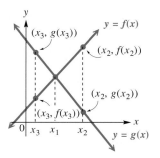

FIGURE 31

Notice that if we choose any x value *less than* x_1, such as x_3 in the figure, the point $(x_3, f(x_3))$ is *below* the point $(x_3, g(x_3))$ and on the same vertical line, indicating that $f(x_3) < g(x_3)$. Similarly, if we choose any x value *greater than* x_1, such as x_2 in the figure, the point $(x_2, f(x_2))$ is *above* the point $(x_2, g(x_2))$, indicating that $f(x_2) > g(x_2)$. This discussion leads to the following extension of the intersection-of-graphs method of solution of equations.

**INTERSECTION-OF-GRAPHS METHOD OF SOLUTION
OF LINEAR INEQUALITIES**

Suppose that f and g are linear functions. The solution set of $f(x) > g(x)$ is the set of all real numbers x such that the graph of f is *above* the graph of g. The solution set of $f(x) < g(x)$ is the set of all real numbers x such that the graph of f is *below* the graph of g.

If an inequality involves one of the symbols $\geq$ or $\leq$, the same method is applied, with the solution of the corresponding equation included in the solution set. This is summarized as follows.

SPECIFYING INTERVALS OF SOLUTION FOR LINEAR INEQUALITIES

If f and g are linear functions, and $f(x) = g(x)$ has a single solution, k, the solution set of

$$f(x) > g(x) \qquad \text{or} \qquad f(x) < g(x)$$

will be of the form (k, ∞) or $(-\infty, k)$, with the endpoint of the interval not included. On the other hand, the solution set of

$$f(x) \geq g(x) \qquad \text{or} \qquad f(x) \leq g(x)$$

will be of the form $[k, \infty)$ or $(-\infty, k]$, with the endpoint of the interval included.

EXAMPLE 6

Providing Graphical Support for an Analytic Solution

The inequality

$$3x - 2(2x + 6) \leq 2(x + 3),$$

solved in Example 5, has solution set $[-6, \infty)$. Support this result graphically.

SOLUTION   Start by entering the left side as y_1 and the right side as y_2.

$$y_1 = 3x - 2(2x + 6)$$
$$y_2 = 2(x + 3)$$

TECHNOLOGICAL NOTE
When supporting the solution set in Example 6 using graphs, the calculator will not determine whether the endpoint of the interval is included or excluded. This must be done by looking at the inequality symbol in the given inequality.

The graph, shown in Figure 32, indicates that the point of intersection of the two lines is $(-6, -6)$. The x-coordinate, -6, gives the included endpoint of the solution set of the inequality. Because the graph of y_1 is *below* the graph of y_2 when x is *greater than* -6, our solution set of $[-6, \infty)$ is supported.   ◼

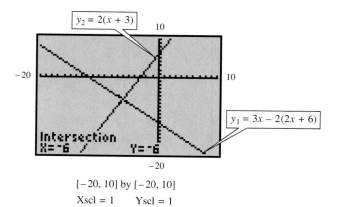

[−20, 10] by [−20, 10]
Xscl = 1   Yscl = 1

FIGURE 32

CAUTION   It is just coincidental that the y-value of the point of intersection of the graphs in Figure 32 is the same as the x-value (that is, -6). When determining the solution set of either an equation or an inequality using the intersection-of-graphs method, we are interested in finding the x-*values* (domain values of the functions) that give the same y-values (range values).

Since there is another method of providing graphical support for the solution of a linear equation (the x-intercept method), it too can be extended to linear inequalities. Suppose that we wish to solve

$$f(x) > g(x).$$

Subtracting $g(x)$ from both sides gives

$$f(x) - g(x) > 0.$$

If we call the expression on the left side $F(x)$, we are interested in solving

$$F(x) > 0.$$

We know that $F(x) = 0$ has as its solution the x-intercept of the graph of $y = F(x)$. Therefore, all solutions of $F(x) > 0$ will be the x-values of the points *above* the point at which the graph intersects the x-axis. Similarly, all solutions of $F(x) < 0$ will be the x-values of the points *below* the point at which the graph intersects the x-axis.

---

x-INTERCEPT METHOD OF SOLUTION OF LINEAR INEQUALITIES

The solution set of $F(x) > 0$ is the set of all real numbers x such that the graph of F is *above* the x-axis. The solution set of $F(x) < 0$ is the set of all real numbers x such that the graph of F is *below* the x-axis.

Figure 33 illustrates this discussion, and summarizes the solution sets for the appropriate inequalities.

TECHNOLOGICAL NOTE
If two functions defined by Y1 and Y2 are already entered into your calculator, you can enter Y3 as Y2 − Y1. Then if you direct the calculator to graph Y3 only, you can solve the equation Y1 = Y2 by finding the x-intercept of Y3. Consult your owner's manual to see how this is accomplished. It will save you a lot of time and effort.

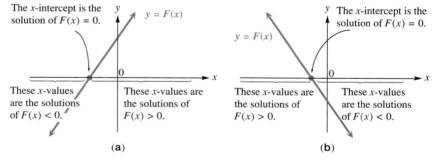

FIGURE 33

EXAMPLE 7

Providing Graphical Support for an Analytic Solution

Let $f(x) = -2(3x + 1)$ and $g(x) = 4(x + 2)$. If we let $y_1 = f(x)$ and $y_2 = g(x)$ and graph $y_1 - y_2$, we obtain the graph shown in Figure 34.

(a) Use the graph to solve the *equation $f(x) = g(x)$*, that is,

$$-2(3x + 1) = 4(x + 2).$$

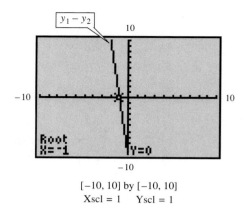

$[-10, 10]$ by $[-10, 10]$
Xscl = 1   Yscl = 1

Figure 34

SOLUTION   The x intercept of $y_1 - y_2$ is -1, so the solution set of this *equation* is $\{-1\}$.

(b) Use the graph to solve the *inequality* $f(x) < g(x)$, that is,

$$-2(3x + 1) < 4(x + 2).$$

SOLUTION   Because the symbol is $<$, we want to find the x-values of the points *below* the x-axis. All such points are to the *right* of $x = -1$, leading to the solution set $(-1, \infty)$.

(c) Use the graph to solve the *inequality* $f(x) > g(x)$, that is,

$$-2(3x + 1) > 4(x + 2).$$

SOLUTION   Because the symbol is $>$, this time we choose the x-values of the points *above* the x-axis. All such points are to the *left* of $x = -1$, leading to the solution set $(-\infty, -1)$.

(d) Use the results of parts (b) and (c) to express the solution sets of

$$-2(3x + 1) \leq 4(x + 2) \qquad \text{and} \qquad -2(3x + 1) \geq 4(x + 2).$$

SOLUTION   Notice here that the only difference is that we now have the symbols that include equality. Therefore, the solution sets are

$$[-1, \infty) \qquad \text{and} \qquad (-\infty, -1], \quad \text{respectively.} \qquad \blacksquare$$

FOR **GROUP DISCUSSION**

One of the authors of this text was once told the following fact by Steele Andrews, a wonderful man who taught him in graduate school:

"Equality is the boundary between less than and greater than."

Use Example 7 to explain this simple, but important, concept.

Solving Quadratic Equations and Inequalities

We will now extend our methods to quadratic equations and inequalities.

QUADRATIC EQUATION IN ONE VARIABLE

An equation that can be written in the form

$$ax^2 + bx + c = 0,$$

where a, b, and c are real numbers with $a \neq 0$, is a **quadratic equation in standard form.**

We are often interested in solving quadratic equations of the form $x^2 = k$, where k is a real number. This type of equation can be solved by factoring using the following sequence of equivalent equations.

$$x^2 = k$$
$$x^2 - k = 0$$
$$(x - \sqrt{k})(x + \sqrt{k}) = 0$$
$$x - \sqrt{k} = 0 \quad \text{or} \quad x + \sqrt{k} = 0$$
$$x = \sqrt{k} \quad \text{or} \quad x = -\sqrt{k}$$

We have proved the following statement, which we will call the square root property for solving quadratic equations. Recall from algebra that $i = \sqrt{-1}$.

SQUARE ROOT PROPERTY FOR SOLVING QUADRATIC EQUATIONS

The solution set of $x^2 = k$ is

a. $\{\pm\sqrt{k}\}$ if $k > 0$
b. $\{0\}$ if $k = 0$
c. $\{\pm i\sqrt{|k|}\}$ if $k < 0$.

As shown in Figure 35, the graph of $y_1 = x^2$ intersects the graph of $y_2 = k$ twice if $k > 0$, once if $k = 0$, and not at all if $k < 0$.

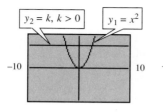
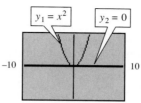
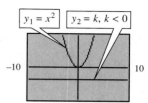

There are two points of intersection if $k > 0$.	There is only one point of intersection (the origin) if $k = 0$.	There are no points of intersection if $k < 0$.
(a)	**(b)**	**(c)**

FIGURE 35

EXAMPLE 8

Using the Square Root Property

Solve each of the following quadratic equations.

(a) $x^2 = 7$

SOLUTION   Since $7 > 0$, there will be two real solutions.

$$x^2 = 7$$
$$x = \pm\sqrt{7}$$

This result may be supported graphically by using the intersection-of-graphs method. If we graph $y_1 = x^2$ and $y_2 = 7$ in a standard viewing window, and then locate the points of intersection, we will find that the x-coordinates are approximately -2.65 and 2.65, which are approximations for $\pm\sqrt{7}$. See Figure 36. The solution set is $\{\pm\sqrt{7}\}$.

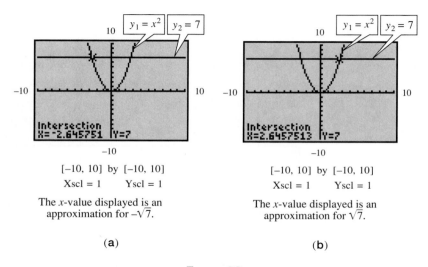

$[-10, 10]$ by $[-10, 10]$
Xscl = 1   Yscl = 1

The x-value displayed is an approximation for $-\sqrt{7}$.

(a)

$[-10, 10]$ by $[-10, 10]$
Xscl = 1   Yscl = 1

The x-value displayed is an approximation for $\sqrt{7}$.

(b)

FIGURE 36

(b) $x^2 = -5$

SOLUTION   There is no real number whose square is -5. However, this equation has two complex imaginary solutions.

$$x^2 = -5$$
$$x = \pm\sqrt{-5}$$
$$x = \pm i\sqrt{5}$$

The solution set is $\{\pm i\sqrt{5}\}$. Notice that the graphs of $y_1 = x^2$ and $y_2 = -5$ do not intersect. This indicates that there are no *real* solutions. See Figure 37.   ■

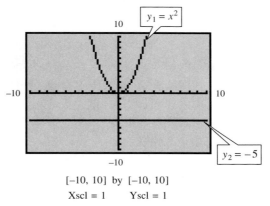

$$[-10, 10] \text{ by } [-10, 10]$$
$$\text{Xscl} = 1 \qquad \text{Yscl} = 1$$

There are no points of intersection,
and thus no *real* solutions.

FIGURE 37

TECHNOLOGICAL NOTE
Programs for the various
makes and models of
graphing calculators are
available from the manu-
facturers, users groups,
etc. You may also be in-
terested in learning to
program yourself. A pro-
gram for the quadratic
formula is almost an es-
sential if you expect to
use your calculator to its
maximum potential. Some
of the more advanced
models even have polyno-
mial solver functions that
will solve polynomial functions of degree 2 and
greater, and thus these
have a "built-in" quadratic
formula.

In algebra we learn that some quadratic equations may be solved by factoring. However, a more powerful method involves the *quadratic formula.* Its derivation can be found in any standard college algebra text.

> **QUADRATIC FORMULA**
>
> The solutions of the quadratic equation $ax^2 + bx + c = 0$, where $a \neq 0$, are
>
> $$x = \frac{-b \pm \sqrt{b^2 - 4ac}}{2a}.$$

CAUTION Notice that the fraction bar in the quadratic formula extends under the $-b$ term in the numerator.

The expression under the radical in the quadratic formula, $b^2 - 4ac$, is called the **discriminant.** The value of the discriminant determines whether the quadratic equation has two real solutions, one real solution, or no real solutions. In the latter case, there will be two imaginary solutions.

EXAMPLE 9

Using the Quadratic Formula

Solve the equation

$$x(x - 2) = 2x - 2$$

using the quadratic formula, and support your solutions graphically using the intersection-of-graphs method.

SOLUTION Before we can apply the quadratic formula, we must rewrite the equation in the form $ax^2 + bx + c = 0$.

$x(x - 2) = 2x - 2$	Given equation
$x^2 - 2x = 2x - 2$	Distributive property
$x^2 - 4x + 2 = 0$	Subtract 2x and add 2 on both sides.

Here $a = 1$, $b = -4$, and $c = 2$. Substitute these values into the quadratic formula to get

$$x = \frac{-b \pm \sqrt{b^2 - 4ac}}{2a}$$

$$= \frac{-(-4) \pm \sqrt{(-4)^2 - 4(1)2}}{2(1)} \qquad a = 1, b = -4, c = 2$$

$$= \frac{4 \pm \sqrt{16 - 8}}{2}$$

$$= \frac{4 \pm \sqrt{8}}{2} \qquad \text{The discriminant, 8,} \\ \text{is positive, so there} \\ \text{are two real solutions.}$$

$$= \frac{4 \pm 2\sqrt{2}}{2} \qquad \sqrt{16 - 8} = \sqrt{8} = 2\sqrt{2}$$

$$= \frac{2(2 \pm \sqrt{2})}{2} \qquad \text{Factor out a 2 in the numerator.}$$

$$= 2 \pm \sqrt{2} \qquad \text{Lowest terms}$$

The solution set is $\{2 + \sqrt{2}, 2 - \sqrt{2}\}$, abbreviated as $\{2 \pm \sqrt{2}\}$.

We can support our solution graphically by considering the graphs of $y_1 = x(x - 2)$ and $y_2 = 2x - 2$. (Note that the original form of our equation is $y_1 = y_2$.) By using the capabilities of the calculator, we can find that the x-coordinates of the points of intersection are approximately .59 and 3.41. (See Figure 38.) These are also approximations of $2 - \sqrt{2}$ and $2 + \sqrt{2}$, supporting our results obtained by the quadratic formula.   ◘

NOTE   For the equation in Example 9,

$$2 - \sqrt{2} \quad \text{and} \quad 2 + \sqrt{2} \qquad \text{are the } \textit{exact} \text{ solutions;}$$

$$.59 \quad \text{and} \quad 3.41 \qquad \text{are } \textit{approximations} \text{ of the exact solutions.}$$

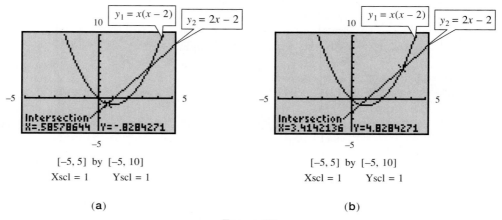

[-5, 5] by [-5, 10]   [-5, 5] by [-5, 10]
Xscl = 1   Yscl = 1   Xscl = 1   Yscl = 1

(a)   (b)

FIGURE 38

We know that the solution set of $P(x) = 0$ is determined by the x-intercepts of the graph of P. The solution set of $P(x) < 0$ consists of all values in the domain of P (that is, x-values) such that the graph of P lies *below* the x-axis. And the solution set of $P(x) > 0$ consists of all values in the domain such that the graph of P lies *above* the x-axis. If P is a quadratic function, then these statements may be summarized in the following diagram.

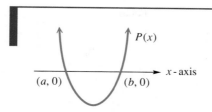

Solution Set of	is
$P(x) = 0$	$\{a, b\}$
$P(x) < 0$	the interval (a, b)
$P(x) > 0$	$(-\infty, a) \cup (b, \infty)$

Solution Set of	is
$P(x) = 0$	$\{a, b\}$
$P(x) < 0$	$(-\infty, a) \cup (b, \infty)$
$P(x) > 0$	the interval (a, b)

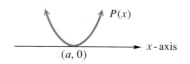

Solution Set of	is
$P(x) = 0$	$\{a\}$
$P(x) < 0$	$\emptyset$
$P(x) > 0$	$(-\infty, a) \cup (a, \infty)$

Solution Set of	is
$P(x) = 0$	$\{a\}$
$P(x) < 0$	$(-\infty, a) \cup (a, \infty)$
$P(x) > 0$	$\emptyset$

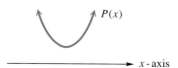

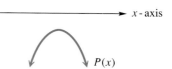

$P(x) = 0$ has no *real* solutions, but two complex imaginary solutions. Real solution set of $P(x) < 0$ is $\emptyset$. Real solution set of $P(x) > 0$ is $(-\infty, \infty)$.

$P(x) = 0$ has no *real* solutions, but two complex imaginary solutions. Real solution set of $P(x) < 0$ is $(-\infty, \infty)$. Real solution set of $P(x) > 0$ is $\emptyset$.

Suppose that the graph of a quadratic polynomial intersects the x-axis in two points. Then the two solutions of the polynomial *equation* divide the real number line (x-axis) into three intervals. Within each interval, the polynomial is either always positive or always negative.

The third line in the solution of the equation in Example 9 indicates that an equivalent equation is $x^2 - 4x + 2 = 0$. Use this information as you read Example 10.

EXAMPLE **10**
Solving a Quadratic Inequality Graphically

Use the graph in Figure 39 and the result of Example 9 to solve the following inequalities for intervals with *exact values* at endpoints.

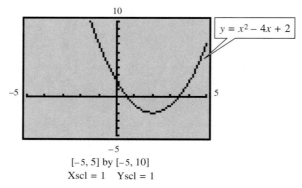

[−5, 5] by [−5, 10]
Xscl = 1   Yscl = 1

The *exact* x-intercepts are $2 - \sqrt{2}$ and
$2 + \sqrt{2}$. The solution set of $x^2 - 4x + 2 \leq 0$
is $[2 - \sqrt{2}, 2 + \sqrt{2}\,]$

FIGURE 39

(a) $x^2 - 4x + 2 \leq 0$

SOLUTION   We found by the quadratic formula that the exact solutions of $x^2 - 4x + 2 = 0$ are $2 - \sqrt{2}$ and $2 + \sqrt{2}$. Since the graph of $y = x^2 - 4x + 2$ lies below or intersects the x-axis between or at these values (as shown in Figure 39), the solution set for this inequality is the closed interval $[2 - \sqrt{2}, 2 + \sqrt{2}]$.

(b) $x^2 - 4x + 2 \geq 0$

SOLUTION   The graph lies above or intersects the x-axis to the left of and including $2 - \sqrt{2}$, and to the right of and including $2 + \sqrt{2}$. Therefore, the solution set is $(-\infty, 2 - \sqrt{2}\,] \cup [2 + \sqrt{2}, \infty)$.   ∎

FOR GROUP DISCUSSION

1.   The function $P(x) = x^2 - x - 20$ has integer zeros. Graph the function in a window that will show both x-intercepts. Then solve each of the following:

  a. $x^2 - x - 20 = 0$ **b.** $x^2 - x - 20 < 0$ **c.** $x^2 - x - 20 > 0$.

2.   The function $P(x) = x^2 + x + 20$ has no real zeros. Graph the function in a window that will show the vertex and the y-intercept. (Why won't the standard window work?) Then solve each of the following:

  a. $x^2 + x + 20 = 0$ **b.** $x^2 + x + 20 < 0$ **c.** $x^2 + x + 20 > 0$.
  (Give only real solutions.)

1.3 EXERCISES

In Exercises 1–6 two linear functions, y_1 and y_2, are graphed in a viewing window with the point of intersection of the graphs given in the display at the bottom. Using the intersection-of-graphs method of graphical solution, give the solution set of $y_1 = y_2$.

1.

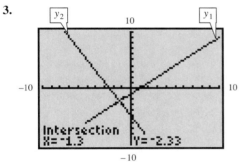

2.

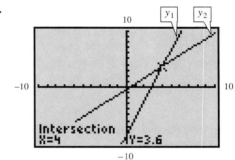

3.

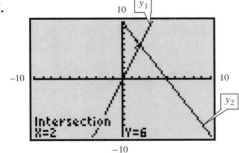

4.

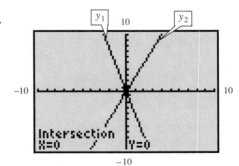

5.

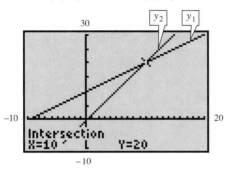

6.

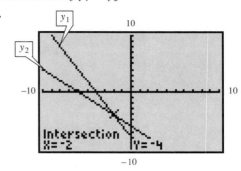

7. In Exercises 1–6 above, give the interpretation of the *y*-value shown at the bottom of the display.

8. If $y_1 = f(x)$ and $y_2 = g(x)$, and the solution set of $y_1 = y_2$ is $\{-4\}$, what is the value of $f(-4) - g(-4)$? How does your answer relate to the *x*-intercept method of graphical solution of equations?

In Exercises 9–12, linear functions y_1 and y_2 have been defined and then the graph of $y_1 - y_2$ has been graphed in an appropriate viewing window. Use the x-intercept method of graphical solution to solve the equation $y_1 = y_2$.

9.

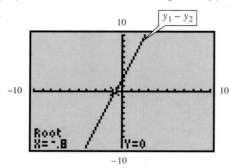

10.

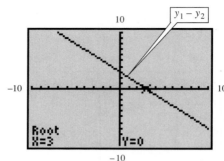

11.

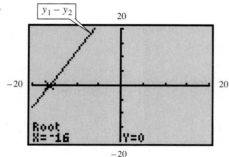

12.

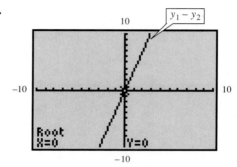

Solve each equation analytically. Check it analytically by direct substitution, and then support your solution graphically.

13. $2x - 5 = x + 7$

14. $9x - 17 = 2x + 4$

15. $.01x + 3.1 = 2.03x - 2.96$

16. $.04x + 2.1 = .02x + 1.92$

17. $-(x + 5) - (2 + 5x) + 8x = 3x - 5$

18. $-(8 + 3x) + 5 = 2x + 3$

19. $\dfrac{2x + 1}{3} + \dfrac{x - 1}{4} = \dfrac{13}{2}$

20. $\dfrac{x - 2}{4} + \dfrac{x + 1}{2} = 1$

Refer to the graphs of the linear functions $y_1 = f(x)$ and $y_2 = g(x)$ in the figure to find the solution set of each equation or inequality in Exercises 21–32. (Remember that the solution sets consist of x-values.)

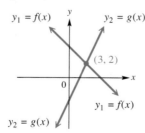

21. $f(x) = g(x)$

22. $y_1 = y_2$

23. $f(x) > g(x)$

24. $y_1 > y_2$

25. $y_1 < y_2$

26. $f(x) < g(x)$

27. $g(x) \geq f(x)$

28. $y_2 \geq y_1$

29. $f(x) - g(x) = 0$

30. $y_1 - y_2 = 0$

31. $g(x) - f(x) = 0$

32. $y_2 - y_1 = 0$

In Exercises 33–36, refer to the graph of the linear function $y = f(x)$ *to solve the inequalities specified in parts (a)–(d). Express solution sets in interval notation.*

33. (a) $f(x) > 0$
  (b) $f(x) < 0$
  (c) $f(x) \geq 0$
  (d) $f(x) \leq 0$

34. (a) $f(x) < 0$
  (b) $f(x) \leq 0$
  (c) $f(x) \geq 0$
  (d) $f(x) > 0$

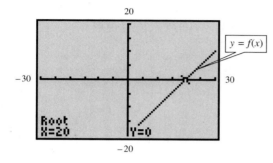

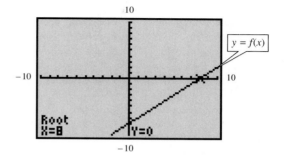

35. (a) $f(x) \leq 0$
  (b) $f(x) > 0$
  (c) $f(x) < 0$
  (d) $f(x) \geq 0$

36. (a) $f(x) < 0$
  (b) $f(x) > 0$
  (c) $f(x) \leq 0$
  (d) $f(x) \geq 0$

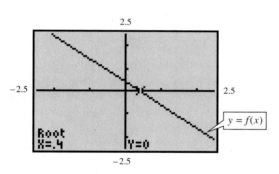

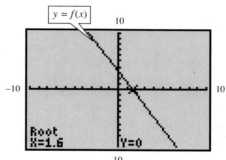

Solve each inequality analytically, giving the solution set using interval notation. Support your answer graphically. (Hint: Once part (a) is done, part (b) follows from the answer to part (a).)

37. (a) $10x + 5 - 7x \geq 8(x + 2) + 4$
  (b) $10x + 5 - 7x < 8(x + 2) + 4$

38. (a) $6x + 2 + 10x > -2(2x + 4) + 10$
  (b) $6x + 2 + 10x \leq -2(2x + 4) + 10$

39. (a) $x + 2(-x + 4) - 3(x + 5) < -4$
  (b) $x + 2(-x + 4) - 3(x + 5) \geq -4$

40. (a) $-11x - (6x - 4) + 5 - 3x \leq 1$
  (b) $-11x - (6x - 4) + 5 - 3x > 1$

41. (a) $\dfrac{1}{3}x - \dfrac{1}{5}x \leq 2$

  (b) $\dfrac{1}{3}x - \dfrac{1}{5}x > 2$

42. (a) $\dfrac{3x}{2} + \dfrac{4x}{7} \geq -5$

  (b) $\dfrac{3x}{2} + \dfrac{4x}{7} < -5$

43. (a) $\dfrac{x - 2}{2} - \dfrac{x + 6}{3} > -4$

  (b) $\dfrac{x - 2}{2} - \dfrac{x + 6}{3} \leq -4$

44. (a) $\dfrac{2x + 3}{5} - \dfrac{3x - 1}{2} < \dfrac{4x + 7}{2}$

  (b) $\dfrac{2x + 3}{5} - \dfrac{3x - 1}{2} \geq \dfrac{4x + 7}{2}$

Find all solutions, both real and imaginary, of the following quadratic equations.
For the equations with real solutions, support your answers graphically.

45. $x^2 = 16$ **46.** $x^2 = 144$ **47.** $3x^2 = 27$ **48.** $2x^2 = 48$

49. $x^2 = -16$ **50.** $x^2 = -100$ **51.** $x^2 = 2$ **52.** $x^2 = 3$

Solve each of the following equations by the quadratic formula. Find all solutions, both real and imaginary. If the equation is not in the form $P(x) = 0$, you will need to write it this way in order to identify a, b, and c. For the equations with real solutions, support your answers graphically by using the x-intercept method, graphing $y = P(x)$, and then locating those intercepts.

53. $x^2 - 8x + 15 = 0$ **54.** $x^2 + 5x - 6 = 0$ **55.** $x^2 - 2x - 4 = 0$

56. $x^2 + 8x + 13 = 0$ **57.** $2x^2 + 2x = -1$ **58.** $9x^2 - 12x = -8$

59. $x(x - 1) = 1$ **60.** $x(x - 3) = 2$ **61.** $x^2 - 5x = x - 7$

62. $11x^2 - 3x + 2 = 4x + 1$ **63.** $4x^2 - 12x = -11$ **64.** $x^2 = 2x - 5$

Exercises 65–72 refer to the graphs of the quadratic functions f, g, and h shown here.

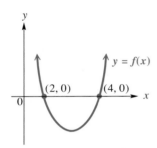   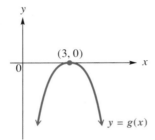   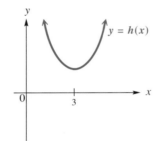

65. What is the solution set of $f(x) = 0$?

66. What is the solution set of $f(x) < 0$?

67. What is the solution set of $f(x) > 0$?

68. What is the solution set of $g(x) = 0$?

69. What is the solution set of $g(x) < 0$?

70. What is the solution set of $g(x) > 0$?

71. Solve $h(x) > 0$.

72. Solve $h(x) < 0$.

*Suppose that $P(x)$ defines a quadratic polynomial. We will call $P(x) = 0$ the **associated quadratic equation** for the inequalities $P(x) > 0$, $P(x) \geq 0$, $P(x) < 0$, and $P(x) \leq 0$. In Exercises 73–80, we give two quadratic inequalities. Solve the associated quadratic equation using the quadratic formula, and then use a calculator-generated graph to express the solution sets of the inequalities given. Give exact values for endpoints.*

73. (a) $x^2 + 4x + 3 \geq 0$ **74. (a)** $x^2 + 6x + 8 < 0$
  (b) $x^2 + 4x + 3 < 0$ **(b)** $x^2 + 6x + 8 \geq 0$

75. (a) $2x^2 - 9x + 4 > 0$ **76. (a)** $3x^2 + 13x + 10 \leq 0$
  (b) $2x^2 - 9x + 4 \leq 0$ **(b)** $3x^2 + 13x + 10 > 0$

77. (a) $-x^2 + 2x + 1 \geq 0$ **78. (a)** $-x^2 - 5x + 2 > 0$
  (b) $-x^2 + 2x + 1 < 0$ **(b)** $-x^2 - 5x + 2 \leq 0$

79. (a) $4x^2 + 3x + 1 \leq 0$ **80. (a)** $-3x^2 + x - 2 > 0$
  (b) $4x^2 + 3x + 1 > 0$ **(b)** $-3x^2 + x - 2 \leq 0$

1.4 ANALYSIS OF GRAPHS OF ELEMENTARY FUNCTIONS AND RELATIONS

Continuity; Increasing and Decreasing Functions ▮ The Identity
Function ▮ The Squaring Function and Symmetry with Respect to the
y-Axis ▮ The Cubing Function and Symmetry with Respect to the
Origin ▮ The Square Root and Cube Root Functions ▮ The Absolute Value
Function ▮ The Relation *x* = *y*² and Symmetry with Respect to the
x-Axis ▮ Even and Odd Functions ▮ Concavity and Extrema

Continuity; Increasing and Decreasing Functions

The graph of a linear function, a straight line, may be drawn by hand over any
interval of its domain without picking the pencil up from the paper. In mathematics
we say that a function with this property is **continuous** over any interval. The
formal definition of continuity requires concepts from calculus, but we can give an
informal definition at this level of study.

INFORMAL DEFINITION OF CONTINUITY

A function is continuous over an interval of its domain if its hand-drawn graph
over that interval can be sketched without lifting the pencil from the paper.

If a function is not continuous at a point, then it may have a point of discontinuity (Figure 40(a)), or it may have a vertical *asymptote* (a vertical line which the
graph does not intersect, as in Figure 40(b)). More will be said about asymptotes
in a later chapter.

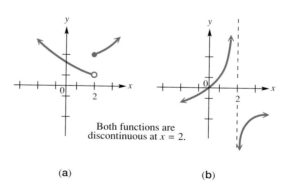

Both functions are
discontinuous at *x* = 2.

(a)   (b)

FIGURE 40

Notice that both graphs in Figure 40 are graphs of functions, since they pass
the vertical line test.

EXAMPLE 1

Determining
Intervals of
Continuity

The figures on the next page show graphs of functions and the descriptions indicate
the intervals of the domain over which they are continuous.

(a) The function in Figure 41 is continuous over the entire domain of real numbers, $(-\infty, \infty)$.

(b)   The function in Figure 42 has a point of discontinuity at $x = 3$. It is continuous over the interval $(-\infty, 3)$ and the interval $(3, \infty)$.

(c)   The function in Figure 43 has a vertical asymptote at $x = -2$, as indicated by the dashed line. It is continuous over the interval $(-\infty, -2)$ and the interval $(-2, \infty)$.   ▯

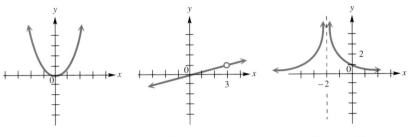

| **FIGURE 41** | **FIGURE 42** | **FIGURE 43** |

If a function is not constant over an interval, then its graph will either rise from left to right or will fall from left to right. We use the words *increasing* and *decreasing* to describe this behavior. For example, a linear function with a positive slope is increasing over its entire domain, while one with a negative slope is decreasing. See Figure 44.

Informally speaking, a function **increases** on an interval of its domain if its graph rises from left to right. It **decreases** on an interval if its graph falls from left to right. It is **constant** on an interval if its graph is horizontal on the interval.

The formal definitions of these concepts follow.

INCREASING, DECREASING, AND CONSTANT FUNCTIONS

Suppose that a function f is defined over an interval I.

a. f increases on I if, whenever $x_1 < x_2, f(x_1) < f(x_2)$;
b. f decreases on I if, whenever $x_1 < x_2, f(x_1) > f(x_2)$;
c. f is constant on I if, for every x_1 and $x_2, f(x_1) = f(x_2)$.

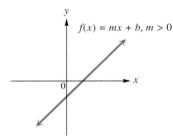

f is *increasing*, since it rises
from left to right.

(a)

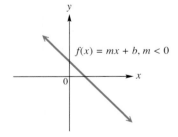

f is *decreasing*, since it falls
from left to right.

(b)

FIGURE 44

Figure 45 illustrates these ideas.

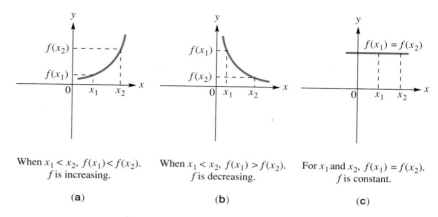

When $x_1 < x_2$, $f(x_1) < f(x_2)$.
f is increasing.

(a)

When $x_1 < x_2$, $f(x_1) > f(x_2)$.
f is decreasing.

(b)

For x_1 and x_2, $f(x_1) = f(x_2)$.
f is constant.

(c)

FIGURE 45

EXAMPLE 2

Determining Intervals Over Which a Function is Increasing, Decreasing, or Constant

Figure 46 shows the graph of a function. Determine the intervals over which the function is increasing, decreasing, or constant.

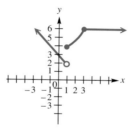

FIGURE 46

SOLUTION In making our determination, we must always ask "What is the *y*-value doing as *x* is getting larger?" For this graph, we see that on the interval $(-\infty, 1)$, the *y*-values are *decreasing;* on the interval $(1, 3)$, the *y*-values are *increasing;* and on the interval $(3, \infty)$, the *y*-values are *constant* (all are 6). Therefore, the function is

decreasing on $(-\infty, 1)$,

increasing on $(1, 3)$,

constant on $(3, \infty)$. ∎

CAUTION A common error involves writing range values when determining intervals like those in Example 2. Remember that we are determining intervals of the domain, and thus are interested in *x*-values for our interval designations.

FOR GROUP DISCUSSION

1. In the standard viewing window of your calculator, enter any linear function $y = mx + b$ with $m > 0$. Now trace the graph from *left to right*. Watch the *y*-values as *x* gets larger. What is happening to *y*? How does this reinforce the concepts presented so far in this section?
2. Repeat this exercise, but with $m < 0$.
3. Repeat this exercise, but with $m = 0$.

The rest of this section is devoted to analysis of several important basic functions that are studied in college algebra. The concepts reviewed will be important as we progress through this book.

The Identity Function

If we let $m = 1$ and $b = 0$ for the general form of the linear function $f(x) = mx + b$, we get the **identity function** $f(x) = x$. This function pairs every real number with itself.

IDENTITY FUNCTION

$f(x) = x$ (Figure 47)

Domain: $(-\infty, \infty)$

Range: $(-\infty, \infty)$

The identity function $f(x) = x$ increases on its entire domain $(-\infty, \infty)$, and is continuous on its entire domain.

x	x
0	0
1	1
-1	-1
2	2
-2	-2

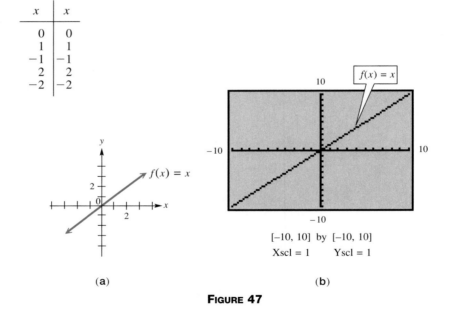

(a)   (b)

FIGURE 47

| FOR **GROUP DISCUSSION** |

Choose *any* viewing window on your graphing calculator and graph $y = x$. Trace, and compare the x and y values as you trace. What do you notice? Why do you think $f(x) = x$ is called the identity function?

The Squaring Function and Symmetry with Respect to the *y*-Axis

We now look at the graph of the simplest degree 2 function, the **squaring function** $f(x) = x^2$. This function pairs every real number with its square. Its graph is called a **parabola.** In the previous section we solved quadratic equations, so the graph of a parabola should be somewhat familiar.

SQUARING FUNCTION

$f(x) = x^2$ (Figure 48)

Domain: $(-\infty, \infty)$

Range: $[0, \infty)$

The squaring function $f(x) = x^2$ decreases on the interval $(-\infty, 0)$ and increases on the interval $(0, \infty)$. It is continuous on its entire domain.

x	x^2
0	0
1	1
-1	1
2	4
-2	4

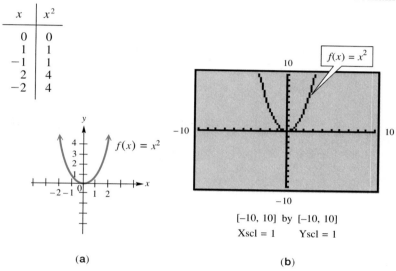

$f(x) = x^2$

$[-10, 10]$ by $[-10, 10]$
Xscl = 1   Yscl = 1

(a)   (b)

FIGURE 48

The point at which the graph changes from decreasing to increasing (the point $(0, 0)$) is called the **vertex** of the parabola.

Notice that if we were able to "fold" the graph of $f(x) = x^2$ along the *y*-axis, the two halves would coincide exactly. In mathematics we refer to this property as symmetry, and we say that the graph of $f(x) = x^2$ is **symmetric with respect to the *y*-axis.** This may be generalized as follows.

SYMMETRY WITH RESPECT TO THE *y*-AXIS

If a function f is defined so that

$$f(x) = f(-x)$$

for all x in its domain, then the graph of f is symmetric with respect to the *y*-axis.

Some *particular* cases illustrating that the graph of $f(x) = x^2$ is symmetric with respect to the y-axis are as follows:

$$f(-4) = f(4) = 16$$
$$f(-3) = f(3) = 9$$
$$f(-2) = f(2) = 4$$
$$f(-1) = f(1) = 1$$
$$f(-0) = f(0) = 0.$$

This pattern holds for any real number x, since $f(-x) = (-x)^2 = x^2 = f(x)$.

The Cubing Function and Symmetry with Respect to the Origin

The function $f(x) = x^3$ is the simplest degree 3 function, and it is an example of a *cubic* function. It pairs each real number with the third power, or cube, of the number.

CUBING FUNCTION

$f(x) = x^3$ (Figure 49)

Domain: $(-\infty, \infty)$

Range: $(-\infty, \infty)$

The cubing function $f(x) = x^3$ increases on its entire domain $(-\infty, \infty)$. It is also continuous on its entire domain $(-\infty, \infty)$.

x	x^3
0	0
1	1
-1	-1
2	8
-2	-8

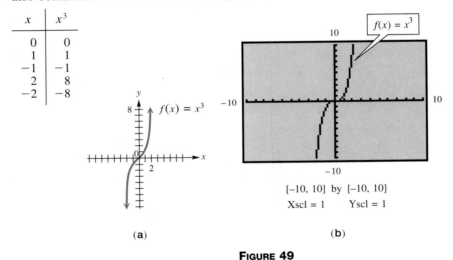

[−10, 10] by [−10, 10]
Xscl = 1   Yscl = 1

(a)   (b)

FIGURE 49

The point at which the graph changes from "opening downward" to "opening upward" (the point $(0, 0)$) is called an **inflection point.**

Notice that if we were able to "fold" the graph of $f(x) = x^3$ along the y-axis and then along the x-axis, forming a "corner" at the origin, the two parts of the graph would coincide exactly. We say that the graph of $f(x) = x^3$ is **symmetric with respect to the origin.** This may be generalized as follows.

> ## SYMMETRY WITH RESPECT TO THE ORIGIN
>
> If a function f is defined so that
> $$f(-x) = -f(x)$$
> for all x in its domain, then the graph of f is symmetric with respect to the origin.

Some *particular* cases illustrating that the graph of $f(x) = x^3$ is symmetric with respect to the origin are as follows:

$$f(-2) = -f(2) = -8$$
$$f(-1) = -f(1) = -1$$
$$f(-0) = -f(0) = -0 = 0.$$

This pattern holds for any real number x, since

$$f(-x) = (-x)^3 = (-1)^3 x^3 = -x^3 = -f(x).$$

EXAMPLE 3
Determining Symmetry Analytically and Supporting it Graphically

(a) Show analytically and support graphically the fact that $f(x) = x^4 - 3x^2 - 8$ has a graph that is symmetric with respect to the y-axis.

SOLUTION We must show that $f(-x) = f(x)$ for any x.

$$f(-x) = (-x)^4 - 3(-x)^2 - 8$$
$$= (-1)^4 x^4 - 3(-1)^2 x^2 - 8$$
$$= x^4 - 3x^2 - 8$$
$$= f(x)$$

This *proves* that there is symmetry with respect to the y-axis. The graph in Figure 50 supports this conclusion, since it appears to have this symmetry. (*Note:* Visual support is not a proof!)

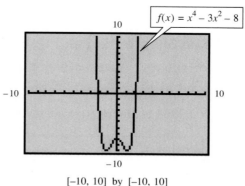

$$f(x) = x^4 - 3x^2 - 8$$

$[-10, 10]$ by $[-10, 10]$
Xscl = 1   Yscl = 1

This graph is symmetric with respect to the y-axis.

FIGURE 50

(b) Show analytically and support graphically the fact that $f(x) = x^3 - 4x$ has a graph that is symmetric with respect to the origin.

SOLUTION   In this case, we must show that $f(-x) = -f(x)$ for any x.

$$f(-x) = (-x)^3 - 4(-x)$$
$$= (-1)^3 x^3 + 4x$$
$$= -x^3 + 4x \qquad *$$
$$= -(x^3 - 4x)$$
$$= -f(x)$$

In the line denoted *, note that the signs of the coefficients are all *opposites* of those in $f(x)$. We completed the argument by factoring out -1, showing that the final result is $-f(x)$.

The graph in Figure 51 supports our conclusion that the graph is symmetric with respect to the origin, for folding it along the y-axis and then along the x-axis would lead to coinciding curves.   ∎

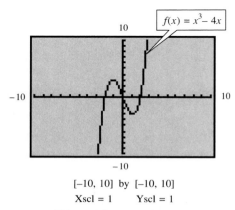

$f(x) = x^3 - 4x$

$[-10, 10]$ by $[-10, 10]$
Xscl = 1   Yscl = 1

This graph is symmetric with
respect to the origin.

FIGURE 51

The Square Root and Cube Root Functions

We now investigate functions that are defined by expressions involving radicals. The first of these is the **square root function,** $f(x) = \sqrt{x}$. Notice that for the function value to be a real number, we must have $x \geq 0$. Thus, the domain is restricted to nonnegative numbers.

SQUARE ROOT FUNCTION

$f(x) = \sqrt{x}$ (Figure 52)

Domain: $[0, \infty)$ Range: $[0, \infty)$

The square root function $f(x) = \sqrt{x}$ increases on $(0, \infty)$. It is also continuous on $[0, \infty)$.

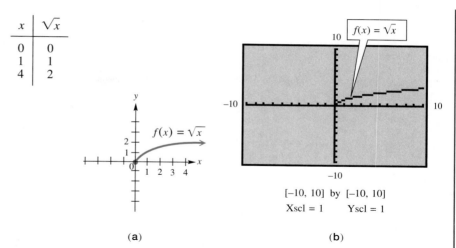

(a)   **(b)**

FIGURE 52

(The definition of rational exponents allows us to also enter $\sqrt{x}$ as $x^{1/2}$ on a calculator.)

The **cube root function,** $f(x) = \sqrt[3]{x}$, differs from the square root function in that *any* real number, positive, zero, or *negative*, has a real cube root, and thus the domain is $(-\infty, \infty)$. Also, when $x > 0$, $\sqrt[3]{x} > 0$, when $x = 0$, $\sqrt[3]{x} = 0$, and when $x < 0$, $\sqrt[3]{x} < 0$. As a result, the range is also $(-\infty, \infty)$.

CUBE ROOT FUNCTION

$f(x) = \sqrt[3]{x}$ (Figure 53)

Domain: $(-\infty, \infty)$ Range: $(-\infty, \infty)$

The cube root function $f(x) = \sqrt[3]{x}$ increases on its entire domain $(-\infty, \infty)$. It is also continuous on $(-\infty, \infty)$.

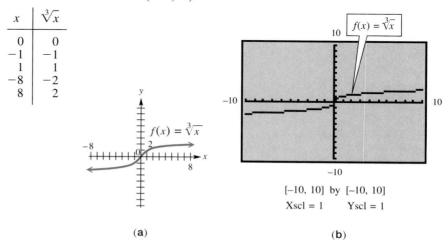

(a)   **(b)**

FIGURE 53

(The definition of rational exponents allows us to also enter $\sqrt[3]{x}$ as $x^{1/3}$ on a calculator.)

TECHNOLOGICAL NOTE
You should become familiar with the command on your particular calculator that allows you to graph the absolute value.

The Absolute Value Function

On a number line, the absolute value of a real number x, denoted $|x|$, represents its undirected distance from the origin, 0. The **absolute value function,** which pairs every real number with its absolute value, is defined as follows:

$$f(x) = |x| = \begin{cases} x & \text{if } x \geq 0 \\ -x & \text{if } x < 0. \end{cases}$$

Notice that this function is defined in two parts. We use $|x| = x$ if x is positive or zero, and we use $|x| = -x$ if x is negative. Since x can be any real number, the domain of the absolute value function is $(-\infty, \infty)$, but since $|x|$ cannot be negative, the range is $[0, \infty)$.

ABSOLUTE VALUE FUNCTION

$f(x) = |x|$ (Figure 54)

Domain: $(-\infty, \infty)$

Range: $[0, \infty)$

The absolute value function $f(x) = |x|$ decreases on the interval $(-\infty, 0)$ and increases on $(0, \infty)$. It is continuous on its entire domain.

x	$|x|$
0	0
1	1
-1	1
2	2
-2	2

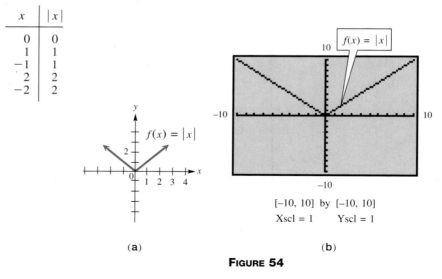

$[-10, 10]$ by $[-10, 10]$
Xscl = 1   Yscl = 1

(a)   (b)

FIGURE 54

FOR **GROUP DISCUSSION**

Based on the discussion so far in this section, answer the following questions.

1. Which functions have graphs that are symmetric with respect to the y-axis?
2. Which functions have graphs that are symmetric with respect to the origin?
3. Which functions have graphs that show neither of these symmetries?
4. Why is it not possible for the graph of a function to be symmetric with respect to the x-axis?

The Relation $x = y^2$ and Symmetry with Respect to the x-Axis

Recall that a function is a relation that satisfies the condition that every domain value is paired with one and only one range value. However, there are cases where we are interested in graphing relations that are not functions, and one of the simplest of these is the relation defined by the equation $x = y^2$. Notice that the table of selected ordered pairs below indicates that this relation has two different y-values for positive values of x.

SELECTED ORDERED PAIRS FOR $x = y^2$

x	y
0	0
1	± 1
4	± 2
9	± 3

two different y-values for the same x-value

If we plot these points and join them with a smooth curve, we find that the graph of $x = y^2$ is a parabola opening to the right. See Figure 55.

    If a graphing calculator is set for the function mode, it is not possible to graph $x = y^2$ directly. (However, if it is set for the *parametric* mode, such a curve is possible with direct graphing.) To overcome this problem, we begin with $y^2 = x$ and take the square root on each side, remembering to choose both the positive and negative square roots of x:

$$x = y^2 \qquad \text{Given equation}$$
$$y^2 = x \qquad \text{Transform so that } y \text{ is on the left.}$$
$$y = \pm\sqrt{x}. \qquad \text{Take square roots.}$$

Now we have $x = y^2$ defined by two *functions*, $y_1 = \sqrt{x}$ and $y_2 = -\sqrt{x}$. Entering both of these into a calculator gives the graph shown in Figure 56.

x	y
0	0
1	± 1
4	± 2
9	± 3

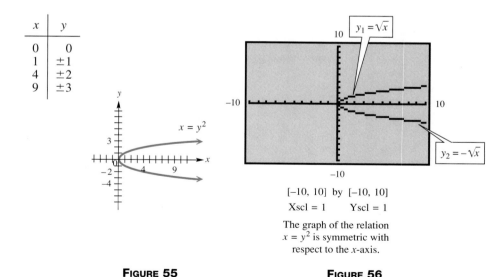

$[-10, 10]$ by $[-10, 10]$
Xscl = 1   Yscl = 1

The graph of the relation $x = y^2$ is symmetric with respect to the x-axis.

FIGURE 55   **FIGURE 56**

    It appears that if we were to fold the graph of $x = y^2$ along the x-axis, the two halves of the parabola would coincide. This is indeed the case, and this graph exhibits symmetry with respect to the x-axis.

> ## SYMMETRY OF A GRAPH WITH RESPECT TO THE x-AXIS
>
> If replacing y with $-y$ in an equation results in the same equation, then the graph is symmetric with respect to the x-axis.

To illustrate this, if we begin with $x = y^2$ and replace y with $-y$, we get

$$x = (-y)^2$$
$$x = (-1)^2 y^2$$
$$x = y^2. \qquad \text{The same equation with which we started}$$

FOR **GROUP DISCUSSION**

With the two functions
$$y_1 = \sqrt{x} \qquad \text{and} \qquad y_2 = -\sqrt{x}$$
graphed on your calculator, trace to any point on y_1 and notice the x-value. Now switch back and forth from y_1 to y_2 at this x-value. What happens to the y-value. Why is this so?

A summary of the types of symmetry just discussed follows.

Type of Symmetry	Example	Basic Fact About Points on the Graph
y-axis symmetry	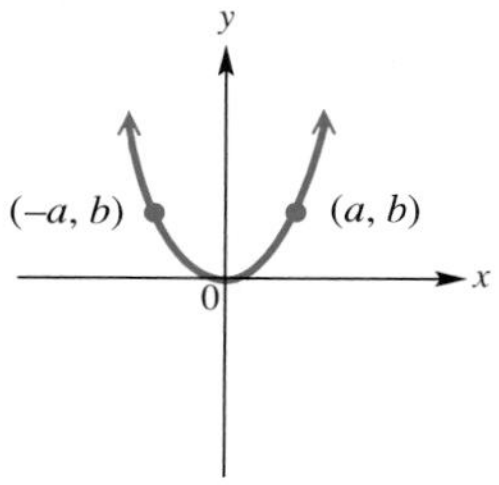	If (a, b) is on the graph, so is $(-a, b)$.
origin symmetry	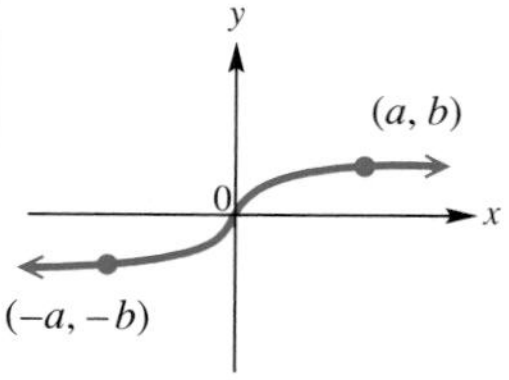	If (a, b) is on the graph, so is $(-a, -b)$.
x-axis symmetry (not possible for a function)	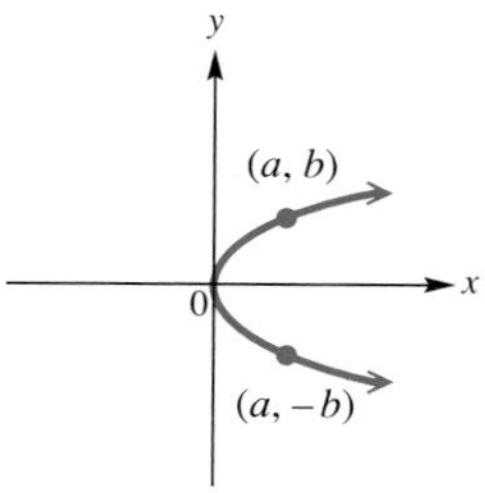	If (a, b) is on the graph, so is $(a, -b)$.

Even and Odd Functions

Closely associated with the concepts of symmetry with respect to the y-axis and symmetry with respect to the origin are the ideas of even and odd functions.

EVEN AND ODD FUNCTIONS

A function f is called an **even function** if $f(-x) = f(x)$ for all x in the domain of f. (Its graph is symmetric with respect to the y-axis.)

A function f is called an **odd function** if $f(-x) = -f(x)$ for all x in the domain of f. (Its graph is symmetric with respect to the origin.)

As an illustration, $f(x) = x^2$ is an even function because

$$f(-x) = (-x)^2 = x^2 = f(x).$$

The function $f(x) = x^3$ is an odd function because

$$f(-x) = (-x)^3 = -x^3 = -f(x).$$

A function may be neither even nor odd; for example, $f(x) = \sqrt{x}$ is neither even nor odd.

EXAMPLE 4

Determining Analytically Whether a Function is Even, Odd, or Neither

Decide if the functions defined as follows are even, odd, or neither.

(a) $f(x) = 8x^4 - 3x^2$

SOLUTION   Replacing x with $-x$ gives

$$f(-x) = 8(-x)^4 - 3(-x)^2 = 8x^4 - 3x^2 = f(x).$$

Since $f(x) = f(-x)$ for each x in the domain of the functon, f is an even function.

(b) $f(x) = 6x^3 - 9x$

SOLUTION   Here

$$f(-x) = 6(-x)^3 - 9(-x) = -6x^3 + 9x = -f(x).$$

This function is odd.

(c) $f(x) = 3x^2 + 2x$

SOLUTION

$$f(-x) = 3(-x)^2 + 2(-x)$$
$$= 3x^2 - 2x$$

Since $f(-x) \neq f(x)$ and $f(-x) \neq -f(x)$, f is neither even nor odd.   ❑

Concavity and Extrema

Recall that for a quadratic function $f(x) = ax^2 + bx + c$, if $a > 0$, the graph is at all times opening upward. If water were to be poured from above, the graph would, in a sense, "hold water." We say that this graph is *concave up* for all values in its domain. On the other hand, if $a < 0$, the graph opens downward at all times, and it would similarly "dispel water" if it were poured from above. In this case, the graph is *concave down* for all values in its domain. See Figure 57.

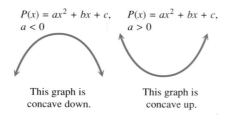

$$P(x) = ax^2 + bx + c, \quad P(x) = ax^2 + bx + c,$$
$$a < 0 \qquad\qquad a > 0$$

This graph is
concave down.

This graph is
concave up.

FIGURE 57

NOTE A *formal* discussion of concavity requires concepts beyond the scope of this text. It is studied more rigorously in calculus.

EXAMPLE 5
Illustrating Concavity

Referring back to the cubing function (Figure 49) as an illustration, we see that the graph is "opening downward" on $(-\infty, 0)$, so it is concave down here. The graph is "opening upward" on $(0, \infty)$, making it concave up on $(0, \infty)$.

TECHNOLOGICAL NOTE
Modern graphing calculators are capable of finding the coordinates of local maxima and local minima, provided an appropriate interval is designated. Refer to your owner's manual to see how your calculator does this.

Notice in Figure 58 that the two graphs have "peaks" and/or "valleys" where the function changes from increasing to decreasing or vice-versa. We first saw this when we studied quadratic functions, noticing that the vertex could be a maximum or minimum point on the graph. In general, the highest point at a "peak" is known as a **local maximum point,** and the lowest point at a "valley" is known as a **local minimum point.** Function values at such points are called **local maxima** (plural of maximum) and **local minima** (plural of minimum). Collectively they are called **extrema** (plural of **extremum**). Figure 58 and the accompanying chart illustrate these ideas for typical graphs.

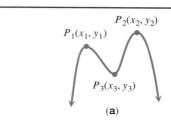

(a)

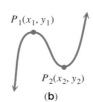

(b)

FIGURE 58

Extreme Point	Specifics	Extreme Point	Specifics
P_1	P_1 is a local maximum point. The function has a local maximum value of y_1 at $x = x_1$.	P_1	P_1 is a local maximum point. The function has a local maximum value of y_1 at $x = x_1$.
P_2	P_2 is a local maximum point. The function has a local maximum value of y_2 at $x = x_2$.	P_2	P_2 is a local minimum point. The function has a local minimum value of y_2 at $x = x_2$.
P_3	P_3 is a local minimum point. The function has a local minimum value of y_3 at $x = x_3$.		

Refer again to Figure 58(a). Notice that the point P_2 is the absolute highest point on the graph, and the range of the function is $(-\infty, y_2]$. We call P_2 the **absolute maximum point** on the graph, and y_2 the **absolute maximum value** of the function. Because the y-values approach $-\infty$, this function has no absolute minimum value. On the other hand, because the graph in Figure 58(b) is that of a function with range $(-\infty, \infty)$, it has neither an absolute maximum nor an absolute minimum.

EXAMPLE 6

Identifying Local
and Absolute
Extrema

Consider the graphs in Figure 59.

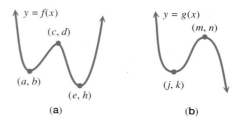

FIGURE 59

(a) Name and classify the local extrema of f.

SOLUTION   The points (a, b) and (e, h) are local minimum points. The point (c, d) is a local maximum.

(b) Name and classify the local extrema of g.

SOLUTION   The point (j, k) is a local minimum and the point (m, n) is a local maximum.

(c) Discuss absolute extrema for f and g.

SOLUTION   The absolute minimum value of function f is the number h, since the range of f is $[h, \infty)$. It has no absolute maximum value. Function g has no absolute extrema, since its range is $(-\infty, \infty)$.   ∎

1.4 EXERCISES

Determine the intervals of the domain over which the given function is continuous.

1.

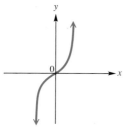

2.

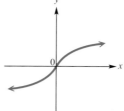

3.

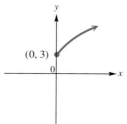

(0, 3)

4.

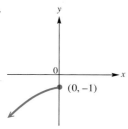

(0, −1)

5.

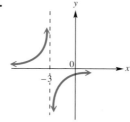

6.

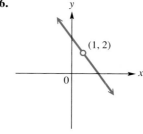

(1, 2)

7.
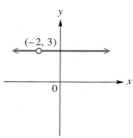

8.
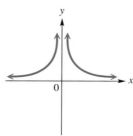

9. Graph the function $y = \frac{x^2 - 9}{x + 3}$ in the standard viewing window of your calculator. At first glance, does this graph seem to be continuous over the entire interval of the domain shown in the window? Now, try to locate the point for which $x = -3$. What happens? Why do you think this happens? (Functions of this kind, called *rational functions,* are studied in college algebra.)

10. Based on your work in Exercise 9, do you think that determination of continuity strictly by observation of a calculator-generated graph is foolproof?

Determine the intervals of the domain over which the given function is **(a)** *increasing,* **(b)** *decreasing, and* **(c)** *constant.*

11.
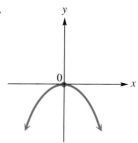

12.
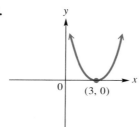

13.
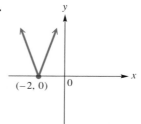

14.
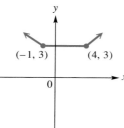

15.
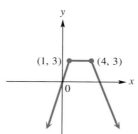

16.
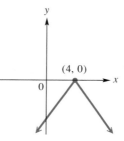

17.
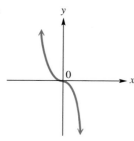

18.
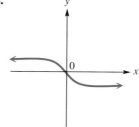

19.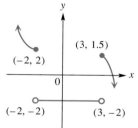

20.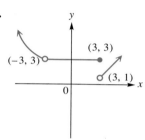

In Exercises 21–28, you are given a function and an interval. Graph the function in the standard viewing window of your calculator, and trace from left to right along a representative portion of the specified interval. Then fill in the blank of this sentence with either increasing or decreasing: *OVER THE INTERVAL SPECIFIED, THIS FUNCTION IS _____ .*

21. $f(x) = x^5$; $(-\infty, \infty)$ **22.** $f(x) = -x^3$; $(-\infty, \infty)$ **23.** $f(x) = x^4$; $(-\infty, 0)$

24. $f(x) = x^4$; $(0, \infty)$ **25.** $f(x) = -|x|$; $(-\infty, 0)$ **26.** $f(x) = -|x|$; $(0, \infty)$

27. $f(x) = \pi x + 3$; $(-\infty, \infty)$ **28.** $f(x) = -\sqrt{2}x - 1$; $(-\infty, \infty)$

Beginning in this exercise set, exercises called **Relating Concepts** will often appear. These exercises are designed to be worked so that you can see the connections between skills learned earlier and skills to be mastered in the section at hand. They will also at times "tie together" concepts from earlier courses or this course with those currently being studied.

*R*elating Concepts

Use your knowledge of slopes of lines to determine whether each function is increasing or decreasing *without actually graphing*. Then confirm your answer by graphing in the standard window.

29. $y = 2.36x - 1.56$ **30.** $y = \sqrt{5}\,x - \sqrt{3}$

31. $y = -\sqrt{6}\,x + .45$ **32.** $y = -.876\,x + \sqrt{5}$

Based on a visual observation, determine whether each graph is symmetric with respect to the following: (**a**) x-axis, (**b**) y-axis, (**c**) origin.

33.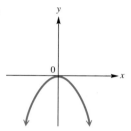

34.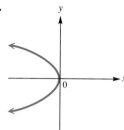

35.

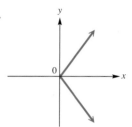

36.

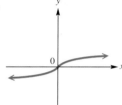

37.

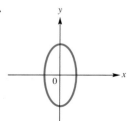

38.

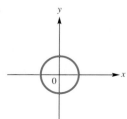

39.

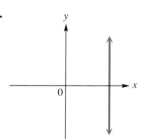

40.

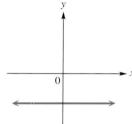

41.

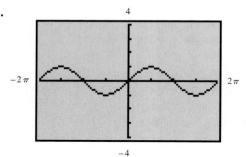

42.

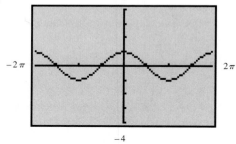

43.

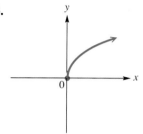

44.

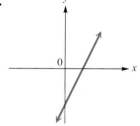

In Exercises 45–50, you are given a calculator-generated graph of a relation that exhibits a type of symmetry. Based on the given information at the bottom of each figure, determine the coordinates of another point that must also lie on the graph.

45. symmetric with respect to the *y*-axis

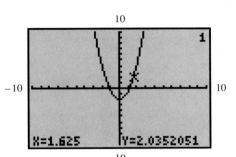

46. symmetric with respect to the *y*-axis

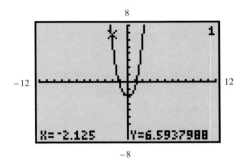

47. symmetric with respect to the *x*-axis

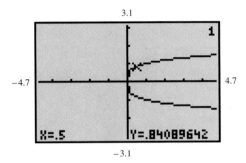

48. symmetric with respect to the *x*-axis

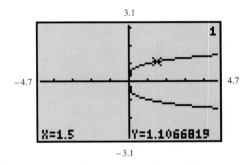

49. symmetric with respect to the origin

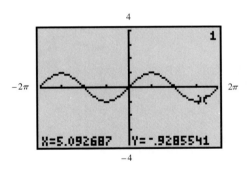

50. symmetric with respect to the origin

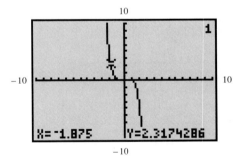

Use the method of Example 3 to determine whether the given function is symmetric with respect to the y-axis, symmetric with respect to the origin, or neither of these. Then graph the function on your calculator to support your conclusion, using the window [−5, 5] *by* [−10, 10].

51. $f(x) = -x^3 + 2x$

52. $f(x) = x^5 - 2x^3$

53. $f(x) = .5x^4 - 2x^2 + 1$

54. $f(x) = .75x^2 + |x| + 1$

55. $f(x) = x^3 - x + 3$

56. $f(x) = x^4 - 5x + 2$

Using the standard viewing windows below (or on a reproduction of them), graph by hand the basic functions described in this section, avoiding the temptation to look back at the figures of the section. Then check your work using your calculator. Finally, use the graphs to answer true *or* false *to the statements in Exercises 57–66.*

$f(x) = x$

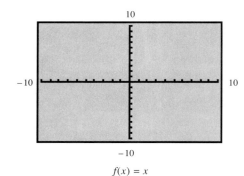

$f(x) = x$

$f(x) = x^2$

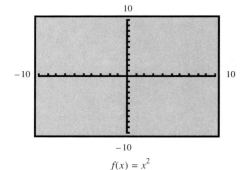

$f(x) = x^2$

$f(x) = x^3$

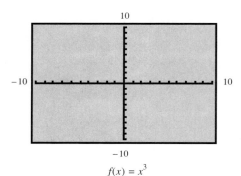

$f(x) = x^3$

$f(x) = \sqrt{x}$

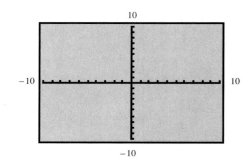

$f(x) = \sqrt{x}$

$f(x) = \sqrt[3]{x}$

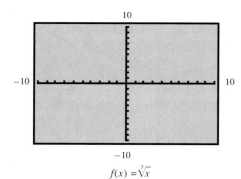

$f(x) = \sqrt[3]{x}$

$f(x) = |x|$

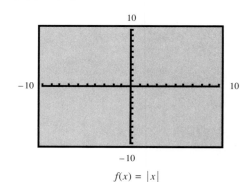

$f(x) = |x|$

57. The range of $f(x) = x^2$ is the same as the range of $f(x) = |x|$.

58. The functions $f(x) = x^2$ and $f(x) = |x|$ increase on the same interval.

59. The functions $f(x) = \sqrt{x}$ and $f(x) = \sqrt[3]{x}$ have the same domain.

60. The function $f(x) = \sqrt[3]{x}$ decreases on its entire domain.

61. The function $f(x) = x$ has its domain equal to its range.

62. The function $f(x) = \sqrt{x}$ is continuous on the interval $(-\infty, 0)$.

63. None of the functions shown decrease on the interval $(0, \infty)$.

64. Both $f(x) = x$ and $f(x) = x^3$ have graphs that are symmetric with respect to the origin.

65. Both $f(x) = x^2$ and $f(x) = |x|$ have graphs that are symmetric with respect to the y-axis.

66. None of the graphs shown are symmetric with respect to the x-axis.

Refer to the function described and determine whether it is even, odd, or neither.

67. the function in Exercise 51   **68.** the function in Exercise 52

69. the function in Exercise 53   **70.** the function in Exercise 54

71. the function in Exercise 55   **72.** the function in Exercise 56

In Exercises 73–82, use your calculator to its maximum capability.
(a) *Graph the function in the window specified.*
(b) *Determine all local minimum points, and tell if any is an absolute minimum point. (Approximate coordinates to the nearest hundredth.)*
(c) *Determine all local maximum points, and tell if any is an absolute maximum point. (Approximate coordinates to the nearest hundredth.)*
(d) *Determine the range. (If an approximation is necessary, give it to the nearest hundredth.)*
(e) *Determine all intercepts. For each function, there is at least one x-intercept that is an integer. For those that are not integers, give an approximation to the nearest hundredth. You should determine the y-intercept analytically.*

73. $y = x^3 - 4x^2 + x + 6;\ [-10, 10]$ by $[-10, 10]$

74. $y = x^3 + x^2 - 22x - 40;\ [-10, 10]$ by $[-100, 10]$

75. $y = -2x^3 - 14x^2 + 2x + 84;\ [-10, 10]$ by $[-100, 100]$

76. $y = -3x^3 + 6x^2 + 39x - 60;\ [-10, 10]$ by $[-100, 100]$

77. $y = x^5 + 4x^4 - 3x^3 - 17x^2 + 6x + 9;\ [-4, 4]$ by $[-20, 20]$

78. $y = -2x^5 + 7x^4 + x^3 - 20x^2 + 4x + 16;\ [-4, 4]$ by $[-20, 20]$

79. $y = 2x^4 + 3x^3 - 17x^2 - 6x - 72;\ [-10, 10]$ by $[-200, 100]$

80. $y = 3x^4 - 33x^2 + 54;\ [-6, 6]$ by $[-100, 100]$

81. $y = -x^6 + 24x^4 - 144x^2 + 256;\ [-6, 6]$ by $[-300, 300]$

82. $y = -3x^6 + 2x^5 + 9x^4 - 8x^3 + 11x^2 + 4;\ [-6, 6]$ by $[-100, 100]$

F*urther Explorations*

For each of the following pairs of TABLES, determine whether the functions in Y_1 are even functions, odd functions, or neither.

1.

X	Y1
0	0
1	1
2	4
3	9
4	16
5	25
6	36

X=0

X	Y1
0	0
-1	1
-2	4
-3	9
-4	16
-5	25
-6	36

X=0

2.

X	Y1	
0	0	
1	1	
2	8	
3	27	
4	64	
5	125	
6	216	
X=0		

X	Y1	
0	0	
-1	-1	
-2	-8	
-3	-27	
-4	-64	
-5	-125	
-6	-216	
X=0		

3.

X	Y1	
0	-5	
1	-3	
2	-1	
3	1	
4	3	
5	5	
6	7	
X=0		

X	Y1	
0	-5	
-1	-3	
-2	-1	
-3	1	
-4	3	
-5	5	
-6	7	
X=0		

4.

X	Y1	
0	5	
1	6	
2	7	
3	8	
4	9	
5	10	
6	11	
X=0		

X	Y1	
0	5	
-1	4	
-2	3	
-3	2	
-4	1	
-5	0	
-6	-1	
X=0		

1.5 TRANSFORMATIONS OF GRAPHS OF FUNCTIONS

Vertical Shifts ▮ Horizontal Shifts ▮ Combinations of Vertical and Horizontal Shifts ▮ Effects of Shifts on Domain and Range ▮ Vertical Stretching ▮ Vertical Shrinking ▮ Reflecting Across an Axis ▮ Combining Transformations of Graphs

In this section we will examine how the graphs of the elementary functions introduced in the previous section can be shifted, stretched, and shrunk in the plane. The basic ideas can then be generalized to apply to the graphs of the circular functions introduced in Chapter 2.

Vertical Shifts

TECHNOLOGICAL NOTE
You can save time when entering the functions in the first Group Discussion. If you let Y1 represent the basic function as shown, you can enter Y2 as Y1 + 3, Y3 as Y1 − 2, and Y4 as Y1 + 5. Then, you need only change Y1 in each function group, since the other functions are defined in terms of Y1.

FOR **GROUP DISCUSSION**

In each group of functions below, we give four related functions. Graph the four functions in the first group (Group A), and then answer the questions below regarding those functions. Then repeat the process for Group B, Group C, and Group D. Use the standard viewing window in each case.

A	B	C	D
$y_1 = x^2$	$y_1 = x^3$	$y_1 = \sqrt{x}$	$y_1 = \sqrt[3]{x}$
$y_2 = x^2 + 3$	$y_2 = x^3 + 3$	$y_2 = \sqrt{x} + 3$	$y_2 = \sqrt[3]{x} + 3$
$y_3 = x^2 - 2$	$y_3 = x^3 - 2$	$y_3 = \sqrt{x} - 2$	$y_3 = \sqrt[3]{x} - 2$
$y_4 = x^2 + 5$	$y_4 = x^3 + 5$	$y_4 = \sqrt{x} + 5$	$y_4 = \sqrt[3]{x} + 5$

1. How does the graph of y_2 compare to the graph of y_1?
2. How does the graph of y_3 compare to the graph of y_1?
3. How does the graph of y_4 compare to the graph of y_1?
4. If $c > 0$, how do you think the graph of $y_1 + c$ would compare to the graph of y_1?
5. If $c > 0$, how do you think the graph of $y_1 - c$ would compare to the graph of y_1?

Choosing your own value of c, support your answers to Items 4 and 5 graphically. (Be sure that your choice is appropriate for the standard window.)

The objective of the preceding group discussion activity was to make conjectures about how the addition or subtraction of a constant c would affect the graph of a function $y = f(x)$. In each case, we obtained a vertical shift, or **translation,** of the graph of the basic function with which we started. Although our observations were based on the graphs of four different elementary functions, they can be generalized to any function.

VERTICAL SHIFTING OF THE GRAPH OF A FUNCTION

If $c > 0$, the graph of $y = f(x) + c$ is obtained by shifting the graph of $y = f(x)$ *upward* a distance of c units. The graph of $y = f(x) - c$ is obtained by shifting the graph of $y = f(x)$ *downward* a distance of c units.

In Figure 60 we give a graphical interpretation of the statement above.

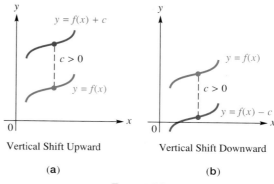

Vertical Shift Upward   Vertical Shift Downward

(a)   (b)

FIGURE 60

EXAMPLE **1**
Recognizing Vertical Shifts on Calculator-Generated Graphs

Figure 61 shows the graphs of four functions. The function labeled y_1 is the function $f(x) = |x|$. The viewing window is $[-10, 10]$ by $[-10, 10]$, with Xscl = 1 and Yscl = 1. Each of y_2, y_3, and y_4 are functions of the form $f(x) + c$ or $f(x) - c$, for $c > 0$. Give the rule for each of these functions.

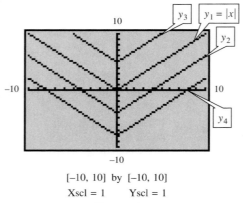

$[-10, 10]$ by $[-10, 10]$
Xscl = 1   Yscl = 1

FIGURE 61

SOLUTION   Because the graph of y_2 lies 4 units below the graph of y_1, we have $y_2 = |x| - 4$. The graph of y_3 is a shift of the graph of y_1 a distance of 5 units upward, so the equation for y_3 is $y_3 = |x| + 5$. Finally, the graph of y_4 is a vertical shift of the graph of y_1 8 units downward, so its equation is $y_4 = |x| - 8$.   ❚

When using a graphing calculator to investigate shifts of graphs, it is important to use an appropriate window; otherwise, the graph may not appear. For example, the graph of $y = |x| + 12$ does not appear in the viewing window $[-10, 10]$ by $[-10, 10]$. Why is this so? There are many windows that would show the graph. (Name one.)

Horizontal Shifts

╔══════════════ FOR **GROUP DISCUSSION** ══════════════╗

In each group of functions below, we give four related functions. Graph the four functions in the first group (Group A), and then answer the questions below regarding those functions. Then repeat the process for Group B, Group C, and Group D. Use the standard window in each case.

A	B	C	D
$y_1 = x^2$	$y_1 = x^3$	$y_1 = \sqrt{x}$	$y_1 = \sqrt[3]{x}$
$y_2 = (x - 3)^2$	$y_2 = (x - 3)^3$	$y_2 = \sqrt{x - 3}$	$y_2 = \sqrt[3]{x - 3}$
$y_3 = (x - 5)^2$	$y_3 = (x - 5)^3$	$y_3 = \sqrt{x - 5}$	$y_3 = \sqrt[3]{x - 5}$
$y_4 = (x + 4)^2$	$y_4 = (x + 4)^3$	$y_4 = \sqrt{x + 4}$	$y_4 = \sqrt[3]{x + 4}$

1.   How does the graph of y_2 compare to the graph of y_1?
2.   How does the graph of y_3 compare to the graph of y_1?
3.   How does the graph of y_4 compare to the graph of y_1?
4.   If $c > 0$, how do you think the graph of $y_5 = f(x - c)$ would compare to the graph of $y_1 = f(x)$?
5.   If $c > 0$, how do you think the graph of $y_5 = f(x + c)$ would compare to the graph of $y_1 = f(x)$?

Choosing your own value of c, support your answers to Items 4 and 5 graphically. Be sure that your choice is appropriate for the standard window.

╚══╝

The results of the above discussion should remind you of the results found earlier. There we saw how graphs of functions can be shifted vertically. Now, we see how they can be shifted *horizontally*. The observations can be generalized as follows.

HORIZONTAL SHIFTING OF THE GRAPH OF A FUNCTION

If $c > 0$, the graph of $y = f(x - c)$ is obtained by shifting the graph of $y = f(x)$ to the *right* a distance of c units. The graph of $y = f(x + c)$ is obtained by shifting the graph of $y = f(x)$ to the *left* a distance of c units.

CAUTION   Errors of interpretation frequently occur when horizontal shifts are involved. In order to determine the direction and magnitude of horizontal shifts, find the value of x that would cause the expression within the parentheses to equal 0. For example, the graph of $f(x) = (x - 5)^2$ would be shifted 5 units to the *right,* because $+5$ would cause $x - 5$ to equal 0. On the other hand, the graph of $f(x) = (x + 4)^2$ would be shifted 4 units to the *left,* because -4 would cause $x + 4$ to equal 0.

Figure 62 illustrates the effect of horizontal shifts of the graph of a function $y = f(x)$.

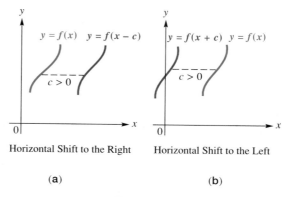

Horizontal Shift to the Right   Horizontal Shift to the Left

(a)   (b)

FIGURE 62

EXAMPLE 2

Recognizing Horizontal Shifts on Calculator-Generated Graphs

Figure 63 shows the graphs of four functions. As in Example 1, the function labeled y_1 is the function $f(x) = |x|$. The viewing window is $[-10, 10]$ by $[-10, 10]$, with

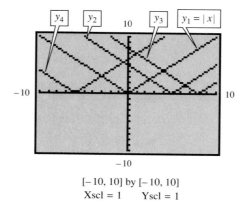

$[-10, 10]$ by $[-10, 10]$
Xscl $= 1$ Yscl $= 1$

FIGURE 63

Xscl = 1 and Yscl = 1. Each of y_2, y_3, and y_4 are functions of the form $f(x - c)$ or $f(x + c)$, where $c > 0$. Give the rule for each of these y's.

SOLUTION   The graph of y_2 is the same as the graph of y_1, but it is shifted 5 units to the *right*. Therefore, we have $y_2 = |x - 5|$. Similarly, since y_3 is the graph of y_1 shifted 8 units to the *right*, $y_3 = |x - 8|$. The graph of y_4 is obtained by shifting that of y_1 6 units to the *left*, so its equation is $y_4 = |x + 6|$.   ◼

Combinations of Vertical and Horizontal Shifts

Now that we have seen how graphs of functions can be shifted vertically and shifted horizontally, it is not difficult to extend these ideas to graphs that are obtained by applying *both* types of translations.

Describe how the graph of $y = |x + 15| - 20$ would be obtained by translating the graph of $y = |x|$. Determine an appropriate viewing window, and support the results by plotting both functions with a graphing calculator.

SOLUTION   The function defined by $y = |x + 15| - 20$ is translated 15 units to the *left* (because of the $|x + 15|$) and 20 units *downward* as compared to the graph of $y = |x|$. Because the point at which the graph changes from decreasing to increasing is now $(-15, -20)$, the standard viewing window is not appropriate. We must choose a window that contains the point $(-15, -20)$ in order to obtain a complete graph. While many such windows are possible, one such window is shown in Figure 64. The display at the bottom of the screen indicates that the point $(-15, -20)$ lies on the graph.   ◼

> **TECHNOLOGICAL NOTE**
> Some later models of graphing calculators have the capability of showing dynamically how transformations of graphs can be made. Read your owner's manual to see if yours has this feature.

> **EXAMPLE 3**
>
> Applying Both Vertical and Horizontal Shifts

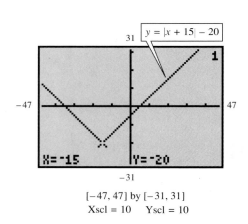

$[-47, 47]$ by $[-31, 31]$
Xscl = 10   Yscl = 10

FIGURE 64

FOR **GROUP DISCUSSION**

For each of the following pairs of functions, describe how the graph of y_2 can be obtained from the graph of y_1 by shifting, and then support your answer by graphing in an appropriate viewing window.

1. $y_1 = x^2$; $y_2 = (x - 12)^2 + 25$ **2.** $y_1 = x^3$; $y_2 = (x + 10)^3 - 15$

3. $y_1 = \sqrt{x}$; $y_2 = \sqrt{x - 20} - 30$ **4.** $y_1 = \sqrt[3]{x}$; $y_2 = \sqrt[3]{x + 20} + 40$

Effects of Shifts on Domain and Range

The domains and ranges of functions may or may not be affected by vertical and horizontal shifts. For example, if the domain of a function is $(-\infty, \infty)$, a horizontal shift will not affect the domain. Similarly, if the range is $(-\infty, \infty)$, a vertical shift will not affect the range. However, if the domain is not $(-\infty, \infty)$, a horizontal shift will affect the domain, and if the range is not $(-\infty, \infty)$, a vertical shift will affect the range. The next example illustrates this.

EXAMPLE 4

Determining Domains and Ranges of Shifted Graphs

The four functions graphed in Figures 65–68 are those discussed in "For Group Discussion" following Example 3. Give the domain and the range of each function.

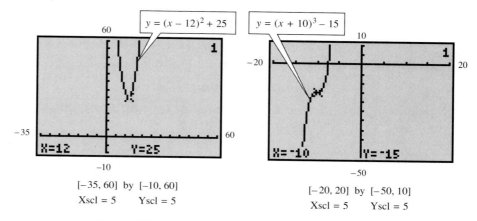

[−35, 60] by [−10, 60]
Xscl = 5   Yscl = 5

FIGURE 65

[−20, 20] by [−50, 10]
Xscl = 5   Yscl = 5

FIGURE 66

SOLUTION   The graph of $y = (x - 12)^2 + 25$ is shown in Figure 65. It is a translation of the graph of $y = x^2$ 12 units to the right and 25 units upward. The original domain $(-\infty, \infty)$ is not affected. However, the range of this function is $[25, \infty)$, because of the vertical translation. The graph in Figure 66, that of $y = (x + 10)^3 - 15$, was obtained by vertical and horizontal shifts of the graph of $y = x^3$, a function that has both domain and range equal to $(-\infty, \infty)$. Neither is affected here, and so the domain and range of $y = (x + 10)^3 - 15$ are also both $(-\infty, \infty)$.

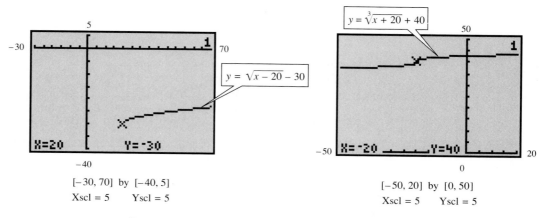

[−30, 70] by [−40, 5]
Xscl = 5   Yscl = 5

FIGURE 67

[−50, 20] by [0, 50]
Xscl = 5   Yscl = 5

FIGURE 68

The function $y = \sqrt{x}$ has domain $[0, \infty)$. The function graphed in Figure 67, $y = \sqrt{x - 20} - 30$, was obtained by shifting the basic graph 20 units to the right, so the new domain is $[20, \infty)$. On the other hand, the original range, $[0, \infty)$ has been affected by the shift of the graph 30 units downward. The new range is $[-30, \infty)$. The situation in Figure 68 is similar to that of Figure 66. A graph with domain and range both $(-\infty, \infty)$, that is, $y = \sqrt[3]{x}$, has been shifted 20 units to the left and 40 units upward. No matter what direction and magnitude these shifts might have been, the domain and the range are both unaffected. They both remain $(-\infty, \infty)$.   ◨

The feature of graphing calculators that allows us to locate a point on a graph helps us support the results obtained in Example 4. You might wish to experiment with yours to confirm those results.

We continue our discussion on how the graphs of functions may be altered. We saw how adding or subtracting a constant can cause a vertical or horizontal shift. Now we will see how multiplication by a constant alters the graph of a function.

Vertical Stretching

FOR **GROUP DISCUSSION**

In each group of functions below, we give four related functions. Graph the four functions in the first group (Group A), and then answer the questions below regarding those functions. Then repeat the process for Group B and Group C. Use the window specified for each group.

A $[-5, 5]$ by $[-5, 20]$	B $[-5, 15]$ by $[-5, 10]$	C $[-20, 20]$ by $[-10, 10]$
$y_1 = x^2$	$y_1 = \sqrt{x}$	$y_1 = \sqrt[3]{x}$
$y_2 = 2x^2$	$y_2 = 2\sqrt{x}$	$y_2 = 2\sqrt[3]{x}$
$y_3 = 3x^2$	$y_3 = 3\sqrt{x}$	$y_3 = 3\sqrt[3]{x}$
$y_4 = 4x^2$	$y_4 = 4\sqrt{x}$	$y_4 = 4\sqrt[3]{x}$

1. How does the graph of y_2 compare to the graph of y_1?
2. How does the graph of y_3 compare to the graph of y_1?
3. How does the graph of y_4 compare to the graph of y_1?
4. If we choose $c > 4$, how do you think the graph of $y_5 = c \cdot y_1$ would compare to the graph of y_4?

Choosing your own value of c, support your answer to Item 4 graphically.

In each group of functions in the preceding activity, we started with an elementary function y_1 and observed how the graphs of functions of the form $y = c \cdot y_1$ compared with y_1 for positive values of c that began at 2 and became progressively larger. In each case, we obtained a *vertical stretch* of the graph of the basic function with which we started. These observations can be generalized to any function.

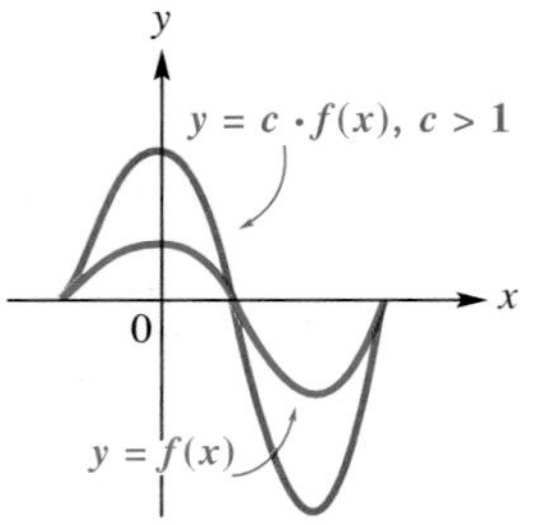

FIGURE 69

VERTICALLY STRETCHING THE GRAPH OF A FUNCTION

If $c > 1$, the graph of $y = c \cdot f(x)$ is obtained by vertically stretching the graph of $y = f(x)$ by a factor of c. In general, the larger the value of c, the greater the stretch.

In Figure 69 we give a graphical interpretation of the statement above.

EXAMPLE 5

Recognizing Vertical
Stretches on
Calculator-Generated
Graphs

Figure 70 shows the graphs of four functions. The function labeled y_1 is the function $f(x) = |x|$. The other three functions, y_2, y_3, and y_4 are defined as follows, but not necessarily in the given order: $2.4|x|$, $3.2|x|$, $4.3|x|$. Determine the correct rule for each graph.

SOLUTION   The values of c here are 2.4, 3.2, and 4.3. The vertical heights of the points with the same x-coordinates on the three graphs will correspond to the magnitudes of these c values. Thus, the graph just above $y_1 = |x|$ will be that of $y = 2.4|x|$, the "highest" graph will be that of $y = 4.3|x|$, and the graph of $y = 3.2|x|$ will lie "between" the others. Therefore, based on our observation of the graphs in the figure, we have $y_2 = 4.3|x|$, $y_3 = 2.4|x|$, and $y_4 = 3.2|x|$.

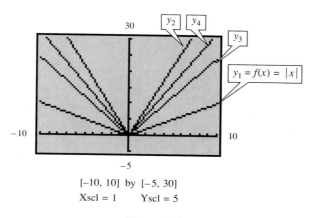

$[-10, 10]$ by $[-5, 30]$
Xscl = 1   Yscl = 5

FIGURE 70

If we were to trace to any point on the graph of y_1 and then move our tracing cursor to the other graphs one by one, we would see that the y-values of the points would be multiplied by the appropriate values of c.   ▯

Vertical Shrinking

FOR **GROUP DISCUSSION**

This discussion parallels the one for vertical stretching given earlier in this section. Follow the same general directions. (*Note:* The fractions $\frac{3}{4}$, $\frac{1}{2}$, and $\frac{1}{4}$ may be entered as their decimal equivalents when plotting the graphs.)

A $[-5, 5]$ by $[-5, 20]$	B $[-5, 15]$ by $[-2, 5]$	C $[-10, 10]$ by $[-2, 10]$
$y_1 = x^2$	$y_1 = \sqrt{x}$	$y_1 = |x|$
$y_2 = \dfrac{3}{4}x^2$	$y_2 = \dfrac{3}{4}\sqrt{x}$	$y_2 = \dfrac{3}{4}|x|$
$y_3 = \dfrac{1}{2}x^2$	$y_3 = \dfrac{1}{2}\sqrt{x}$	$y_3 = \dfrac{1}{2}|x|$
$y_4 = \dfrac{1}{4}x^2$	$y_4 = \dfrac{1}{4}\sqrt{x}$	$y_4 = \dfrac{1}{4}|x|$

(continued on p. 74)

1. How does the graph of y_2 compare to the graph of y_1?
2. How does the graph of y_3 compare to the graph of y_1?
3. How does the graph of y_4 compare to the graph of y_1?
4. If we choose $0 < c < \frac{1}{4}$, how do you think the graph of $y_5 = c \cdot y_1$ would compare to the graph of y_4? Provide support by choosing such a value of c.

In this group discussion, we began with an elementary function y_1 and observed the graphs of $y = c \cdot y_1$, where we began with $c = \frac{3}{4}$ and chose progressively smaller positive values of c. In each case, the graph of y_1 was *vertically shrunk*. These observations, like the ones for vertical stretching, can be generalized to any function.

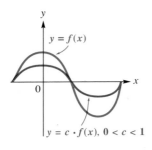

FIGURE 71

> **VERTICALLY SHRINKING THE GRAPH OF A FUNCTION**
> If $0 < c < 1$, the graph of $y = c \cdot f(x)$ is obtained by vertically shrinking the graph of $y = f(x)$ by a factor of c. In general, the smaller the value of c, the greater the shrink.

Figure 71 shows a graphical interpretation of vertical shrinking. Notice that the x-intercepts are not affected by a vertical shrink.

Figure 72 shows the graphs of four functions. The function labeled y_1 is the function $f(x) = x^3$. The other three functions, y_2, y_3, and y_4 are defined as follows, but not necessarily in the given order: $.5x^3$, $.3x^3$, and $.1x^3$. Determine the correct rule for each graph.

EXAMPLE 6

Recognizing Vertical
Shrinks on
Calculator-Generated
Graphs

SOLUTION   The smaller the positive value of c, where $0 < c < 1$, the more of a shrink toward the x-axis will there be. Since we have $c = .5, .3,$ and $.1$, the function rules must be as follows: $y_2 = .1x^3$, $y_3 = .5x^3$, and $y_4 = .3x^3$.   ◨

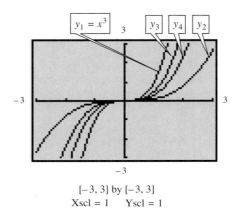

$[-3, 3]$ by $[-3, 3]$
Xscl $= 1$ Yscl $= 1$

FIGURE 72

Reflecting Across an Axis

In the previous section and so far in this one, we have seen how graphs can be transformed by shifting, stretching, and shrinking. We will now examine how graphs can be reflected across an axis.

FOR GROUP DISCUSSION

In each pair of functions, we give two related functions. Graph $y_1 = f(x)$ and $y_2 = -f(x)$ in the standard viewing window, and then answer the questions below for each pair.

A
$y_1 = x^2$
$y_2 = -x^2$

B
$y_1 = |x|$
$y_2 = -|x|$

C
$y_1 = \sqrt{x}$
$y_2 = -\sqrt{x}$

D
$y_1 = x^3$
$y_2 = -x^3$

With respect to the x-axis,

1. how does the graph of y_2 compare with the graph of y_1?
2. how would the graph of $y = -\sqrt[3]{x}$ compare with the graph of $y = \sqrt[3]{x}$, based on your answer to Item 1? Confirm your answer by actual graphing.

Now, in each pair of functions, we give two related functions. Graph $y_1 = f(x)$ and $y_2 = f(-x)$ in the standard viewing window, and then answer the questions below for each pair.

A
$y_1 = \sqrt{x}$
$y_2 = \sqrt{-x}$

B
$y_1 = \sqrt{x - 3}$
$y_2 = \sqrt{-x - 3}$

C
$y_1 = \sqrt[3]{x + 4}$
$y_2 = \sqrt[3]{-x + 4}$

With respect to the y-axis,

3. how does the graph of y_2 compare with the graph of y_1?
4. how would the graph of $y = \sqrt[3]{-x}$ compare with the graph of $y = \sqrt[3]{x}$, based on your answer to Item 3? Confirm your answer by actual graphing.

Based upon the preceding group discussion, we can see how the graph of a function can be reflected across an axis. The results of that discussion are now formally summarized.

REFLECTING THE GRAPH OF A FUNCTION ACROSS AN AXIS

For a function $y = f(x)$,

(a) the graph of $y = -f(x)$ is a reflection of the graph of f across the x-axis.

(b) the graph of $y = f(-x)$ is a reflection of the graph of f across the y-axis.

Figure 73 shows how the reflections described above affect the graph of a function in general.

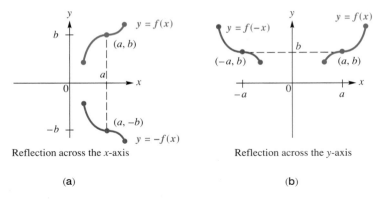

Reflection across the x-axis

Reflection across the y-axis

(a)

(b)

FIGURE 73

Figure 74 shows the graph of a function $y = f(x)$.

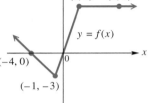

FIGURE 74

(a) Sketch the graph of $y = -f(x)$.

SOLUTION We must reflect the graph across the x-axis. This means that if a point (a, b) lies on the graph of $y = f(x)$, then the point $(a, -b)$ must lie on the graph of $y = -f(x)$. Using the labeled points to assist us, we find the graph of $y = -f(x)$ in Figure 75.

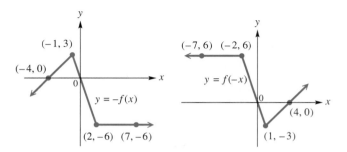

FIGURE 75 **FIGURE 76**

(b) Sketch the graph of $y = f(-x)$.

SOLUTION Here we must reflect the graph across the y-axis, meaning that if a point (a, b) lies on the graph of $y = f(x)$, then the point $(-a, b)$ must lie on the graph of $y = f(-x)$. Again using the labeled points to guide us, we obtain the graph of $y = f(-x)$ as shown in Figure 76.

To illustrate how reflections appear on calculator-generated graphs, observe the two graphs shown in Figures 77 and 78. We use a higher degree polynomial function (covered in detail in college algebra courses) here in Figure 77. The graph of y_2 is a reflection of the graph of y_1 across the x-axis. The display at the bottom of the screen shows that the point $(-2, 23)$ is on the graph of y_1. Therefore, if we were to use the capabilities of the calculator, we should locate the point $(-2, -23)$ on the graph of y_2. (*Note:* The function y_1 is defined as $y_1 = x^4 - 2x + 3$. You might wish to verify the above statement.)

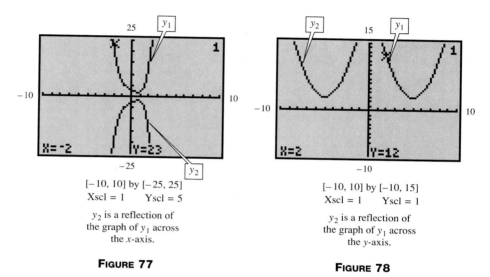

$[-10, 10]$ by $[-25, 25]$
Xscl = 1   Yscl = 5

y_2 is a reflection of
the graph of y_1 across
the x-axis.

FIGURE 77

$[-10, 10]$ by $[-10, 15]$
Xscl = 1   Yscl = 1

y_2 is a reflection of
the graph of y_1 across
the y-axis.

FIGURE 78

Figure 78 illustrates a reflection across the y-axis. The graph of y_2 is a reflection of the graph of y_1 across the y-axis. Notice that the point $(2, 12)$ lies on the graph of y_1. What point must lie on the graph of y_2? (To verify your answer, graph $y_1 = (x - 5)^2 + 3$. So we have $y_2 = (-x - 5)^2 + 3$. Now see if you were correct.)

Combining Transformations of Graphs

The graphs of $y_1 = x^2$ and $y_2 = -2x^2$ are shown in the same viewing window in Figure 79.

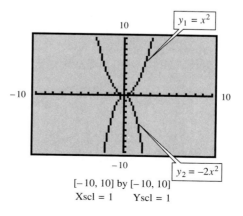

$[-10, 10]$ by $[-10, 10]$
Xscl = 1   Yscl = 1

FIGURE 79

Notice that in terms of the types of transformations we have studied, the graph of y_2 is obtained by vertically stretching the graph of y_1 by a factor of 2 and then reflecting across the x-axis. Thus we have a case of a combination of transformations. As you might expect, we can create an infinite number of functions by vertically stretching or shrinking, shifting left or right, and reflecting across an axis. The next example investigates examples of this type of function.

EXAMPLE **8**

Describing a Combination of Transformations of a Graph

(a)   Describe how the graph of $y = -3(x - 4)^2 + 5$ can be obtained by transforming the graph of $y = x^2$.

SOLUTION   The fact that we have $(x - 4)^2$ in our function indicates that the graph of $y = x^2$ must be shifted 4 units to the *right*. Since the coefficient of $(x - 4)^2$ is -3 (a negative number with absolute value greater than 1), the graph is stretched vertically by a factor of 3 and then reflected across the x-axis. The constant $+5$ indicates that the graph is finally shifted up 5 units. Figure 80 shows the graph of both $y = x^2$ and $y = -3(x - 4)^2 + 5$.

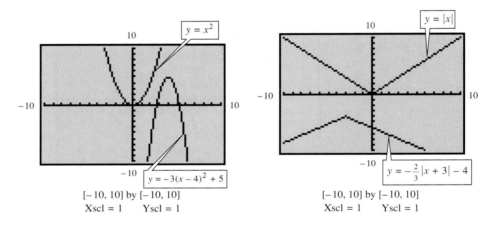

$y = x^2$

$y = -3(x - 4)^2 + 5$

[$-10, 10$] by [$-10, 10$]
Xscl = 1   Yscl = 1

FIGURE 80

$y = |x|$

$y = -\frac{2}{3}|x + 3| - 4$

[$-10, 10$] by [$-10, 10$]
Xscl = 1   Yscl = 1

FIGURE 81

(b)   Give the equation of the function that would be obtained by starting with the graph of $y = |x|$, shifting 3 units to the left, vertically shrinking the graph by a factor of $\frac{2}{3}$, reflecting across the x-axis, and shifting the graph 4 units down, in this order.

SOLUTION   Shifting 3 units to the left means that $|x|$ is transformed to $|x + 3|$. Vertically shrinking by a factor of $\frac{2}{3}$ means multiplying $|x + 3|$ by $\frac{2}{3}$, and reflecting across the x-axis changes $\frac{2}{3}$ to $-\frac{2}{3}$. Finally, shifting 4 units down means subtracting 4. Putting this all together leads to the following equation:

$$y = -\frac{2}{3}|x + 3| - 4.$$

The graphs of both $y = |x|$ and our new function are shown in Figure 81.   ❚

NOTE   The order in which the transformations are made is important. If the same transformations are made in a different order, a different equation can result.

1.5 EXERCISES

Exercises 1–25 are grouped in "fives." For each group of five functions, match the correct graph A, B, C, D, *or* E *to the function without using your calculator. You should use the concepts developed in this section to work these exercises based on visual observation. Then, after you have answered each group of five, use your calculator to check your answers. Every graph in these groups is plotted in the standard viewing window.*

1. $y = x^2 - 3$
2. $y = (x - 3)^2$
3. $y = (x + 3)^2$
4. $y = (x - 3)^2 + 2$
5. $y = (x + 3)^2 + 2$

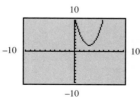
A

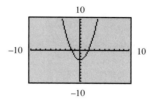
B

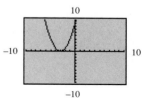
C

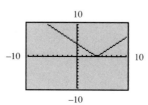
D

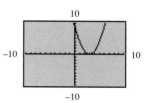
E

6. $y = |x| + 4$
7. $y = |x + 4|$
8. $y = |x - 4|$
9. $y = |x + 4| - 3$
10. $y = |x - 4| - 3$

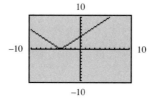
A

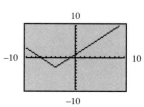
B

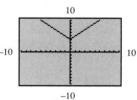
C

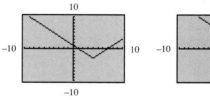
D

E

11. $y = \sqrt{x} + 6$

12. $y = \sqrt{x + 6}$

13. $y = \sqrt{x - 6}$

14. $y = \sqrt{x + 2} - 4$

15. $y = \sqrt{x - 2} - 4$

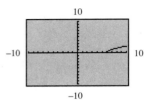

A   B   C

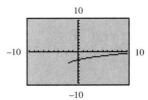
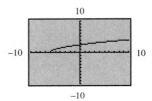

D   E

16. $y = \sqrt[3]{x} + 5$

17. $y = \sqrt[3]{x + 5}$

18. $y = \sqrt[3]{x - 4} + 2$

19. $y = \sqrt[3]{x + 4} + 2$

20. $y = \sqrt[3]{x - 4} - 2$

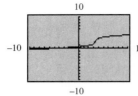
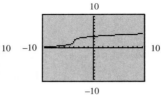
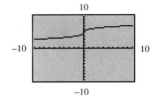

A   B   C

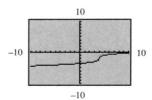

D   E

21. $y = x^3 + 3$

22. $y = (x - 3)^3$

23. $y = (x + 3)^3$

24. $y = (x + 2)^3 - 4$

25. $y = (x - 2)^3 - 4$

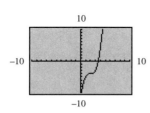
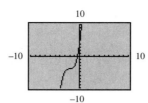
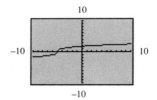

A   B   C

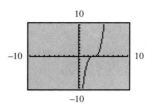
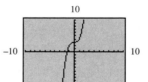

D   E

26. In which quadrant does the vertex of the graph of $y = (x - h)^2 + k$ lie, if $h < 0$ and $k < 0$?

*In Exercises 27–32, use the results of the specified earlier exercises and corresponding graphs to determine (**a**) the domain and (**b**) the range of the given function.*

27. $y = |x + 4| - 3$ (Exercise 9)

28. $y = |x - 4| - 3$ (Exercise 10)

29. $y = \sqrt{x - 2} - 4$ (Exercise 15)

30. $y = \sqrt{x + 2} - 4$ (Exercise 14)

31. $y = \sqrt[3]{x + 5}$ (Exercise 17)

32. $y = \sqrt[3]{x} + 5$ (Exercise 16)

The concepts introduced in this section can be applied to functions whose graphs may not be familiar to you. Given here is the graph of a function whose graph is the top half of a circle. We will call it $y = f(x)$. Each tick mark represents 1 unit.

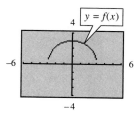

Now match the function specified with the appropriate graph from the choices A, B, C, or D.

33. $y = f(x) + 1$

34. $y = f(x + 1)$

35. $y = f(x - 1)$

36. $y = f(x) - 1$

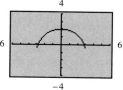

A

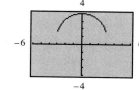

B

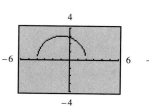

C

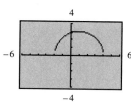

D

Given the graph shown below, sketch by hand the graph of the function described, indicating how the three points labeled on the original graph have been translated.

37. $y = f(x) + 2$

38. $y = f(x) - 2$

39. $y = f(x + 2)$

40. $y = f(x - 2)$

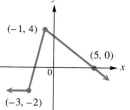

Suppose that h and k are both positive numbers. Match the equation with the correct graph in Exercises 41–44.

41. $y = (x - h)^2 - k$

42. $y = (x + h)^2 - k$

43. $y = (x + h)^2 + k$

44. $y = (x - h)^2 + k$

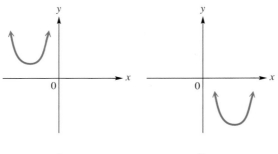

A

B

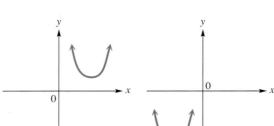

C

D

*R*elating Concepts

*Each function in Exercises 45–50 is a translation of one of the basic functions $y = x^2$, $y = x^3$, $y = \sqrt{x}$, $y = \sqrt[3]{x}$, or $y = |x|$. Using the concepts of increasing and decreasing functions discussed in Section 1.4, determine the interval of the domain over which the function is (**a**) increasing and (**b**) decreasing.*

45.

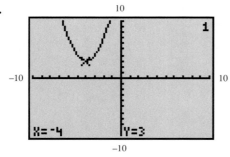

46.

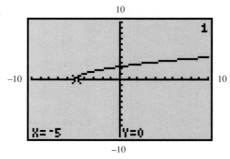

47.

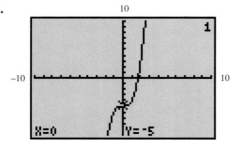

48.

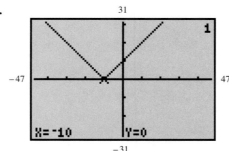

49. $y = \sqrt[3]{x - 12} + 18$ **50.** $y = (x - 10)^2 + 13$

Draw a rough sketch by hand of the graphs of y_1, y_2, and y_3. Do not plot points. In each case, y_2 and y_3 can be graphed by one or more of these: a vertical and/or horizontal shift of the graph of y_1, a vertical stretch or shrink of the graph of y_1, or a reflection of the graph of y_1 across an axis. After you have made your sketches, check by graphing them in an appropriate viewing window of your calculator.

51. $y_1 = x$, $y_2 = x + 3$, $y_3 = x - 3$ **52.** $y_1 = x^2$, $y_2 = x^2 - 1$, $y_3 = x^2 + 1$

53. $y_1 = x^3$, $y_2 = x^3 + 4$, $y_3 = x^3 - 4$ **54.** $y_1 = \sqrt{x}$, $y_2 = \sqrt{x} + 6$, $y_3 = \sqrt{x} - 6$

55. $y_1 = |x|$, $y_2 = |x - 3|$, $y_3 = |x + 3|$ **56.** $y_1 = \sqrt[3]{x}$, $y_2 = \sqrt[3]{x - 4}$, $y_3 = \sqrt[3]{x + 4}$

57. $y_1 = |x|$, $y_2 = |x| - 3$, $y_3 = |x| + 3$ **58.** $y_1 = \sqrt[3]{x}$, $y_2 = \sqrt[3]{x} - 4$, $y_3 = \sqrt[3]{x} + 4$

59. $y_1 = x^2$, $y_2 = (x - 2)^2 + 1$, $y_3 = -(x + 2)^2$ **60.** $y_1 = x^2$, $y_2 = -(x + 3)^2 - 2$, $y_3 = -(x - 4)^2$

61. $y_1 = |x|$, $y_2 = -2|x - 1| + 1$, $y_3 = -\dfrac{1}{2}|x| - 4$ **62.** $y_1 = |x|$, $y_2 = -|x + 1| - 4$, $y_3 = -|x - 1|$

In Exercises 63–68, fill in the blanks with the appropriate responses. (Remember that the vertical stretch or shrink factor is positive.)

63. The graph of $y = -4x^2$ can be obtained from the graph of $y = x^2$ by vertically stretching by a factor of _____ and reflecting across the _____ -axis.

64. The graph of $y = -6\sqrt{x}$ can be obtained from the graph of $y = \sqrt{x}$ by vertically stretching by a factor of _____ and reflecting across the _____ -axis.

65. The graph of $y = -\frac{1}{4}|x + 2| - 3$ can be obtained from the graph of $y = |x|$ by shifting horizontally _____ units to the _____ , vertically shrinking by a factor of _____ , reflecting across the _____-axis, and shifting vertically _____ units in the _____ direction.

66. The graph of $y = -\frac{2}{5}|-x| + 6$ can be obtained from the graph of $y = |x|$ by reflecting across the _____ -axis, vertically shrinking by a factor of _____ , reflecting a second time across the _____ -axis, and shifting vertically _____ units in the _____ direction.

67. The graph of $y = 6\sqrt[3]{(x - 3)}$ can be obtained from the graph of $y = \sqrt[3]{x}$ by shifting horizontally _____ units to the _____ and stretching vertically by a factor of _____ .

68. The graph of $y = .5\sqrt[3]{(x + 2)}$ can be obtained from the graph of $y = \sqrt[3]{x}$ by shifting horizontally _____ units to the _____ and vertically shrinking by a factor of _____ .

Give the equation of the function whose graph is described.

69. The graph of $y = x^2$ is vertically shrunk by a factor of $\frac{1}{2}$, and the resulting graph is shifted 7 units downward.

70. The graph of $y = x^3$ is vertically stretched by a factor of 3. This graph is then reflected across the x-axis. Finally, the graph is shifted 8 units upward.

71. The graph of $y = \sqrt{x}$ is shifted 3 units to the right. This graph is then vertically stretched by a factor of 4.5. Finally, the graph is shifted 6 units downward.

72. The graph of $y = \sqrt[3]{x}$ is shifted 2 units to the left. This graph is then vertically stretched by a factor of 1.5. Finally, the graph is shifted 8 units upward.

In each of Exercises 73–78, the figure shows the graph of a function $y = f(x)$. Sketch by hand the graphs of the functions in parts (a), (b), and (c), and answer the question of part (d).

73. (a) $y = -f(x)$ **(b)** $y = f(-x)$ **(c)** $y = 2f(x)$ **(d)** What is $f(0)$?

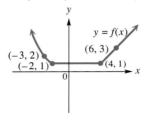

74. (a) $y = -f(x)$ **(b)** $y = f(-x)$ **(c)** $y = 3f(x)$ **(d)** What is $f(4)$?

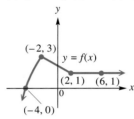

75. (a) $y = -f(x)$ **(b)** $y = f(-x)$ **(c)** $y = f(x + 1)$
(d) What are the x-intercepts of $y = f(x - 1)$?

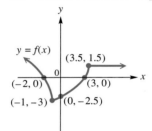

76. (a) $y = -f(x)$ **(b)** $y = f(-x)$ **(c)** $y = -2f(x)$
(d) On what interval of the domain is $f(x) < 0$?

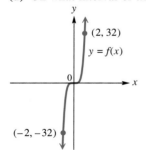

77. (a) $y = -f(x)$ **(b)** $y = f(-x)$ **(c)** $y = .5f(x)$
(d) What symmetry does the graph of $y = f(x)$ exhibit?

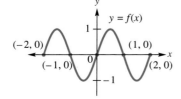

78. (a) $y = -f(x)$ **(b)** $y = f(-x)$ **(c)** $y = 3f(x)$
   (d) What symmetry does the graph of $y = f(x)$ exhibit?

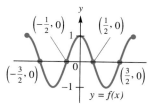

79. If r is an x-intercept of the graph of $y = f(x)$, what statement can be made about an x-intercept of the graph of each of the following? (*Hint:* Draw a picture.)
   (a) $y = -f(x)$ **(b)** $y = f(-x)$ **(c)** $y = -f(-x)$

80. If b is the y-intercept of the graph of $y = f(x)$, what statement can be made about the y-intercept of the graph of each of the following? (*Hint:* Draw a picture.)
   (a) $y = -f(x)$ **(b)** $y = f(-x)$ **(c)** $y = 5f(x)$ **(d)** $y = -3f(x)$

The following sketch shows an example of a function $y = f(x)$ that increases on the interval (a, b) of its domain.

Use this graph as a visual aid, and apply the concepts of reflection introduced in this section to answer each of the following questions. (Make your own sketch if you wish.)

81. Does the function $y = -f(x)$ increase or decrease on the interval (a, b)?

82. Does the function $y = f(-x)$ increase or decrease on the interval $(-b, -a)$?

83. Does the function $y = -f(-x)$ increase or decrease on the interval $(-b, -a)$?

84. If $c > 0$, does the graph of $y = -c \cdot f(x)$ increase or decrease on the interval (a, b)?

85. If the graph of the function $y = f(x)$ is symmetric with respect to the y-axis, what can be said about the symmetry of **(a)** $y = f(-x)$ and **(b)** $y = -f(x)$?

86. If the graph of the function $y = f(x)$ is symmetric with respect to the origin, what can be said about the symmetry of **(a)** $y = f(-x)$ and **(b)** $y = -f(x)$?

Chapter 1   SUMMARY

Ordered pairs of real numbers are plotted in a rectangular coordinate system, and this can be done either by hand or with graphing calculator technology. The window used in graphing a set of ordered pairs with a graphing calculator can drastically alter the picture we view. Ordered pairs of real numbers are called relations, with the set of all first components forming the domain of the relation, and the set of all second components forming the range. A function is a special kind of relation that pairs with each element in its domain one and only one element in its range. One of the simplest, yet most important, kinds of functions is the linear function. Its graph is a straight line, and its equation can always be written in the form $f(x) = ax + b$. The coefficient of x, represented here by a but often represented by m, is the slope of the line, and b represents the y-intercept.

Linear equations are solved analytically by using the properties that allow us to transform equations into simpler equations that have the same solution set. The solution, or root, of a linear equation of the form $f(x) = 0$ is represented by the x-intercept of the associated linear function f. Alternatively, the solution of a linear equation of the form $f(x) = g(x)$ is represented by the x-coordinate of the point of intersection of the graphs of f and g. Linear inequalities are solved analytically using a method similar to solving linear equations. The solution set of the linear inequality $f(x) > 0$ consists of the domain values of the points on the graph of f that lie above the x-axis, while that of $f(x) < 0$ consists of the domain values of the points on the graph of f that lie below the x-axis. Alternatively, the solution set of $f(x) > g(x)$ consists of all domain values common to both f and g for which the graph of f lies above the graph of g. A similar statement can be made for $f(x) < g(x)$, with the word *above* replaced with *below*.

By the square root property the solution of an equation of the form $x^2 = k$ is $x = \sqrt{k}$ or $x = -\sqrt{k}$. If k is negative, the solutions are imaginary numbers. By the quadratic formula, the solution of any quadratic equation $ax^2 + bx + c = 0$ is

$$x = \frac{-b \pm \sqrt{b^2 - 4ac}}{2a}.$$

Quadratic inequalities are solved graphically in a manner similar to the one for solving linear inequalities.

Several important elementary algebraic functions are introduced in this chapter. A student preparing to study trigonometry should be familiar with the graphs of the following algebraic functions: identity, squaring, cubing, square root, cube root, and absolute value. The concepts of symmetry with respect to the y-axis, symmetry with respect to the origin, continuity, and increasing, decreasing, and constant functions are introduced, and these ideas are important in the study of functions. The relation defined by $x = y^2$ is also presented, and its graph exhibits symmetry with respect to the x-axis. A function may be classified as even, odd, or neither, according to the definitions found in Section 1.4. A graph is concave upward in an interval if it "holds water"; it is concave downward if it "dispels water." Local extrema of a function are the maximum or minimum values of the function in some region. Some functions also have an absolute maximum and/or minimum.

The graphs of the basic functions can be altered by vertical and horizontal shifting, vertical stretching and shrinking, and reflections across an axis. These concepts, introduced in Section 1.5, will be used later in this book when we examine the graphs of the circular (trigonometric) functions.

Key Terms

SECTION 1.1

natural numbers
whole numbers
number line
rational numbers
irrational numbers
real numbers
square root
cube root
coordinate of a point
   (on a number line)
coordinate system
a is less than (greater than) b
rectangular (cartesian)
   coordinate system
origin
x-axis
y-axis

coordinate $(x$-$y)$ plane
quadrants
coordinates of a point
   (in the plane)
viewing window
scale

SECTION 1.2

set-builder notation
interval notation
infinity symbol (∞)
compound inequality
relation
domain
range
function
vertical line test
function notation $[f(x)$ notation$]$

SECTION 1.3

analytic method
visual support
linear equation
equivalent equations
addition and multiplication
    properties of equality
conditional equation
intersection-of-graphs method of
    graphical solution
x-intercept method of graphical
    solution
inequality
equivalent inequalities
properties of inequality
linear inequality
quadratic equation in standard
    form
quadratic inequality
quadratic formula

SECTION 1.4

continuity
asymptote
increasing
decreasing
constant
identity function
squaring function
quadratic function
parabola
symmetry with respect to the y-axis
symmetry with respect to the x-axis

cubing function
cubic
symmetry with respect to the origin
inflection point
square root function
cube root function
absolute value function
even function
odd function
extreme point (extremum, extrema)
local minimum point
local minimum value
local maximum point
local maximum value
absolute maximum
absolute maximum value
absolute minimum
absolute minimum value
concavity
concave up
concave down

SECTION 1.5

vertical shift
horizontal shift
vertical stretch
vertical shrink
reflection across the x-axis
reflection across the y-axis

 Chapter 1   **REVIEW EXERCISES**

Refer to the graphs of the linear functions $y_1 = f(x)$ and $y_2 = g(x)$ in the figure at right to match the solution set in the columns on the right with the equation or inequality on the left. Choices may be used once, more than once, or not at all.

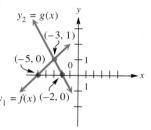

1. $f(x) = g(x)$

2. $f(x) > g(x)$

3. $f(x) < g(x)$

4. $g(x) \geq f(x)$

5. $y_2 - y_1 = 0$

6. $f(x) < 0$

7. $g(x) > 0$

8. $y_2 - y_1 < 0$

A. $(-\infty, -3]$

B. $(-\infty, -3)$

C. $\{3\}$

D. $\{2\}$

E. $\{(3, 2)\}$

F. $\{-5\}$

G. $\{-2\}$

H. $\{0\}$

I. $\{-3\}$

J. $[-3, \infty)$

K. $(-3, \infty)$

L. $[-3]$

M. $(-\infty, -5)$

N. $(-5, \infty)$

O. $(-\infty, -2)$

P. $(-2, \infty)$

Solve each equation using analytic methods.

9. $5[3 + 2(x - 6)] = 3x + 1$

10. $\dfrac{x}{4} - \dfrac{x + 4}{3} = -2$

Exercises 11–13 refer to the linear function

$$f(x) = 5\pi x + (\sqrt{3})x - 6.24(x - 8.1) + (\sqrt[3]{9})x.$$

11. Solve the equation $f(x) = 0$ using graphical methods. Give the solution to the nearest hundredth. Then give an explanation of how you went about solving this equation graphically.

12. Refer to the graph, and give the solution set of $f(x) < 0$.

13. Refer to the graph, and give the solution set of $f(x) \geq 0$.

14. What is the solution set of $f(x) > 0$, based on the screen shown?

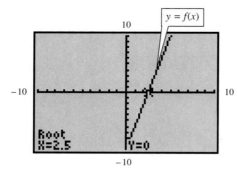

15. What is the solution set of $f(x) - g(x) = 0$, based on the screen shown?

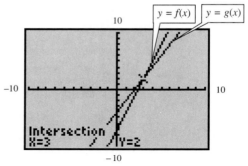

16. If f is a linear function and the solution set of $f(x) \geq 0$ is $(-\infty, -3]$, what is the solution set of $f(x) < 0$?

Match the equation with the graph that most closely resembles its graph.

17. $y = \sqrt{x} + 2$

18. $y = \sqrt{x + 2}$

19. $y = 2\sqrt{x}$

20. $y = -2\sqrt{x}$

21. $y = \sqrt[3]{x} - 2$

22. $y = \sqrt[3]{x - 2}$

23. $y = 2\sqrt[3]{x}$

24. $y = -2\sqrt[3]{x}$

A.
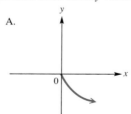

B.
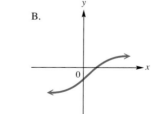

C.
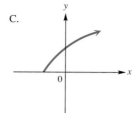

D.
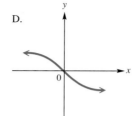

E.
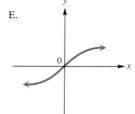

F.
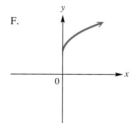

G.
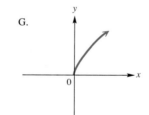

H.
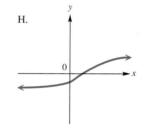

Consider the function whose graph is shown here.

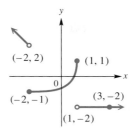

Give the interval(s) over which the function

25. is continuous.

26. increases.

27. decreases.

28. is constant.

29. (a) What is the domain of the function graphed for Exercises 25–28?
  (b) What is the range of the function graphed for Exercises 25–28?

30. Refer to the function defined and determine whether it is even, odd, or neither.

  (a) $f(x) = .2x^3 - 3x + 1$ **(b)** $f(x) = x^4 - \dfrac{1}{2}x^2 - 2$

Determine whether the given relation has x-axis symmetry, y-axis symmetry, origin symmetry, or none of these symmetries. (More than one choice is possible.) Also, if the relation is a function, determine whether it is an even function, an odd function, or neither.

31.

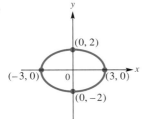

32. $F(x) = x^3 - 6$

33. $y = |x| + 4$

34. $f(x) = \sqrt{x - 5}$

35. $y^2 = x - 5$

36. $f(x) = 3x^4 + 2x^2 + 1$

37. Use the terminology of Section 1.5 to describe how the graph of $y = -3(x + 4)^2 - 8$ can be obtained from the graph of $y = x^2$.

38. Find the rule for the function whose graph is obtained by reflecting the graph of $y = \sqrt{x}$ across the y-axis, then reflecting across the x-axis, shrinking vertically by a factor of $\frac{2}{3}$, and finally translating 4 units upward.

Give the solution set of the equation or inequality, based on the graphs of $y = f(x)$ and $y = g(x)$.

39. $f(x) = g(x)$

40. $f(x) < g(x)$

41. $f(x) \geq g(x)$

42. $f(x) \geq 1$

43. $f(x) \leq 0$

44. $g(x) < 0$

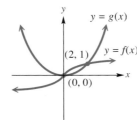

45. Give the solution set of each of the following.
   (a) $2x^2 - 6x - 8 = 0$ **(b)** $2x^2 - 6x - 8 > 0$ **(c)** $2x^2 - 6x - 8 \leq 0$

46. The graph of $P(x) = 2x^2 - 6x - 8$ is shown here. Explain how the graph supports your solution sets in Exercise 45.

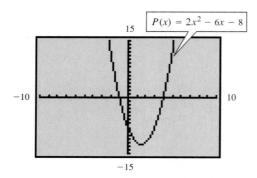

$P(x) = 2x^2 - 6x - 8$

47. Discuss the concavity of the graph in Exercise 46.

Consider the function $P(x) = 3x^2 - x - 1$ for Exercises 48–50.

48. Graph the function in the standard window of your calculator, and use the root-finding capabilities to solve the equation $P(x) = 0$. Then use the quadratic formula to express solutions as exact values.

49. Use your answer to Exercise 48 and the graph of P to solve **(a)** $P(x) > 0$ and **(b)** $P(x) < 0$.

50. Use the capabilities of your calculator to find the coordinates of the local and absolute minimum of the graph. Express coordinates using exact values.

Graph the function $P(x) = -2x^5 + 15x^4 - 21x^3 - 32x^2 + 60x$ in the window $[-8, 8]$ by $[-100, 200]$ to obtain a complete graph. Then use your calculator and the concepts of this chapter to answer Exercises 51–55.

51. How many local maxima does this function have?

52. One local minimum lies on the x-axis and has an integer as its x-value. What are the coordinates of this point?

53. What is the range of P?

54. The graph of P has a local minimum with a negative x-value. Use your calculator to find its coordinates. Express them to the nearest hundredth.

55. Consider the polynomial function $P(x) = -x^4 + 3x^3 + 3x^2 - 7x - 6$. The factored form of the polynomial is $(-x + 2)(x - 3)(x + 1)^2$. Graph the polynomial in the window $[-5, 5]$ by $[-10, 10]$, and solve the equation or inequality.
   (a) $P(x) = 0$ **(b)** $P(x) > 0$ **(c)** $P(x) < 0$

56. Refer to the graphs in Exercises 45–48 in Section 1.5, and give the equations of the functions.

*T*he Trigonometric and Circular Functions and Their Graphs

2.1 GEOMETRIC FOUNDATIONS

Terminology and Basic Concepts ▌ Degree Measure ▌ Angles in Standard
Position ▌ Radian Measure ▌ The Arc Length Formula

Terminology and Basic Concepts

A line may be drawn through the two distinct points *A* and *B*. This line is called **line
AB.** The portion of the line between *A* and *B*, including points *A* and *B* themselves,
is **segment AB.** The portion of line *AB* that starts at *A* and continues through *B*, and
on past *B*, is called **ray AB.** Point *A* is the endpoint of the ray. (See Figure 1.)

An **angle** is formed by rotating a ray around its endpoint. The ray in its initial
position is called the **initial side** of the angle, while the ray in its location after the
rotation is the **terminal side** of the angle. The endpoint of the ray is the **vertex** of
the angle. Figure 2 shows the initial and terminal sides of an angle with vertex *A*.

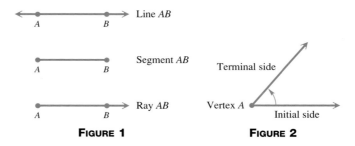

FIGURE 1   **FIGURE 2**

If the rotation of the terminal side is counterclockwise, the angle is **positive.** If the rotation is clockwise, the angle is **negative.** Figure 3 shows two angles, one positive and one negative.

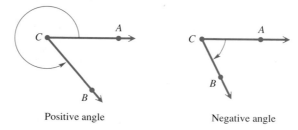

Positive angle   Negative angle

FIGURE 3

An angle can be named by using the name of its vertex. For example, either angle in Figure 3 can be called angle *C*. Alternatively, an angle can be named using three letters, with the vertex letter in the middle. For example, either angle also could be named angle *ACB* or angle *BCA*.

Degree Measure

There are two systems in common use for measuring the size of angles. The most common unit of measure is the **degree.** Degree measure was developed by the Babylonians four thousand years ago. To use degree measure, we assign 360 degrees to a complete rotation of a ray. In Figure 4, notice that the terminal side of the angle corresponds to its initial side when it makes a complete rotation.

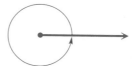

A complete rotation of a ray gives an angle whose measure is 360°.

FIGURE 4

One degree, written 1°, represents $\frac{1}{360}$ of a rotation. Therefore, 90° represents $\frac{90}{360} = \frac{1}{4}$ of a complete rotation, and 180° represents $\frac{180}{360} = \frac{1}{2}$ of a complete rotation. Angles of measure 90° and 180° are shown in Figure 5.

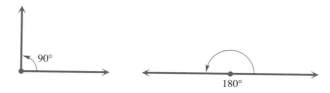

90°   180°

FIGURE 5

Angles are named as shown in the following chart.

Types of Angles

Name	Angle Measure	Example
Acute angle	Between 0° and 90°	60° 82°
Right angle	Exactly 90°	90°
Obtuse angle	Between 90° and 180°	97° 138°
Straight angle	Exactly 180°	180°

If the sum of the measures of two angles is 90°, the angles are called **complementary angles,** or **complements.** Two angles that have a sum of 180° are called **supplementary angles,** or **supplements.**

Consider the angle shown in Figure 6. The angle is formed by the rotation of a ray about its endpoint A. The measure of the rotation is 35°. We will often use notation such as "$A = 35°$" to indicate that the measure of the angle A is 35°.

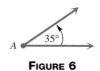

FIGURE 6

Traditionally, portions of a degree have been measured with minutes and seconds. One **minute,** written $1'$, is $\frac{1}{60}$ of a degree.

$$1' = \frac{1°}{60} \quad \text{or} \quad 60' = 1°$$

One **second,** $1''$, is $\frac{1}{60}$ of a minute.

$$1'' = \frac{1'}{60} = \frac{1°}{3600} \quad \text{or} \quad 60'' = 1' \quad \text{or} \quad 3600'' = 1°$$

The measure 12° 42′ 38″ represents 12 degrees, 42 minutes, 38 seconds.

The next example shows how to perform calculations with degrees, minutes, and seconds.

Perform each calculation.

(a) $51° \ 29' + 32° \ 46'$

SOLUTION Add the degrees and the minutes separately.

$$51° \ 29' + 32° \ 46' = (51° + 32°) + (29' + 46') = 83° \ 75'$$

Since $75' = 60' + 15' = 1° \ 15'$, the sum is written

$$83° \ 75' = 83° + (1° \ 15') = 84° \ 15'.$$

(b) $90° - 73° \ 12'$

SOLUTION Write $90°$ as $89° \ 60'$. Then

$$90° - 73° \ 12' = 89° \ 60' - 73° \ 12' = 16° \ 48'. \quad \blacksquare$$

NOTE Many modern graphing calculators have the capability to add and subtract angles given in degrees, minutes, and seconds. Likewise, they are also able to convert from degrees, minutes, seconds to *decimal* degrees (see the explanation immediately following) and vice versa. Read your owner's manual for details on these capabilities.

Because calculators are an integral part of our world today, it is now common to measure angles in **decimal degrees.** For example, $12.4238°$ represents

$$12.4238° = 12 \frac{4238°}{10,000}.$$

The next example shows how to change between decimal degrees and degrees, minutes, and seconds.

(a) Convert $74° \ 8' \ 14''$ to decimal degrees. Round to the nearest thousandth of a degree.

SOLUTION Since $1' = \frac{1°}{60}$ and $1'' = \frac{1°}{3600}$,

$$74° \ 8' \ 14'' = 74° + \frac{8°}{60} + \frac{14°}{3600}$$
$$= 74° + .1333° + .0039°$$
$$= 74.137° \text{ (rounded).}$$

(b) Convert $34.817°$ to degrees, minutes, and seconds.

SOLUTION

$$34.817° = 34° + .817°$$
$$= 34° + (.817)(60') \qquad \text{1 degree = 60 minutes}$$
$$= 34° + 49.02'$$
$$= 34° + 49' + .02'$$
$$= 34° + 49' + (.02)(60'') \qquad \text{1 minute = 60 seconds}$$
$$= 34° + 49' + 1'' \text{ (rounded)}$$
$$= 34° \ 49' \ 1'' \quad \blacksquare$$

Angles in Standard Position

An angle is in **standard position** if its vertex is at the origin and its initial side is along the positive x-axis. The two angles in Figure 7 are in standard position. An angle in standard position is said to lie in the quadrant in which its terminal side lies. For example, an acute angle is in quadrant I and an obtuse angle is in quadrant II. Angles in standard position having their terminal sides along the x-axis or y-axis, such as angles with measures 90°, 180°, 270°, and so on, are called **quadrantal angles.**

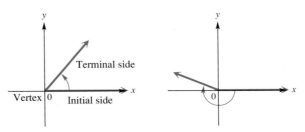

FIGURE 7

A complete rotation of a ray results in an angle of measure 360°. But there is no reason why the rotation need stop at 360°. By continuing the rotation, angles of measure larger than 360° can be produced. The angles in Figure 8(a) have measures 60° and 420°. These two angles have the same initial side and the same terminal side, but different amounts of rotation. Angles that have the same initial side and the same terminal side are called **coterminal angles.** As shown in Figure 8(b), angles with measures 110° and 830° are coterminal.

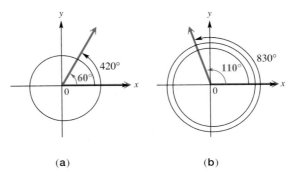

(a)   (b)

FIGURE 8

EXAMPLE **3**

Finding Measures of Coterminal Angles

Find the angles of smallest possible positive measure coterminal with the following angles.

(a) 908°

SOLUTION   Add or subtract 360° as many times as needed to get an angle with measure greater than 0° but less than 360°. Since $908° - 2 \cdot 360° = 908° - 720° = 188°$, an angle of 188° is coterminal with an angle of 908°. See Figure 9.

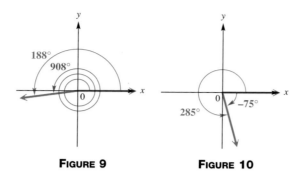

FIGURE 9   **FIGURE 10**

(b) $-75°$

SOLUTION   Use a rotation of $360° + (-75°) = 285°$. See Figure 10.   ◘

Sometimes it is necessary to find an expression that will generate all angles coterminal with a given angle. For example, suppose that we wish to do this for a $60°$ angle. Since any angle coterminal with $60°$ can be obtained by adding an appropriate integer multiple of $360°$ to $60°$, we can let n represent any integer, and the expression

$$60° + n \cdot 360°$$

will represent all such coterminal angles. The table below shows a few of these.

Value of n	Angle Coterminal with $60°$
2	$60° + 2 \cdot 360° = 780°$
1	$60° + 1 \cdot 360° = 420°$
0	$60° + 0 \cdot 360° = 60°$ (the angle itself)
-1	$60° + (-1) \cdot 360° = -300°$

Radian Measure

Degree measure is used primarily in applications of trigonometry. For more theoretical work in mathematics, angles are measured using a more natural system of measurement: radians. To see how an angle is measured in radians, consider Figure 11. The angle θ (θ is the Greek letter *theta*) is in standard position and a circle of radius r is shown with its center at the origin. Angle θ is called a **central angle,** because its vertex is at the center of the circle. As shown in the figure, angle θ intercepts an arc of length s on the circle. If $s = r$, then θ is said to have a measure of one radian. In general, radian measure is defined as follows.

RADIAN MEASURE

Suppose that a circle has radius $r > 0$. Let θ be a central angle of the circle. If θ intercepts an arc of length s on the circle, then the radian measure of θ is given by the formula

$$\theta = \frac{s}{r}.$$

Radian measure can be thought of as a "pure number" (since the units for r and s "cancel") and consequently no symbol for radian measure is needed (as opposed to degree measure, where ° is needed). Therefore, if no unit of measure is indicated for an angle, radian measure is understood.

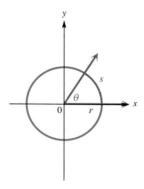

When $r = s$, θ measures
1 radian.

FIGURE 11

The circumference of a circle, the distance around the circle, is given by $C = 2\pi r$, where r is the radius of the circle. The formula $C = 2\pi r$ shows that the radius can be laid off 2π times around a circle. Therefore, an angle of 360°, which corresponds to a complete circle, intercepts an arc equal in length to 2π times the radius of the circle. Because of this, an angle of 360° has a measure of 2π radians:

$$360° = 2\pi \text{ radians}.$$

An angle of 180° is half the size of an angle of 360°, so an angle of 180° has half the radian measure of an angle of 360°.

$$180° = \frac{1}{2}(2\pi) \text{ radians}$$

DEGREE / RADIAN RELATIONSHIP

$$180° = \pi \text{ radians}$$

Dividing both sides of $180° = \pi$ radians by π leads to

$$1 \text{ radian} = \frac{180°}{\pi},$$

or approximately,

$$1 \text{ radian} \approx \frac{180°}{3.1415927} \approx 57.295779° \approx 57°\ 17'\ 45''.$$

Since $180° = \pi$ radians, dividing both sides by 180 gives

$$1° = \frac{\pi}{180} \text{ radians},$$

or, approximately,

$$1° \approx \frac{3.1415927}{180} \text{ radians} \approx .01745329 \text{ radians}.$$

Angle measures can be converted back and forth between degrees and radians by using one of several methods described below.

CONVERTING BETWEEN DEGREES AND RADIANS

1. Proportion: $\dfrac{\text{Radian measure}}{\pi} = \dfrac{\text{Degree measure}}{180}$

2. Formulas:

From	To	Multiply by
Radians	Degrees	$\frac{180°}{\pi}$
Degrees	Radians	$\frac{\pi}{180°}$

3. If a radian measure involves a multiple of π, replace π with $180°$, and simplify in order to convert to degrees.

EXAMPLE 4

Converting Degrees to Radians

Convert each degree measure to radians.

(a) $45°$

SOLUTION   By the proportion method,

$$\frac{\text{Radian measure}}{\pi} = \frac{45}{180}.$$

Multiply both sides by π.

$$\text{Radian measure} = \frac{45\pi}{180} = \frac{\pi}{4}$$

To use the formula method, multiply by $\frac{\pi}{180°}$.

$$45° = 45°\left(\frac{\pi}{180°}\right) = \frac{45\pi}{180} = \frac{\pi}{4} \text{ radians}$$

(b) $240°$

SOLUTION   Using the formula, $240° = 240°\left(\dfrac{\pi}{180°}\right) = \dfrac{4\pi}{3}$ radians.   ◻

EXAMPLE 5

Converting Radians to Degrees

Convert each of the following radian measures to degrees.

(a) $\dfrac{9\pi}{4}$

SOLUTION   Using the proportion method,

$$\frac{\dfrac{9\pi}{4}}{\pi} = \frac{x}{180°}$$

$$\frac{9}{4} = \frac{x}{180°}$$

$$x = \frac{9}{4}(180°) = \mathbf{405°}.$$

(b) $\dfrac{11\pi}{3}$

SOLUTION   Using the formula, we multiply this radian measure by $\frac{180°}{\pi}$ to find the corresponding degree measure.

$$\frac{11\pi}{3} \cdot \frac{180°}{\pi} = \left(\frac{1980\pi}{3\pi}\right)^{°} = 660°$$

(c) $\dfrac{-5\pi}{6}$

SOLUTION   Fractional multiples of π often appear as radian measures. To convert to degrees, we may replace π with $180°$ and simplify.

$$\frac{-5\pi}{6} \text{ radians} = \frac{-5(180°)}{6} = -150°$$

(d)   4.2 (Write the result correct to the nearest minute.)

SOLUTION

$$4.2 \text{ radians} = 4.2\left(\frac{180°}{\pi}\right) \qquad \text{Multiply by } \tfrac{180°}{\pi}.$$

$$= \frac{4.2(180°)}{\pi}$$

$$\approx 240.642274°$$

$$= 240° + .642274°$$

$$= 240° + .642274(60')$$

$$\approx 240° + 38' + 32.186''$$

$$\approx 240° \ 39'. \quad \blacksquare$$

TECHNOLOGICAL NOTE
If your calculator does not have the capability to convert between radians and degrees, you may wish to write a program that will perform these conversions for you.

NOTE   Check your owner's manual to see if your graphing calculator has the capability of converting from radians to degrees and vice versa.

Understanding the distinction between degree measure and radian measure is essential. To see why, look at Figures 12 and 13, which show angles measuring 30 radians and 30 degrees. These angles are not at all the same.

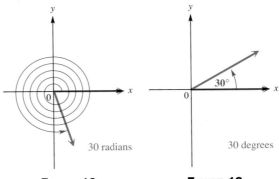

30 radians   30 degrees

FIGURE 12   **FIGURE 13**

Earlier we saw how all angles coterminal with a given degree-measured angle can be expressed by adding multiples of 360°. Similarly, if we wish to express all radian-measured angles coterminal with $\frac{\pi}{3}$, for example, we add an integer multiple of 2π to $\frac{\pi}{3}$: $\frac{\pi}{3} + 2n\pi$. The table below shows a few of these angles.

Value of n	Angle Coterminal with $\frac{\pi}{3}$
2	$\frac{\pi}{3} + 2\,(2\pi) = \frac{13\pi}{3}$
1	$\frac{\pi}{3} + 1\,(2\pi) = \frac{7\pi}{3}$
−1	$\frac{\pi}{3} + (-1)\,(2\pi) = -\frac{5\pi}{3}$
−2	$\frac{\pi}{3} + (-2)\,(2\pi) = -\frac{11\pi}{3}$

The Arc Length Formula

The relationship for radian measure, $\theta = \frac{s}{r}$, can also be expressed in an alternative form which is useful in finding the length of an arc of a circle.

LENGTH OF ARC

The length s of the arc intercepted on a circle of radius r by a central angle of measure θ radians is given by the product of the radius and the radian measure of the angle, or

$$s = r\theta, \quad \theta \text{ in radians.}$$

This formula is a good example of the usefulness of radian measure. To see why, try to write the equivalent formula for an angle measured in degrees.

CAUTION When applying the formula $s = r\theta$, the value of θ *must be* expressed in *radians.*

EXAMPLE 6

Finding Arc Length Using $s = r\theta$

A circle has a radius of 18.2 centimeters. Find the length of the arc intercepted by a central angle having each of the following measures.

(a) $\dfrac{3\pi}{8}$ radians

SOLUTION Here $r = 18.2$ cm and $\theta = \frac{3\pi}{8}$. Since $s = r\theta$,

$$s = 18.2 \left(\frac{3\pi}{8} \right) \text{ centimeters}$$

$$s = \frac{54.6\pi}{8} \text{ centimeters} \qquad \text{The exact answer}$$

or $\qquad s \approx 21.4$ centimeters. $\qquad$ Calculator approximation

(b) 144°

SOLUTION   The formula $s = r\theta$ requires that θ be measured in radians. First, convert θ to radians.

$$144° = 144°\left(\frac{\pi}{180°}\right) \text{ radians} \qquad \text{Multiply by } \tfrac{\pi}{180°}.$$

$$144° = \frac{4\pi}{5} \text{ radians}$$

Now

$$s = 18.2\left(\frac{4\pi}{5}\right) \text{ centimeters} \qquad \text{Use } s = r\theta.$$

$$s = \frac{72.8\pi}{5} \text{ centimeters,}$$

or

$$s \approx 45.7 \text{ centimeters.} \quad \blacksquare$$

<table>
<tr><td>EXAMPLE 7</td></tr>
<tr><td>Finding the Distance Between Two Cities Using Latitudes</td></tr>
</table>

TECHNOLOGICAL NOTE
See the Further Explorations discussion at the end of the exercises for this section for an interesting alternate approach to solving problems like the one in Example 7, employing the TABLE feature of the graphing calculator.

Reno, Nevada, is approximately due north of Los Angeles. The latitude of Reno is 40° N, while that of Los Angeles is 34° N. (The N in 34° N means *north* of the equator.) If the radius of the earth is 6400 kilometers, find the north-south distance between the two cities.

SOLUTION   Latitude gives the measure of a central angle with vertex at the earth's center whose initial side goes through the earth's equator and whose terminal side goes through the given location. As shown in Figure 14 the central angle for Reno and Los Angeles is 6°. The distance between the two cities can thus be found by the formula $s = r\theta$, after 6° is first converted to radians.

$$6° = 6°\left(\frac{\pi}{180°}\right) = \frac{\pi}{30} \text{ radians}$$

The distance between the two cities is

$$s = r\theta$$

$$s = 6400\left(\frac{\pi}{30}\right) \text{ kilometers} \qquad r = 6400,\ \theta = \tfrac{\pi}{30}$$

$$\approx 670 \text{ kilometers.} \quad \blacksquare$$

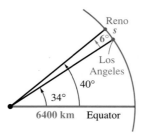

FIGURE 14

2.1 EXERCISES

To the Student: Calculator Considerations

You should read your owner's manual to determine whether your calculator has the capability of converting from degrees, minutes, and seconds to decimal degrees, and vice versa. Also, find out whether it has a function to convert between degree measure and radian measure.

The real number π is used extensively in trigonometry. Learn where the π key is located. Also, be aware that when exact values involving π are required, such as $\frac{\pi}{3}$ and $\frac{3\pi}{4}$, decimal approximations given by the calculator are not acceptable.

Find (a) the complement and (b) the supplement of each of the following angles.

1. $30°$ **2.** $60°$ **3.** $45°$ **4.** $15°$

5. $\dfrac{\pi}{6}$ **6.** $\dfrac{\pi}{3}$ **7.** $\dfrac{\pi}{4}$ **8.** $\dfrac{\pi}{12}$

9. What is the degree measure of the angle that is its own complement?

10. What is the radian measure of the angle that is its own supplement?

Find the angle of smallest positive measure that is coterminal with the given angle.

11. $-40°$ **12.** $-98°$ **13.** $450°$ **14.** $539°$

15. $-\dfrac{\pi}{4}$ **16.** $-\dfrac{\pi}{3}$ **17.** $-\dfrac{3\pi}{2}$ **18.** $-\pi$

Convert each of the following degree measures to radians. Leave answers as multiples of π.

19. $30°$ **20.** $60°$ **21.** $45°$ **22.** $90°$

Convert each of the following radian measures to degrees.

23. $\dfrac{2\pi}{3}$ **24.** $\dfrac{5\pi}{6}$ **25.** $-\dfrac{11\pi}{3}$ **26.** $-\dfrac{7\pi}{4}$

Convert each angle measure to decimal degrees. Use a calculator, and round to the nearest thousandth of a degree.

27. $20°\ 54'$ **28.** $38°\ 42'$ **29.** $91°\ 35'\ 54''$

30. $34°\ 51'\ 35''$ **31.** $-274°\ 18'\ 59''$ **32.** $-165°\ 51'\ 09''$

Convert each angle measure to degrees, minutes, and seconds. Round seconds to whole units. Use a calculator.

33. $31.4296°$ **34.** $59.0854°$ **35.** $89.9004°$

36. $102.3771°$ **37.** $-178.5994°$ **38.** $-122.6853°$

Give an expression that generates all angles coterminal with the given angle. Let n represent any integer. Also, give the quadrant of all such angles.

39. $30°$ **40.** $45°$ **41.** $230°$

42. $135°$ **43.** $270°$ **44.** $-90°$

Give an expression that generates all angles coterminal with the given angle. Let n represent any integer. Also, give the quadrant of all such angles.

45. $\dfrac{\pi}{4}$ **46.** $\dfrac{\pi}{6}$ **47.** $\dfrac{3\pi}{4}$ **48.** $-\dfrac{7\pi}{6}$

Sketch the following angles in standard position. Draw an arrow representing the correct amount of rotation. Find the measure of two other angles, one positive and one negative, that are coterminal with each angle. Give the quadrant of the angle.

49. $75°$ **50.** $122°$ **51.** $-52°$

52. $-159°$ **53.** $\dfrac{5\pi}{3}$ **54.** $-\dfrac{\pi}{2}$

Find the length of the arc intercepted by a central angle θ in a circle of radius r. Give calculator approximations in your answers.

55. $r = 12.3$ cm, $\theta = \dfrac{2\pi}{3}$ **56.** $r = .892$ cm, $\theta = \dfrac{11\pi}{10}$

57. $r = 253$ m, $\theta = \dfrac{2\pi}{5}$ **58.** $r = 120$ mm, $\theta = \dfrac{\pi}{9}$

59. $r = 4.82$ m, $\theta = 60°$ **60.** $r = 71.9$ cm, $\theta = 135°$

Find the distance in kilometers between the pair of cities whose latitudes are given. Assume that the cities are on a north-south line, and that the radius of the earth is 6400 kilometers. Round answers to the nearest hundred.

61. Madison, South Dakota, 44° N, and Dallas, Texas, 33° N

62. Charleston, South Carolina, 33° N, and Toronto, Ontario, 43° N

63. Panama City, Panama, 9° N, and Pittsburgh, Pennsylvania, 40° N

64. Farmersville, California, 36° N, and Penticton, British Columbia, 49° N

65. New York City, New York, 41° N, and Lima, Peru, 12° S

66. Halifax, Nova Scotia, 45° N, and Buenos Aires, Argentina, 34° S

*F**urther Explorations***

You can use the TABLE feature of your graphing calculator to work problems like Exercises 61–66. In the TABLE in the figure, the *x* values are decimal approximations of increments (ΔTbl) of $\frac{\pi}{180}$.

Recall: to convert degrees to radians, multiply the degree measure by $\frac{\pi}{180°}$.

The expression in Y_1 gives the degree measure of the radian measure listed in X. This means that ΔTbl is equivalent to increments of one degree. The expression in Y_2 uses the formula for distance between points on a circle ($s = r\theta$) to find the north-south distance in kilometers between the given measure of latitude and the equator. X must be in radians for the formula $s = r\theta$ to work.

X	Y₁	Y₂
0	0	0
.01745	1	111.7
.03491	2	223.4
.05236	3	335.1
.06981	4	446.8
.08727	5	558.51
.10472	6	670.21

X=0

EXAMPLE   Find the vertical distance between 2° and 5°.

SOLUTION   From the TABLE in the figure, we see that moving 5° north (or south) from the equator is equivalent to traveling 558.51 kilometers, and 2° north (or south) is equivalent to traveling 223.4 kilometers. Then the north-south distance from 2° to 5° is the difference between these two measurements: 558.51 km − 223.4 km ≈ 335.1 km. This can also be confirmed in the TABLE by noting that 5° − 2° = 3° and the north-south distance along the earth's surface, equivalent to 3°, is 335.1 km.

1. If your graphing calculator has a TABLE feature, find expressions for Y_1 and Y_2 that will duplicate the TABLE in the figure.

2. Support your answers to Exercises 61−66 numerically, using the TABLE feature and the expressions in Exercise 1.

2.2 THE TRIGONOMETRIC FUNCTIONS AND BASIC IDENTITIES

Introduction to the Trigonometric Functions ▮ The Reciprocal Identities ▮ The Pythagorean and Quotient Identities

Introduction to the Trigonometric Functions

The study of trigonometry covers the six trigonometric functions defined in this section. To define these six basic functions, we start with an angle θ in standard position. Choose any point P having coordinates (x, y) on the terminal side of angle θ. (The point P must not be the vertex of the angle.) See Figure 15.

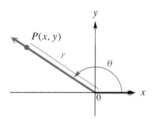

FIGURE 15

The distance r from $P(x, y)$ to the origin, $(0, 0)$, can be found from the distance formula.*

$$r = \sqrt{(x - 0)^2 + (y - 0)^2}$$
$$r = \sqrt{x^2 + y^2}$$

Notice that $r > 0$, since distance is never negative.

The six trigonometric functions of angle θ, called **sine, cosine, tangent, cotangent, secant,** and **cosecant,** are defined in terms of x, y, and r. We use the standard abbreviations in these definitions.

---

* From algebra, the distance from (x_1, y_1) to (x_2, y_2) is $\sqrt{(x_2 - x_1)^2 + (y_2 - y_1)^2}$.

> ## TRIGONOMETRIC FUNCTIONS
>
> Let (x, y) be a point other than the origin on the terminal side of an angle θ in standard position. The distance from the point to the origin is $r = \sqrt{x^2 + y^2}$. The six trigonometric functions of θ are:
>
> $$\sin \theta = \frac{y}{r} \qquad\qquad \csc \theta = \frac{r}{y} \quad (y \neq 0)$$
>
> $$\cos \theta = \frac{x}{r} \qquad\qquad \sec \theta = \frac{r}{x} \quad (x \neq 0)$$
>
> $$\tan \theta = \frac{y}{x} \quad (x \neq 0) \qquad \cot \theta = \frac{x}{y} \quad (y \neq 0).$$

NOTE Because of the restrictions on the denominators in the definitions of tangent, cotangent, secant, and cosecant, some angles will have undefined function values. This will be discussed in more detail later.

EXAMPLE 1

Finding the Function Values of an Angle

The terminal side of an angle α in standard position goes through the point $(8, 15)$. Find the values of the six trigonometric functions of angle α.

SOLUTION Figure 16 shows angle α. Since $r = \sqrt{x^2 + y^2}$,

$$r = \sqrt{8^2 + 15^2}$$
$$= \sqrt{64 + 225}$$
$$= \sqrt{289}$$
$$= 17.$$

The values of the six trigonometric functions of angle α can now be found with the definitions given above.

$$\sin \alpha = \frac{y}{r} = \frac{15}{17} \qquad \csc \alpha = \frac{r}{y} = \frac{17}{15}$$

$$\cos \alpha = \frac{x}{r} = \frac{8}{17} \qquad \sec \alpha = \frac{r}{x} = \frac{17}{8}$$

$$\tan \alpha = \frac{y}{x} = \frac{15}{8} \qquad \cot \alpha = \frac{x}{y} = \frac{8}{15} \qquad \blacksquare$$

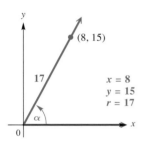

$x = 8$
$y = 15$
$r = 17$

FIGURE 16

EXAMPLE 2

Finding the Function Values of an Angle

The terminal side of angle β in standard position goes through $(-3, -4)$. Find the values of the six trigonometric functions of β.

SOLUTION As shown in Figure 17, $x = -3$ and $y = -4$. The value of r is

$$r = \sqrt{(-3)^2 + (-4)^2}$$
$$r = \sqrt{25}$$
$$r = 5.$$

(Remember that $r > 0$.) Then by the definitions of the trigonometric functions,

$$\sin \beta = \frac{-4}{5} = -\frac{4}{5} \qquad \csc \beta = \frac{5}{-4} = -\frac{5}{4}$$

$$\cos \beta = \frac{-3}{5} = -\frac{3}{5} \qquad \sec \beta = \frac{5}{-3} = -\frac{5}{3}$$

$$\tan \beta = \frac{-4}{-3} = \frac{4}{3} \qquad \cot \beta = \frac{-3}{-4} = \frac{3}{4}. \qquad \blacksquare$$

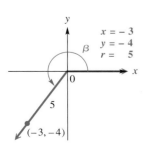

$x = -3$
$y = -4$
$r = 5$

FIGURE 17

The choice of the point on the terminal side is arbitrary. It can be shown that the trigonometric functions of an angle can be found by using *any* point (other than the origin) on the terminal side of the angle.

If the equation of the ray forming the terminal side of an angle in standard position is known, the trigonometric function values can be found as shown in the next example.

EXAMPLE 3

Finding the Function Values of an Angle

Find the six trigonometric function values of the angle θ in standard position, if the terminal side of θ is defined by $x + 2y = 0$, $x \geq 0$.

SOLUTION   The angle is shown in Figure 18. We can use *any* point except $(0, 0)$ on the terminal side of θ to find the trigonometric function values, so if we let $x = 2$, for example, we can find the corresponding value of y.

$$x + 2y = 0, \; x \geq 0$$
$$2 + 2y = 0 \qquad \text{Arbitrarily choose } x = 2.$$
$$2y = -2$$
$$y = -1$$

The point $(2, -1)$ lies on the terminal side, and the corresponding value of r is $r = \sqrt{2^2 + (-1)^2} = \sqrt{5}$. Now use the definitions of the trigonometric functions.

$$\sin \theta = \frac{y}{r} = \frac{-1}{\sqrt{5}} = -\frac{\sqrt{5}}{5} \qquad \csc \theta = \frac{r}{y} = \frac{\sqrt{5}}{-1} = -\sqrt{5}$$

$$\cos \theta = \frac{x}{r} = \frac{2}{\sqrt{5}} = \frac{2\sqrt{5}}{5} \qquad \sec \theta = \frac{r}{x} = \frac{\sqrt{5}}{2}$$

$$\tan \theta = \frac{y}{x} = \frac{-1}{2} = -\frac{1}{2} \qquad \cot \theta = \frac{x}{y} = \frac{2}{-1} = -2 \qquad ∎$$

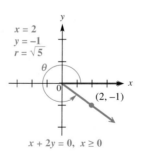

$x = 2$
$y = -1$
$r = \sqrt{5}$

$(2, -1)$

$x + 2y = 0, \; x \geq 0$

FIGURE 18

FOR **GROUP DISCUSSION**

Either individually or with a group of students, rework Example 3 using a different value for x. Find the corresponding y-value, and then show that the six trigonometric function values you obtain are the same as the ones shown above.

In the definition of the trigonometric functions, r is the distance from the origin to the point (x, y). Distance is never negative, so $r > 0$. If we choose a point (x, y) in quadrant I, then both x and y will be positive. Since $r > 0$, all six of the fractions used in the definitions of the trigonometric functions will be positive, so that the values of all six functions will be positive in quadrant I.

A point (x, y) in quadrant II has $x < 0$ and $y > 0$. This makes the values of sine and cosecant positive for quadrant II angles, while the other four functions take on negative values. Similar results can be obtained for the other quadrants, as summarized in the following chart.

SIGNS OF FUNCTION VALUES

θ in quadrant	$\sin \theta$	$\cos \theta$	$\tan \theta$	$\cot \theta$	$\sec \theta$	$\csc \theta$
I	+	+	+	+	+	+
II	+	−	−	−	−	+
III	−	−	+	+	−	−
IV	−	+	−	−	+	−

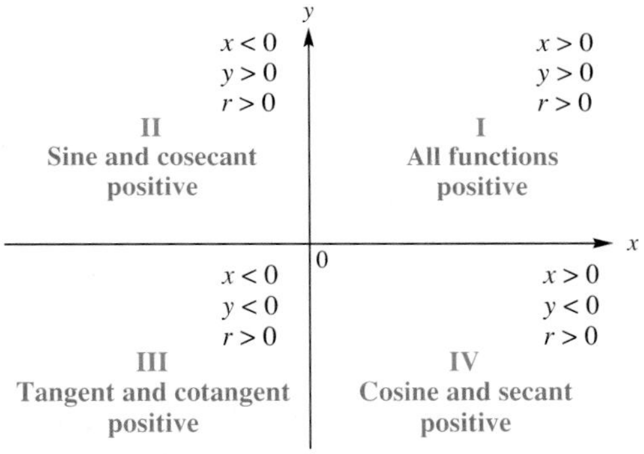

$x < 0$
$y > 0$
$r > 0$
II
Sine and cosecant positive

$x > 0$
$y > 0$
$r > 0$
I
All functions positive

$x < 0$
$y < 0$
$r > 0$
III
Tangent and cotangent positive

$x > 0$
$y < 0$
$r > 0$
IV
Cosine and secant positive

EXAMPLE 4

Identifying the Quadrant of an Angle

Identify the quadrant (or quadrants) for any angle θ that satisfies $\sin \theta > 0$, $\tan \theta < 0$.

SOLUTION   Since $\sin \theta > 0$ in quadrants I and II, while $\tan \theta < 0$ in quadrants II and IV, both conditions are met only in quadrant II.   ∎

Recall that a quadrantal angle has its terminal side coinciding with an axis. If the terminal side of an angle in standard position lies along the y-axis, any point on this terminal side has x-coordinate 0. Similarly, any angle with terminal side on the x-axis has y-coordinate 0 for any point on the terminal side. Since the values of x and y appear in the denominators of some of the trigonometric functions, and since a fraction is undefined if its denominator is 0, some of the trigonometric function values of quadrantal angles will be undefined.

EXAMPLE **5**

Finding
Trigonometric
Function Values of a
Quadrantal Angle

Find the values of the six trigonometric functions for an angle of 90°.

SOLUTION   First, select any point on the terminal side of a 90° angle. Let us select the point $(0, 1)$, as shown in Figure 19. Here $x = 0$ and $y = 1$. Verify that $r = 1$. Then, by the definition of the trigonometric functions,

$$\sin 90° = \frac{1}{1} = 1 \qquad \csc 90° = \frac{1}{1} = 1$$

$$\cos 90° = \frac{0}{1} = 0 \qquad \sec 90° = \frac{1}{0} \text{(undefined)}$$

$$\tan 90° = \frac{1}{0} \text{(undefined)} \qquad \cot 90° = \frac{0}{1} = 0. \quad \blacksquare$$

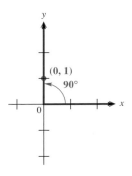

FIGURE 19

In Example 5 we chose the point on the terminal side of the angle that is 1 unit from the origin. However, we could have chosen any point on the terminal side other than the origin itself. We will see that the idea of choosing the point 1 unit from the origin leads us to the "unit circle" concept, an important tool in the approach to *circular functions*, discussed later in this chapter.

FOR **GROUP DISCUSSION**

Refer to Example 5 and discuss how you would go about finding the trigonometric function values of an angle of $\frac{\pi}{2}$ radians.

The conditions under which the trigonometric function values of quadrantal angles are undefined are summarized here.

If the terminal side of a quadrantal angle lies along the y-axis, the tangent and secant functions are undefined since $x = 0$ on the y-axis. If it lies along the x-axis, the cotangent and cosecant functions are undefined since $y = 0$ on the x-axis.

Since the most commonly used quadrantal angles are

$$0° = 0 \text{ radians},$$

$$90° = \frac{\pi}{2} \text{ radians},$$

$$180° = \pi \text{ radians},$$

$$270° = \frac{3\pi}{2} \text{ radians},$$

$$360° = 2\pi \text{ radians},$$

the values of the functions of these angles are summarized in the following table. This table is for reference only; you should be able to reproduce it quickly.

QUADRANTAL ANGLE FUNCTION VALUES

θ	$\sin \theta$	$\cos \theta$	$\tan \theta$	$\cot \theta$	$\sec \theta$	$\csc \theta$
$0° = 0$	0	1	0	Undefined	1	Undefined
$90° = \frac{\pi}{2}$	1	0	Undefined	0	Undefined	1
$180° = \pi$	0	-1	0	Undefined	-1	Undefined
$270° = \frac{3\pi}{2}$	-1	0	Undefined	0	Undefined	-1
$360° = 2\pi$	0	1	0	Undefined	1	Undefined

FOR **GROUP DISCUSSION**

1. Learn how to put your calculator in the *degree mode.* Then verify the sine, cosine, and tangent values for the quadrantal angles shown above.
2. Learn how to put your calculator in the *radian mode.* Then repeat Item 1.

The Reciprocal Identities

The definitions of the trigonometric functions considered earlier in this section were written so that functions directly across from one another are reciprocals. Since $\sin \theta = \frac{y}{r}$ and $\csc \theta = \frac{r}{y}$,

$$\sin \theta = \frac{1}{\csc \theta} \quad \text{and} \quad \csc \theta = \frac{1}{\sin \theta}.$$

Also, $\cos \theta$ and $\sec \theta$ are reciprocals, as are $\tan \theta$ and $\cot \theta$. In summary, we have the **reciprocal identities** that hold for any angle θ that does not lead to a zero denominator.

RECIPROCAL IDENTITIES

$$\sin \theta = \frac{1}{\csc \theta} \qquad \csc \theta = \frac{1}{\sin \theta}$$

$$\cos \theta = \frac{1}{\sec \theta} \qquad \sec \theta = \frac{1}{\cos \theta}$$

$$\tan \theta = \frac{1}{\cot \theta} \qquad \cot \theta = \frac{1}{\tan \theta}$$

NOTE When studying identities, be aware that various forms exist. These forms are obtained by algebraic manipulation. For example,

$$\sin \theta = \frac{1}{\csc \theta}$$

can also be written

$$\csc \theta = \frac{1}{\sin \theta} \qquad \text{and} \qquad (\sin \theta)(\csc \theta) = 1.$$

You should become familiar with all forms of these identities.

EXAMPLE 6

Using the Reciprocal Identities

Find each function value.

(a) $\cos \theta$, if $\sec \theta = \frac{5}{3}$

SOLUTION Since $\cos \theta = \frac{1}{\sec \theta}$,

$$\cos \theta = \frac{1}{\frac{5}{3}} = \frac{3}{5}.$$

(b) $\sin \theta$, if $\csc \theta = -\frac{\sqrt{12}}{2}$

SOLUTION

$$\sin \theta = \frac{1}{-\dfrac{\sqrt{12}}{2}}$$

$$= \frac{-2}{\sqrt{12}}$$

$$= \frac{-2}{2\sqrt{3}} \qquad \sqrt{12} = \sqrt{4 \cdot 3} = 2\sqrt{3}$$

$$= \frac{-1}{\sqrt{3}}$$

$$= \frac{-\sqrt{3}}{3} \qquad \text{Multiply by } \frac{\sqrt{3}}{\sqrt{3}} \text{ to rationalize the denominator.} \quad \blacksquare$$

The reciprocal identities are necessary in evaluating the secant, cosecant, and cotangent function values of an angle on a calculator, because calculators do not have keys specifically marked for these functions. For example, to verify the earlier statement that sec $180° = -1$, we would first be sure that the calculator is in degree mode. Then we would enter $\frac{1}{\cos 180°}$ and obtain the result -1.

The Pythagorean and Quotient Identities

As shown in Figure 20, if the point (x, y) lies on the terminal side of θ in standard position, by the Pythagorean theorem,

$$x^2 + y^2 = r^2.$$

Dividing both sides of this equation by r^2 gives

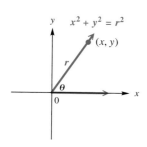

FIGURE 20

$$\frac{x^2}{r^2} + \frac{y^2}{r^2} = \frac{r^2}{r^2},$$

or

$$\left(\frac{x}{r}\right)^2 + \left(\frac{y}{r}\right)^2 = 1.$$

Since $\sin \theta = \frac{y}{r}$ and $\cos \theta = \frac{x}{r}$, this result becomes

$$(\sin \theta)^2 + (\cos \theta)^2 = 1,$$

or, as it is usually written,

$$\sin^2 \theta + \cos^2 \theta = 1.$$

Starting with $x^2 + y^2 = r^2$ and dividing through by x^2 gives

$$\frac{x^2}{x^2} + \frac{y^2}{x^2} = \frac{r^2}{x^2}$$

$$1 + \left(\frac{y}{x}\right)^2 = \left(\frac{r}{x}\right)^2$$

$$1 + (\tan \theta)^2 = (\sec \theta)^2$$

or

$$\tan^2 \theta + 1 = \sec^2 \theta.$$

On the other hand, dividing through by y^2 leads to

$$1 + \cot^2 \theta = \csc^2 \theta.$$

These three identities are called the **Pythagorean identities** since the original equation that led to them, $x^2 + y^2 = r^2$, comes from the Pythagorean theorem.

PYTHAGOREAN IDENTITIES

$$\sin^2 \theta + \cos^2 \theta = 1 \qquad \tan^2 \theta + 1 = \sec^2 \theta$$
$$1 + \cot^2 \theta = \csc^2 \theta$$

As before, we have given only one form of each identity. However, algebraic transformations can be made to get equivalent identities. For example, by subtracting $\sin^2 \theta$ from both sides of $\sin^2 \theta + \cos^2 \theta = 1$ we get the equivalent identity

$$\cos^2 \theta = 1 - \sin^2 \theta.$$

You should be able to transform these identities quickly, and also recognize their equivalent forms.

Recall that $\sin \theta = \frac{y}{r}$ and $\cos \theta = \frac{x}{r}$. Consider the quotient of $\sin \theta$ and $\cos \theta$, where $\cos \theta \neq 0$.

$$\frac{\sin \theta}{\cos \theta} = \frac{\dfrac{y}{r}}{\dfrac{x}{r}} = \frac{y}{r} \div \frac{x}{r} = \frac{y}{r} \cdot \frac{r}{x} = \frac{y}{x} = \tan \theta$$

Similarly, it can be shown that $\frac{\cos \theta}{\sin \theta} = \cot \theta$, for $\sin \theta \neq 0$. Thus we have two more identities, called the **quotient identities.**

QUOTIENT IDENTITIES

$$\frac{\sin \theta}{\cos \theta} = \tan \theta \qquad \frac{\cos \theta}{\sin \theta} = \cot \theta$$

By using the identities discussed so far, it is possible to find all trigonometric function values of an angle in standard position if we know one function value and the quadrant of the angle. The final example of this section illustrates how this is done.

EXAMPLE 7

Finding Other Function Values Given One Value and the Quadrant

Find all other trigonometric function values of α, if $\cos \alpha = -\frac{\sqrt{3}}{4}$ and α is in quadrant II.

SOLUTION   One way to begin is to use $\sin^2 \alpha + \cos^2 \alpha = 1$, and replace $\cos \alpha$ with $-\frac{\sqrt{3}}{4}$.

$$\sin^2 \alpha + \left(-\frac{\sqrt{3}}{4}\right)^2 = 1 \qquad \text{Replace } \cos \alpha \text{ with } -\frac{\sqrt{3}}{4}.$$

$$\sin^2 \alpha + \frac{3}{16} = 1$$

$$\sin^2 \alpha = \frac{13}{16} \qquad \text{Subtract } \frac{3}{16}.$$

$$\sin \alpha = \pm \frac{\sqrt{13}}{4} \qquad \text{Take square roots.}$$

Since α is in quadrant II, $\sin \alpha > 0$, and

$$\sin \alpha = \frac{\sqrt{13}}{4}.$$

To find $\tan \alpha$, use the quotient identity $\tan \alpha = \frac{\sin \alpha}{\cos \alpha}$.

$$\tan \alpha = \frac{\sin \alpha}{\cos \alpha} = \frac{\dfrac{\sqrt{13}}{4}}{\dfrac{-\sqrt{3}}{4}} = \frac{\sqrt{13}}{4} \cdot \frac{4}{-\sqrt{3}} = \frac{\sqrt{13}}{-\sqrt{3}}$$

Rationalize the denominator as follows.

$$\frac{\sqrt{13}}{-\sqrt{3}} = \frac{\sqrt{13}}{-\sqrt{3}} \cdot \frac{\sqrt{3}}{\sqrt{3}}$$

$$= \frac{\sqrt{39}}{-3} = -\frac{\sqrt{39}}{3}$$

Therefore, $\tan \alpha = -\dfrac{\sqrt{39}}{3}$.

The reciprocal identities can be used to find the remaining three function values:

$$\sec \alpha = \frac{1}{\cos \alpha} = \frac{1}{-\dfrac{\sqrt{3}}{4}} = -\frac{4}{\sqrt{3}} = \frac{-4\sqrt{3}}{3}$$

$$\csc \alpha = \frac{1}{\sin \alpha} = \frac{1}{\dfrac{\sqrt{13}}{4}} = \frac{4}{\sqrt{13}} = \frac{4\sqrt{13}}{13}$$

$$\cot \alpha = \frac{1}{\tan \alpha} = \frac{1}{-\dfrac{\sqrt{39}}{3}} = -\frac{3}{\sqrt{39}} = \frac{-3\sqrt{39}}{39} = \frac{-\sqrt{39}}{13}. \quad ▮$$

CAUTION   When working a problem like the one in Example 7, remember that there are usually several different ways to approach it. Furthermore, one of the most common errors involves making an incorrect sign choice for a square root. Always pay close attention to the quadrant of the angle so that the correct sign choice can be made.

2.2 EXERCISES

To the Student: Calculator Considerations

Be sure that you know how to use the reciprocal key in conjunction with the sine, cosine, and tangent keys so that cosecant, secant, and cotangent values can be found. Also be aware of how to find powers of trigonometric functions; while we usually write $\sin^2 \theta$, for example, it may be necessary to enter this as $(\sin \theta)^2$ on your calculator.

Of utmost importance is knowing how to put the calculator in the degree or radian mode. Learn how to do this.

To test yourself, verify these facts (to be explained in the next section): $\csc 30° = 2$; $\sin^2 30° = .25$; $\cot^2 \left(\frac{\pi}{6}\right) = 3$.

Sketch an angle θ in standard position such that θ has the smallest positive measure, and the given point is on the terminal side of θ.

1. $(-3, 4)$ **2.** $(-4, -3)$ **3.** $(5, -12)$ **4.** $(-12, -5)$

Suppose that θ is in standard position and the given point is on the terminal side of θ. Give the exact values of the six trigonometric functions of θ. If a function is not defined, say so.

5. $(-3, 4)$ **6.** $(-4, -3)$ **7.** $(5, -12)$ **8.** $(-12, -5)$

9. $(-7, 24)$ **10.** $(24, 7)$ **11.** $(0, 2)$ **12.** $(0, -9)$

13. $(-4, 0)$ **14.** $(8, 0)$ **15.** $(\sqrt{5}, -2)$ **16.** $(-\sqrt{7}, \sqrt{2})$

17. $(1, 1)$ **18.** $(-4, -4)$ **19.** $(1, \sqrt{3})$ **20.** $(-2\sqrt{3}, -2)$

21. For any nonquadrantal angle θ, $\sin \theta$ and $\csc \theta$ will have the same sign. Explain why this is so.

22. If $\cot \theta$ is undefined, what is the value of $\tan \theta$?

23. How is the value of r interpreted geometrically in the definitions of the sine, cosine, secant, and cosecant functions?

24. If the terminal side of an angle β is in quadrant III, what is the sign of each of the trigonometric function values of β?

Suppose that the point (x, y) is in the indicated quadrant. Decide whether the given ratio is positive or negative. (Hint: It may be helpful to draw a sketch.)

25. II, $\dfrac{y}{r}$ **26.** II, $\dfrac{x}{r}$ **27.** III, $\dfrac{y}{r}$ **28.** III, $\dfrac{x}{r}$

29. III, $\dfrac{y}{x}$ **30.** III, $\dfrac{x}{y}$ **31.** IV, $\dfrac{x}{r}$ **32.** IV, $\dfrac{y}{r}$

33. IV, $\dfrac{y}{x}$ **34.** IV, $\dfrac{x}{y}$ **35.** III, $\dfrac{r}{x}$ **36.** II, $\dfrac{r}{y}$

In Exercises 37–42, an equation with a restriction on x is given. This is an equation of the terminal side of an angle θ in standard position. Sketch the smallest positive such angle θ, and find the values of the six trigonometric functions of θ.

37. $y = -2x, \quad x \geq 0$ **38.** $y = -\dfrac{3}{5}x, \quad x \geq 0$ **39.** $y = \dfrac{4}{7}x, \quad x \leq 0$

40. $y = -6x, \quad x \leq 0$ **41.** $y = -\dfrac{5}{3}x, \quad x \leq 0$ **42.** $y = \dfrac{6}{5}x, \quad x \geq 0$

Use the appropriate reciprocal identity to find each function value. In Exercises 43–50, give exact values, and in Exercises 51–54, give calculator approximations.

43. $\sin \theta$, if $\csc \theta = 3$ **44.** $\cos \alpha$, if $\sec \alpha = -2.5$ **45.** $\cot \beta$, if $\tan \beta = -\dfrac{1}{5}$

46. $\sin \alpha$, if $\csc \alpha = \sqrt{15}$ **47.** $\csc \alpha$, if $\sin \alpha = \dfrac{\sqrt{2}}{4}$ **48.** $\sec \beta$, if $\cos \beta = -\dfrac{1}{\sqrt{7}}$

49. $\tan \theta$, if $\cot \theta = -\dfrac{\sqrt{5}}{3}$ **50.** $\cot \theta$, if $\tan \theta = \dfrac{\sqrt{11}}{5}$ **51.** $\sin \theta$, if $\csc \theta = 1.42716321$

52. $\cos \alpha$, if $\sec \alpha = 9.80425133$ **53.** $\tan \alpha$, if $\cot \alpha = .43900273$ **54.** $\csc \theta$, if $\sin \theta = -.37690858$

55. Can a given angle γ satisfy both $\sin \gamma > 0$ and $\csc \gamma < 0$? Explain.

56. One form of a particular reciprocal identity is

$$\tan \theta = \frac{1}{\cot \theta}.$$

Give two other equivalent forms of this identity.

57. What is wrong with the following statement? $\tan 90° = \dfrac{1}{\cot 90°}$

58. If an angle θ has undefined cotangent, which other function value of θ is undefined as well?

Identify the quadrant or quadrants for the angles satisfying the following conditions.

59. $\sin \alpha > 0, \cos \alpha < 0$ **60.** $\cos \beta > 0, \tan \beta > 0$ **61.** $\sec \theta < 0, \csc \theta < 0$

62. $\tan \gamma > 0$, $\cot \gamma > 0$ **63.** $\sin \beta < 0$, $\cos \beta > 0$ **64.** $\cos \beta > 0$, $\sin \beta > 0$

65. $\tan \omega < 0$, $\cot \omega < 0$ **66.** $\csc \theta < 0$, $\cos \theta < 0$ **67.** $\sin \alpha > 0$

68. $\cos \beta < 0$ **69.** $\tan \theta > 0$ **70.** $\csc \alpha < 0$

Use identities to find the indicated function value. Use a calculator in Exercises 79–82.

71. $\cos \theta$, if $\sin \theta = \dfrac{2}{3}$, with θ in quadrant II   **72.** $\tan \alpha$, if $\sec \alpha = 3$, with α in quadrant IV

73. $\csc \beta$, if $\cot \beta = -\dfrac{1}{2}$, with β in quadrant IV   **74.** $\sin \alpha$, if $\cos \alpha = -\dfrac{1}{4}$, with α in quadrant II

75. $\sec \theta$, if $\tan \theta = \dfrac{\sqrt{7}}{3}$, with θ in quadrant III   **76.** $\tan \theta$, if $\cos \theta = \dfrac{1}{3}$, with θ in quadrant IV

77. $\sin \theta$, if $\sec \theta = 2$, with θ in quadrant IV   **78.** $\cos \beta$, if $\csc \beta = -4$, with β in quadrant III

79. $\cot \alpha$, if $\csc \alpha = -3.5891420$, with α in quadrant III

80. $\sin \beta$, if $\cot \beta = 2.40129813$, with β in quadrant I

81. $\tan \beta$, if $\sin \beta = .49268329$, with β in quadrant II

82. $\csc \alpha$, if $\tan \alpha = .98244655$, with α in quadrant III

Find all six trigonometric function values for each of the following angles. Use a calculator in Exercises 91 and 92.

83. $\cos \alpha = -\dfrac{3}{5}$, with α in quadrant III   **84.** $\tan \alpha = -\dfrac{15}{8}$, with α in quadrant II

85. $\sin \beta = \dfrac{7}{25}$, with β in quadrant II   **86.** $\cot \gamma = \dfrac{3}{4}$, with γ in quadrant III

87. $\csc \theta = 2$, with θ in quadrant II   **88.** $\tan \beta = \sqrt{3}$, with β in quadrant III

89. $\cot \alpha = \dfrac{\sqrt{3}}{8}$, with $\sin \alpha > 0$ **90.** $\sin \beta = \dfrac{\sqrt{5}}{7}$, with $\tan \beta > 0$

91. $\sin \alpha = .164215$, with α in quadrant II   **92.** $\cot \theta = -1.49586$, with θ in quadrant IV

Further Explorations

The figure shows a TABLE for $Y_1 = \cos \theta$ and $Y_2 = \sin \theta$, with $\Delta\text{Tbl} = 5°$. *Note:* $\cos \theta$ and $\sin \theta$ must be entered as $\cos x$ and $\sin x$ in the $Y =$ menu of the graphing calculator.

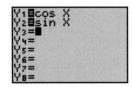 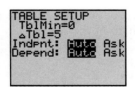

1. Use the TABLE on your graphing calculator and the up and/or down arrows to scroll through the TABLE to find the angle(s) between $0°$ and $360°$ in which $|\cos \theta| = |\sin \theta|$. (Be sure your calculator is set for DEGREE MODE.)

2. Sketch by hand the terminal sides of each of the angles in Further Explorations Exercise 1. Find the equation for the lines containing each of these terminal sides.

3. Use the equations from Further Explorations Exercise 2 and the definitions of $\cos \theta$ and $\sin \theta$ ($\cos \theta = \frac{x}{r}$ and $\sin \theta = \frac{y}{r}$) to explain why $|\sin \theta| = |\cos \theta|$ for the angles found in Further Explorations Exercise 1.

2.3 EVALUATING TRIGONOMETRIC FUNCTIONS

Function Values of Special Angles ▮ Cofunction Identities ▮ Calculator Approximations ▮ Reference Angles

Function Values of Special Angles

We begin this section by deriving the exact values of the trigonometric functions of three special angles: $30°$ ($\frac{\pi}{6}$ radians), $45°$ ($\frac{\pi}{4}$ radians), and $60°$ ($\frac{\pi}{3}$ radians). This will be done in a purely analytic manner.

To find the function values of a $45°$ angle, we note that the ray $y = x$, $x \geq 0$, bisects the first quadrant and is therefore the terminal side of a $45°$ angle in standard position. See Figure 21. If we choose the point $A(1, 1)$ on this terminal side, the distance r from the origin O to A is

$$r = \sqrt{1^2 + 1^2} = \sqrt{2}.$$

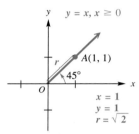

FIGURE 21

Using the definitions of the trigonometric functions, we have

$$\cos 45° = \frac{x}{r} = \frac{1}{\sqrt{2}} = \frac{\sqrt{2}}{2} \qquad \sec 45° = \frac{r}{x} = \frac{\sqrt{2}}{1} = \sqrt{2}$$

$$\sin 45° = \frac{y}{r} = \frac{1}{\sqrt{2}} = \frac{\sqrt{2}}{2} \qquad \csc 45° = \frac{r}{y} = \frac{\sqrt{2}}{1} = \sqrt{2}$$

$$\tan 45° = \frac{y}{x} = \frac{1}{1} = 1 \qquad \cot 45° = \frac{x}{y} = \frac{1}{1} = 1.$$

To derive the exact function values for a $30°$ angle, observe Figure 22. Angle SOP is a $30°$ angle in standard position. Let P be the point (x, y) on the terminal side of this angle that is 1 unit from the origin. Now let angle SOQ be a $-30°$

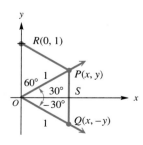

FIGURE 22

angle in standard position, with the Q designating the point on its terminal side 1 unit from the origin. The coordinates of Q are $(x, -y)$, since Q is a reflection of P across the x-axis.

If we join P and Q with a line segment, we obtain the triangle POQ. The measure of angle POQ is 60°, since $30° + |-30°| = 60°$. Triangle POQ is an isosceles triangle since OP and OQ each measure 1 unit, and consequently angles OPQ and OQP each measure 60°. Thus triangle POQ is also equilangular, from which it follows that it is equilateral, with the length of PQ equal to 1.

Let R be the point $(0, 1)$. Angle ROP measures $90° - 30° = 60°$, and by reasoning similar to that above, the length of RP is 1, the same as PQ. We now apply the distance formula.

$$RP = PQ$$
$$\sqrt{(x - 0)^2 + (y - 1)^2} = \sqrt{(x - x)^2 + (y - (-y))^2}$$

$x^2 + y^2 - 2y + 1 = (2y)^2$	Square both sides.
$-2y + 2 = 4y^2$	$x^2 + y^2 = 1$
$0 = 4y^2 + 2y - 2$	
$0 = 2y^2 + y - 1$	Divide by 2.

Use the quadratic formula and the fact that $y > 0$ to obtain

$$y = \frac{-1 + \sqrt{1^2 - 4(2)(-1)}}{2(2)} = \frac{-1 + \sqrt{9}}{4} = \frac{-1 + 3}{4} = \frac{2}{4} = \frac{1}{2}.$$

Since $x^2 + y^2 = 1$, we solve for x as follows.

$$x^2 + \left(\frac{1}{2}\right)^2 = 1$$

$$x^2 + \frac{1}{4} = 1$$

$$x^2 = \frac{3}{4}$$

$$x = \frac{\sqrt{3}}{2} \qquad (x > 0)$$

Now we apply the definitions of the trigonometric functions to obtain the exact values for the functions of 30°.

$$\cos 30° = \frac{x}{r} = \frac{\frac{\sqrt{3}}{2}}{1} = \frac{\sqrt{3}}{2} \qquad\qquad \sec 30° = \frac{r}{x} = \frac{1}{\frac{\sqrt{3}}{2}} = \frac{2}{\sqrt{3}} = \frac{2\sqrt{3}}{3}$$

$$\sin 30° = \frac{y}{r} = \frac{\frac{1}{2}}{1} = \frac{1}{2} \qquad\qquad \csc 30° = \frac{r}{y} = \frac{1}{\frac{1}{2}} = 2$$

$$\tan 30° = \frac{y}{x} = \frac{\frac{1}{2}}{\frac{\sqrt{3}}{2}} = \frac{1}{\sqrt{3}} = \frac{\sqrt{3}}{3} \qquad\qquad \cot 30° = \frac{x}{y} = \frac{\frac{\sqrt{3}}{2}}{\frac{1}{2}} = \sqrt{3}$$

In Exercises 93–98, we show a similar argument for the functions of a 60° angle. There, we find that the point 1 unit from the origin and on the terminal side of a 60° angle in standard position has coordinates $(\frac{1}{2}, \frac{\sqrt{3}}{2})$. Thus,

$$\cos 60° = \frac{x}{r} = \frac{\frac{1}{2}}{1} = \frac{1}{2} \qquad \sec 60° = \frac{r}{x} = \frac{1}{\frac{1}{2}} = 2$$

$$\sin 60° = \frac{y}{r} = \frac{\frac{\sqrt{3}}{2}}{1} = \frac{\sqrt{3}}{2} \qquad \csc 60° = \frac{r}{y} = \frac{1}{\frac{\sqrt{3}}{2}} = \frac{2}{\sqrt{3}} = \frac{2\sqrt{3}}{3}$$

$$\tan 60° = \frac{y}{x} = \frac{\frac{\sqrt{3}}{2}}{\frac{1}{2}} = \sqrt{3} \qquad \cot 60° = \frac{x}{y} = \frac{\frac{1}{2}}{\frac{\sqrt{3}}{2}} = \frac{1}{\sqrt{3}} = \frac{\sqrt{3}}{3}.$$

The following chart summarizes the *exact* trigonometric function values of 30°, 45°, and 60° angles and their radian equivalents.

FUNCTION VALUES OF SPECIAL ANGLES

θ	$\sin \theta$	$\cos \theta$	$\tan \theta$	$\cot \theta$	$\sec \theta$	$\csc \theta$
$30° = \dfrac{\pi}{6}$	$\dfrac{1}{2}$	$\dfrac{\sqrt{3}}{2}$	$\dfrac{\sqrt{3}}{3}$	$\sqrt{3}$	$\dfrac{2\sqrt{3}}{3}$	2
$45° = \dfrac{\pi}{4}$	$\dfrac{\sqrt{2}}{2}$	$\dfrac{\sqrt{2}}{2}$	1	1	$\sqrt{2}$	$\sqrt{2}$
$60° = \dfrac{\pi}{3}$	$\dfrac{\sqrt{3}}{2}$	$\dfrac{1}{2}$	$\sqrt{3}$	$\dfrac{\sqrt{3}}{3}$	2	$\dfrac{2\sqrt{3}}{3}$

NOTE   It is not difficult to reproduce this chart if you learn the values of sin 30°, cos 30°, and sin 45°. Then the rest of the chart can be completed by using the reciprocal identities, the quotient identities, and the cofunction identities. (See the material that immediately follows.)

Cofunction Identities

Notice the following pattern in the chart above.

$$\sin 30° = \cos 60° \qquad \csc 30° = \sec 60° \qquad \sin 45° = \cos 45°$$
$$\cos 30° = \sin 60° \qquad \sec 30° = \csc 60° \qquad \tan 45° = \cot 45°$$
$$\tan 30° = \cot 60° \qquad \cot 30° = \tan 60° \qquad \sec 45° = \csc 45°$$

The angles 30° and 60° are complementary, and 45° is its own complement. The function pairs

sine, cosine

tangent, cotangent

and   **secant, cosecant**

are called **cofunctions.** The pattern suggests that cofunction values of complementary angles are equal, and this can be shown to be true in general. We now state the cofunction identities.

COFUNCTION IDENTITIES

If A is an angle measured in degrees,

$$\sin A = \cos(90° - A) \qquad \csc A = \sec(90° - A)$$
$$\cos A = \sin(90° - A) \qquad \sec A = \csc(90° - A)$$
$$\tan A = \cot(90° - A) \qquad \cot A = \tan(90° - A).$$

If A is an angle measured in radians,

$$\sin A = \cos\left(\frac{\pi}{2} - A\right) \qquad \csc A = \sec\left(\frac{\pi}{2} - A\right)$$

$$\cos A = \sin\left(\frac{\pi}{2} - A\right) \qquad \sec A = \csc\left(\frac{\pi}{2} - A\right)$$

$$\tan A = \cot\left(\frac{\pi}{2} - A\right) \qquad \cot A = \tan\left(\frac{\pi}{2} - A\right).$$

EXAMPLE **1**

Writing Functions in Terms of Cofunctions

Write each of the following in terms of cofunctions.

(a) $\cos 52°$

SOLUTION Since $\cos A = \sin(90° - A)$,

$$\cos 52° = \sin(90° - 52°) = \sin 38°.$$

(b) $\tan \dfrac{\pi}{6}$

SOLUTION Since $\tan A = \cot(\frac{\pi}{2} - A)$,

$$\tan \frac{\pi}{6} = \cot\left(\frac{\pi}{2} - \frac{\pi}{6}\right) = \cot \frac{\pi}{3} \qquad ◻$$

FOR GROUP DISCUSSION

1. With your calculator in degree mode, find values for $\cos 52°$ and $\sin 38°$. Refer to Example 1(a), and discuss your results. (The displays are actually approximations for these functions, and we will investigate them further later in this section.)
2. With your calculator in radian mode, find values for $\tan \frac{\pi}{6}$ and $\cot \frac{\pi}{3}$. Refer to Example 1(b), and discuss your results.

TECHNOLOGICAL NOTE
One of the latest models of graphing "calculators," the TI-92, is more accurately described as a hand-held computer. It *is*, unlike less sophisticated models, able to return exact values of the trigonometric function values of 30°, 45°, and 60°, and their radian equivalents.

Calculator Approximations

While it is interesting to see how exact trigonometric function values can be found for certain special angles, in most applications of trigonometry we rely on calculator approximations for function values.

FOR **GROUP DISCUSSION**

1. With your calculator in degree mode, find the value of sin 30°. Verify that it corresponds to the exact value of sin 30° given in the chart above.
2. With your calculator in radian mode, find the value of tan $\frac{\pi}{4}$. Verify that it corresponds to the exact value of tan $\frac{\pi}{4}$ given in the chart above.
3. With your calculator in degree mode, find an approximation for cos 30°. Then use the computational capabilities to find an approximation for the exact value given for cos 30° in the chart above. Verify that they correspond.

CAUTION   One of the most common errors involving calculator use in trigonometry is working in the incorrect angle measure mode. Be sure that your calculator is in the proper mode before calculating a function value.

EXAMPLE **2**

Finding Calculator Approximations of Trigonometric Functions

The following approximations of trigonometric function values can be obtained with a calculator. (The number of displayed digits will vary among calculator models.)

(a)   sin 49°12′ ≈ .7569950557

SOLUTION   It may be necessary to convert to decimal degrees before calculating. Be sure that the calculator is in degree mode.

(b)   cot 3 ≈ −7.015252551

SOLUTION   Calculators usually do not have keys for the secant, cosecant, and cotangent functions. To find this approximation, first put the calculator in radian mode, find tan 3, and then find the reciprocal of the result. Here we are using the identity cot $\theta = \dfrac{1}{\tan \theta}$.

(c)   sin(−147°) ≈ −.544639035

SOLUTION   Notice that this function value is negative because −147° is a quadrant III angle, and the sine function is negative in quadrant III.   ∎

TECHNOLOGICAL NOTE
Students often confuse the symbols for the inverse trigonometric functions with the reciprocal functions. For example, $\sin^{-1} x$ represents an angle whose sine is x, and not the reciprocal of sin x (which is csc x). In order to find reciprocal function values, you must use the function in conjunction with the reciprocal function of the calculator.

Sometimes it is necessary to find an angle measure when we know a trigonometric function value of that angle, and information about the quadrant in which the angle lies. A complete discussion of inverse trigonometric functions is necessary to understand *why* a calculator will give an inverse function value in a particular quadrant. We will examine inverse trigonometric functions in more detail in the next chapter. For now, we simply present an introduction to the use of inverse trigonometric functions with calculators. (You should locate these functions on your particular model. They are usually designated $\sin^{-1}$, $\cos^{-1}$, and $\tan^{-1}$.)

EXAMPLE **3**

Using Inverse Trigonometric Functions to Find Angles

(a)   Use a calculator to find an angle θ in degrees that satisfies sin θ ≈ .9677091705.

SOLUTION   With the calculator in degree mode, we find that an angle θ having sine value .9677091705 is 75.4°. (While there are infinitely many such angles, the calculator only gives this one.) We write this result as $\sin^{-1}$ .9677091705 ≈ 75.4°.

(b)   Use a calculator to find an angle θ in radians that satisfies tan θ ≈ .25.

SOLUTION   With the calculator in radian mode, we find $\tan^{-1}$ .25 ≈ .2449786631.   ∎

Reference Angles

Associated with every non-quadrantal angle in standard position is a positive acute angle called its reference angle. A **reference angle** for an angle θ, written θ', is the positive acute angle made by the terminal side of angle θ and the *x*-axis. Figure 23 shows several angles θ (each less than one complete counterclockwise revolution) in quadrants II, III, and IV, respectively, with the reference angle θ' also shown. In quadrant I, θ and θ' are the same. If an angle θ is negative or has measure greater than 360°, its reference angle is found by first finding its coterminal angle that is between 0° and 360°, and then using the diagrams in Figure 23.

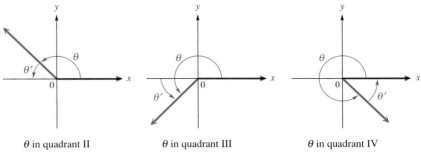

θ in quadrant II θ in quadrant III θ in quadrant IV

FIGURE 23

CAUTION   A very common error is to find the reference angle by using the terminal side of θ and the *y*-axis. *The reference angle is always found with reference to the x-axis.*

EXAMPLE **4**
Finding Reference Angles

Find the reference angles for the following three angles.

(a)   218°

SOLUTION   As shown in Figure 24, the positive acute angle made by the terminal side of this angle and the *x*-axis is $218° - 180° = 38°$. For $\theta = 218°$, the reference angle $\theta' = 38°$.

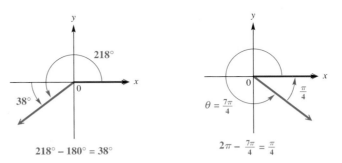

$218° - 180° = 38°$ $2\pi - \frac{7\pi}{4} = \frac{\pi}{4}$

FIGURE 24   **FIGURE 25**

(b) $\dfrac{7\pi}{4}$

SOLUTION   As shown in Figure 25, the reference angle (the positive acute angle made by the terminal side of this angle and the *x*-axis) is $2\pi - \frac{7\pi}{4} = \frac{\pi}{4}$. Note that the angle $-\frac{\pi}{4}$, which has the same terminal ray as $\frac{7\pi}{4}$, also has a reference angle of $\frac{\pi}{4}$.

(c)   1387°

SOLUTION   First find a coterminal angle between 0° and 360°. Divide 1387° by 360° to get a quotient of about 3.9. Begin by subtracting 360° three times (because of the 3 in 3.9):

$$1387° - 3 \cdot 360° = 307°. \qquad \text{307° is in quadrant IV.}$$

The reference angle for 307° (and thus for 1387°) is $360° - 307° = 53°$.   ∎

   The preceding example suggests the following table for finding the reference angle θ' for any angle θ between 0° and 360° or 0 and 2π.

REFERENCE ANGLES FOR θ IN (0°, 360°) OR (0, 2π)		
θ in Quadrant	**θ' Is**	**Example**
I	θ	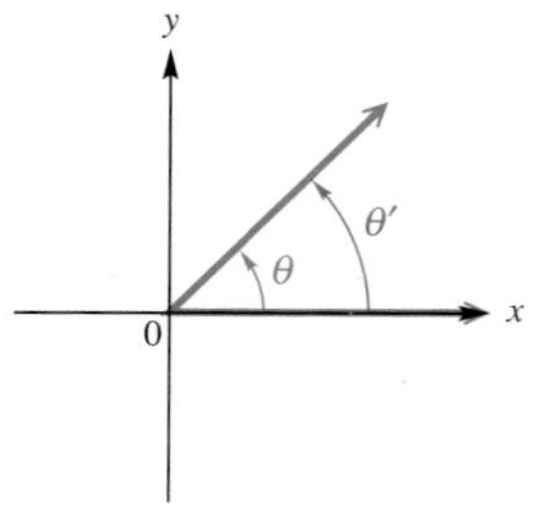
II	$180° - \theta$ $\pi - \theta$	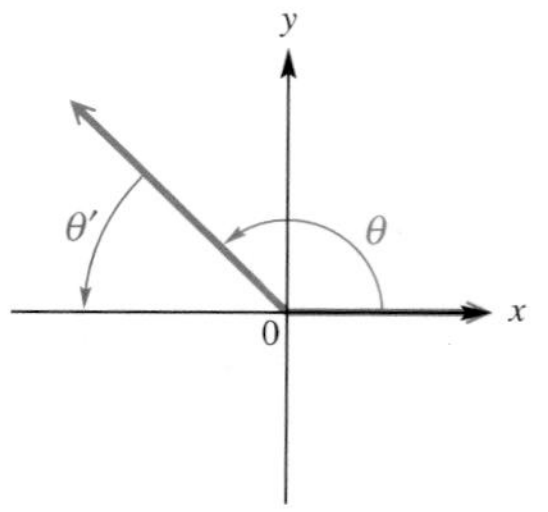
III	$\theta - 180°$ $\theta - \pi$	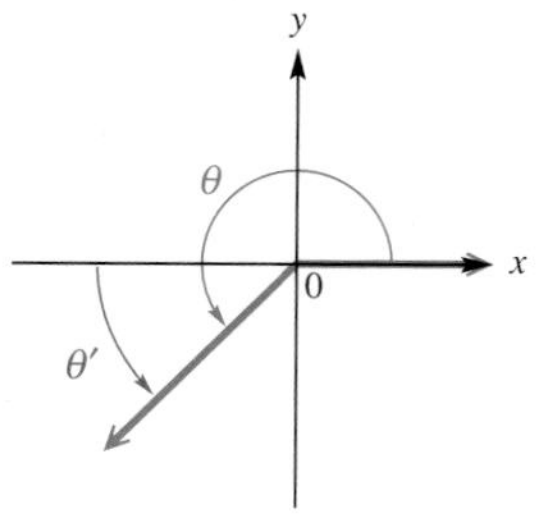
IV	$360° - \theta$ $2\pi - \theta$	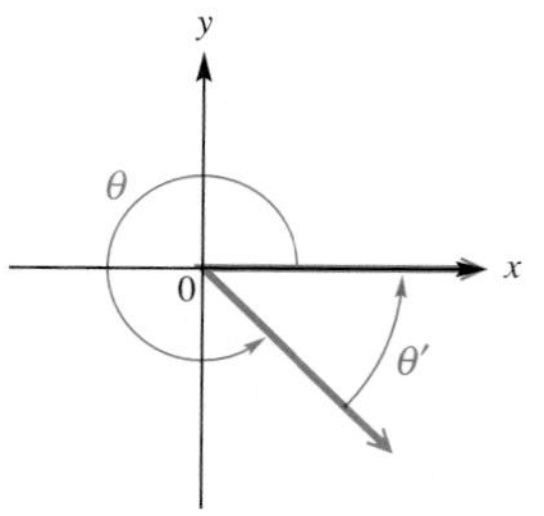

If we wish to find a trigonometric function value of a non-quadrantal angle using reference angles, we use the following guidelines. (Similar guidelines hold for radian measure.)

FINDING TRIGONOMETRIC FUNCTION VALUES FOR ANY NON-QUADRANTAL ANGLE

1. If $\theta > 360°$, or if $\theta < 0°$, find a coterminal angle by adding or subtracting 360° as many times as needed to get an angle of at least 0° but less than 360°.
2. Find the reference angle θ'.
3. Find the necessary values of the trigonometric functions for the reference angle θ'.
4. Find the correct signs for the values found in Step 3. (Use the table of signs in Section 2.2.) This result gives the value of the trigonometric functions for angle θ.

EXAMPLE 5

Finding Trigonometric Function Values Using Reference Angles

Use reference angles to find the exact value of each of the following.

(a) $\cos(-240°)$

SOLUTION   The reference angle is 60°, as shown in Figure 26. Since the cosine is negative in quadrant II,

$$\cos(-240°) = -\cos 60° = -\frac{1}{2}.$$

(b) $\tan \frac{7\pi}{6}$

SOLUTION   The reference angle is $\frac{\pi}{6}$, as shown in Figure 27. The tangent is positive in quadrant III, so,

$$\tan \frac{7\pi}{6} = +\tan \frac{\pi}{6} = \frac{\sqrt{3}}{3}.$$

(c) $\csc 675°$

SOLUTION   Begin by subtracting 360° to get a coterminal angle between 0° and 360°, obtaining $675° - 360° = 315°$. As shown in Figure 28, the reference angle is $360° - 315° = 45°$. An angle of 315° is in quadrant IV, so the cosecant is negative. Therefore,

$$\csc 675° = -\csc 45° = -\sqrt{2}. \quad \blacksquare$$

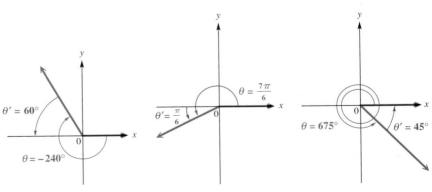

FIGURE 26   **FIGURE 27**   **FIGURE 28**

The ideas discussed in this section can be reversed to find the measures of certain angles, given a trigonometric function value and an interval in which the angle must lie. We are most often interested in the interval [0°, 360°) or [0, 2π).

EXAMPLE 6
Finding Angle Measures Given an Interval and a Function Value (Degree Measure)

Find all values of θ, if θ is in the interval [0°, 360°) and cos $\theta = \frac{-\sqrt{2}}{2}$.

SOLUTION   Since cosine here is negative, θ must lie in either quadrant II or III. Since the absolute value of cos θ is $\frac{\sqrt{2}}{2}$, the reference angle θ' must be 45°. The two possible angles θ are sketched in Figure 29.

The quadrant II angle θ must equal 180° − 45° = 135°, and the quadrant III angle θ must equal 180° + 45° = 225°.   ∎

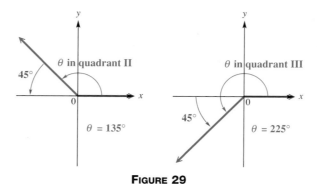

FIGURE 29

EXAMPLE 7
Finding Angle Measures Given an Interval and a Function Value (Radian Measure)

Find two angles in the interval [0, 2π) that satisfy cos $\theta \approx .3623577545$.

SOLUTION   With the calculator in radian mode, we find that one such θ is

$$\cos^{-1} .3623577545 \approx 1.2.$$

Since $1.2 < \frac{\pi}{2}$, θ is in quadrant I. We must find *another* value of θ that satisfies the given condition. This other value of θ will have its reference angle θ' equal to 1.2, and must be in quadrant IV, since the angle given by the calculator is in quadrant

I and cosine is also positive in quadrant IV. The other value of θ, then, is

$$2\pi - 1.2 \approx 5.083185307.$$

Verify this result by showing that

$$\cos 5.083185307 \approx .3623577545. \quad \blacksquare$$

2.3 EXERCISES

To the Student: Calculator Considerations

Be sure that you know how to put your calculator in degree or radian mode. Also, be aware of the difference between inverse trigonometric functions (such as $\sin^{-1} x$) and reciprocals of trigonometric functions (such as $\csc \theta$, which is equal to $\frac{1}{\sin \theta}$).

Find the six trigonometric function values of the angle θ shown in standard position.

1.

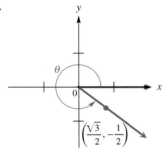

2.

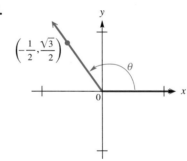

3.

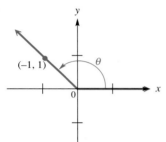

4.

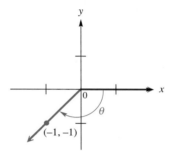

5.

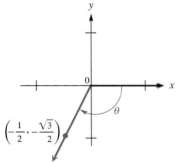

6.

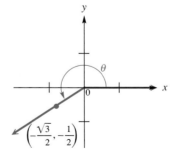

Write each of the following in terms of the cofunction.

7. $\cot 73°$ **8.** $\sec 39°$ **9.** $\sin 38° \, 29'$ **10.** $\tan 25° \, 43'$

11. $\cos \dfrac{\pi}{5}$ **12.** $\sin \dfrac{\pi}{3}$ **13.** $\tan .5$ **14.** $\csc .3$

15. A student was asked to give the exact value of sin 45°. Using his calculator, he gave the answer .7071067812. The teacher did not give him credit. What was the teacher's reason for this?

16. A student was asked to give an approximate value of sin 45. With her calculator in degree mode, she gave the value .7071067812. The teacher did not give her credit. What was her error?

For each of the following (a) give the exact value, (b) state whether the exact value is rational or irrational, and (c) if the exact value is irrational, use your calculator in two ways to support your answer in part (a) by finding a decimal approximation.

17. tan 30°   **18.** cot 30°   **19.** sin 30°   **20.** cos 30°

21. sec 30°   **22.** csc 30°   **23.** csc 45°   **24.** sec 45°

25. cos 45°   **26.** sin 45°   **27.** cot 45°   **28.** tan 45°

29. $\sin \dfrac{\pi}{3}$ **30.** $\cos \dfrac{\pi}{3}$ **31.** $\tan \dfrac{\pi}{3}$ **32.** $\cot \dfrac{\pi}{3}$

33. $\sec \dfrac{\pi}{3}$ **34.** $\csc \dfrac{\pi}{3}$

Use a calculator to find a decimal approximation for each function value. Give as many digits as your calculator displays.

35. sin 38° 40′   **36.** tan 29° 30′   **37.** cos 251° 10′

38. cot 512° 20′   **39.** sec(−108° 20′)   **40.** csc(−29° 30′)

41. tan .4538   **42.** sin .6109   **43.** csc 1.3875

44. cos(−3.0602)   **45.** sin(−17.5784)

46. (a) cos(−17.5784)
(b) Based on your answers in Exercises 45 and 46(a), in what quadrant is an angle of −17.5784 radians?

Complete the following table with exact trigonometric function values using the methods of this section.

	θ	$\sin \theta$	$\cos \theta$	$\tan \theta$	$\cot \theta$	$\sec \theta$	$\csc \theta$
47.	30°	$\frac{1}{2}$	$\frac{\sqrt{3}}{2}$	___	___	$\frac{2\sqrt{3}}{3}$	2
48.	45°	___	___	1	1	___	___
49.	60°	___	$\frac{1}{2}$	$\sqrt{3}$	___	2	___
50.	120°	$\frac{\sqrt{3}}{2}$	___	$-\sqrt{3}$	___	___	$\frac{2\sqrt{3}}{3}$
51.	135°	$\frac{\sqrt{2}}{2}$	$-\frac{\sqrt{2}}{2}$	___	___	$-\sqrt{2}$	$\sqrt{2}$
52.	150°	___	$-\frac{\sqrt{3}}{2}$	$-\frac{\sqrt{3}}{3}$	___	___	2
53.	210°	$-\frac{1}{2}$	___	$\frac{\sqrt{3}}{3}$	$\sqrt{3}$	___	−2
54.	240°	$-\frac{\sqrt{3}}{2}$	$-\frac{1}{2}$	___	___	−2	$-\frac{2\sqrt{3}}{3}$

For each of the following (a) write the function in terms of a function of the reference angle, (b) give the exact value, and (c) use a calculator to show that the decimal value or approximation for the given function is the same as the decimal value or approximation for your answer in part (b).

55. $\sin \dfrac{7\pi}{6}$ **56.** $\cos \dfrac{5\pi}{3}$ **57.** $\tan \dfrac{3\pi}{4}$

58. $\sin \dfrac{5\pi}{3}$ **59.** $\cos \dfrac{7\pi}{6}$ **60.** $\tan \dfrac{4\pi}{3}$

Find all values of θ, if θ is in the interval [0°, 360°) and has the given function value.

61. $\sin \theta = \dfrac{1}{2}$ **62.** $\cos \theta = \dfrac{\sqrt{3}}{2}$ **63.** $\tan \theta = \sqrt{3}$ **64.** $\sec \theta = \sqrt{2}$

65. $\cos \theta = -\dfrac{1}{2}$ **66.** $\cot \theta = -\dfrac{\sqrt{3}}{3}$ **67.** $\sin \theta = -\dfrac{\sqrt{3}}{2}$ **68.** $\cos \theta = -\dfrac{\sqrt{2}}{2}$

Find all values of θ if θ is in the interval [0°, 360°) and has the given function value.
Give approximations to as many decimal places as your calculator displays.

69. $\cos \theta \approx .68716510$ **70.** $\cos \theta \approx .96476120$ **71.** $\sin \theta \approx .41298643$

72. $\sin \theta \approx .63898531$ **73.** $\tan \theta \approx .87692035$ **74.** $\tan \theta \approx 1.2841996$

Find two angles in the interval [0, 2π) that satisfy the given equation. Give calculator
approximations to as many digits as your calculator will display.

75. $\tan \theta \approx .21264138$ **76.** $\cos \theta \approx .78269876$ **77.** $\sin \theta \approx .99184065$

78. $\cot \theta \approx .29949853$ **79.** $\csc \theta \approx 1.0219553$ **80.** $\cos \theta \approx .92728460$

Relating Concepts

In a square window of your calculator that gives a good picture of the first quadrant, graph the line $y = \sqrt{3}x$ with $x \geq 0$. Then trace to any *point on the line. For example, see the figure below. What we see is a simulated view of an angle in standard position, with terminal side in quadrant I. Store the values of x and y in convenient memory locations. For this group of exercises, we call them x_1 and y_1. Work the exercises in order.*

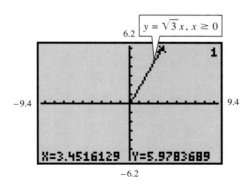

81. Find the value of $\sqrt{x_1^2 + y_1^2}$ and store it in a convenient memory location. (We will call it r.) What does this number mean geometrically?

82. With your calculator in degree mode, find $\tan^{-1}\left(\frac{y_1}{x_1}\right)$.

83. With your calculator in degree mode, find $\sin^{-1}\left(\frac{y_1}{r}\right)$.

84. With your calculator in degree mode, find $\cos^{-1}\left(\frac{x_1}{r}\right)$.

85. Your answers in Exercises 82–84 should all be the same. How does it relate to the angle formed on your screen?

86. Find the value of $\frac{y_1}{x_1}$. Now square it. What do you get? What is the exact value of $\frac{y_1}{x_1}$?

87. Look at the equation of the line you graphed, and make a conjecture: The _____ of a line passing through the origin is equal to the _____ of the angle it forms with the positive x-axis.

88. Find the value of $\left(\frac{x_1}{r}\right)^2 + \left(\frac{y_1}{r}\right)^2$. What identity does this illustrate?

89. Find $\csc 60°$. Then find the value of $\frac{r}{y_1}$. Do they agree?

90. Graph $y_2 = \sqrt{1 - x^2}$ as a second curve in the same viewing window. This is one half of a circle centered at the origin with radius 1. Now use the intersection feature of your calculator to determine the x- and y-coordinates of the points of intersection of the two graphs. What are they?

91. Find a calculator value for cos 60°. How does it compare to the x-coordinate of the point you found in Exercise 90? Why is this so?

92. Find a calculator approximation for sin 60°. How does it compare to the y-coordinate of the point you found in Exercise 90? Why is this so?

Refer to the figure shown here. Work Exercises 93–98 in sequence, and verify that the exact function values for a 60° angle as given in this section are correct.

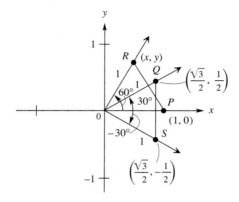

93. Write an equation in x and y based on the fact that $PR = QS$.

94. Solve the equation found in Exercise 93. (Use $x^2 + y^2 = 1$ to eliminate y.)

95. Why must the solution of the equation in Exercise 94 be a positive number?

96. Use $x^2 + y^2 = 1$ to solve for y, remembering $y > 0$.

97. What are the values of x, y, and r you can use to find the trigonometric function values of 60°?

98. Verify that the trigonometric function values of 60° correspond to the ones given in this section.

2.4 THE UNIT CIRCLE AND THE CIRCULAR FUNCTIONS

The Unit Circle ❚ Defining the Circular Functions ❚ A Brief Discussion of Parametric Equations ❚ The Negative Number Identities

The Unit Circle

In algebra we learn that a circle is a set of points (x, y) in a plane, each of which is a fixed distance r (called the *radius*) from a fixed point P (called the *center*). Suppose that the center of a circle is the origin $(0, 0)$ and its radius is 1. Then by the distance formula,

$$\sqrt{(x - 0)^2 + (y - 0)^2} = 1$$
$$\sqrt{x^2 + y^2} = 1$$
$$x^2 + y^2 = 1 \qquad \text{Square both sides.}$$

This final equation is the standard form for the *unit circle.* See Figure 30. The unit circle is a valuable tool in studying the trigonometric functions from a different point of view, as we shall see in this section.

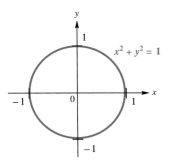

The Unit Circle

FIGURE 30

Defining the Circular Functions

In our work so far, we have defined the six *trigonometric* functions in such a way that the domain of each function was a set of *angles* in standard position. These angles can be measured either in degrees or in radians. In theoretical work, it is usually necessary to modify the trigonometric functions so that their domains consist of sets of *real numbers* rather than angles. Because the interpretations of these functions are based on the graph of the unit circle $x^2 + y^2 = 1$, we refer to them as **circular functions.**

To define the values of the circular functions for any real number s, we use the unit circle, shown in Figure 31(a) and (b). The calculator-generated graph in Figure 31(b) was obtained by using a square viewing window and graphing $x^2 + y^2 = 1$ as $y_1 = \sqrt{1 - x^2}$ and $y_2 = -\sqrt{1 - x^2}$. We start at the point $(1, 0)$ and measure an arc of length s along the circle as in Figure 31(a). If $s > 0$, the arc is measured in a counterclockwise direction, and if $s < 0$, the direction is clockwise. (If $s = 0$, then no arc is measured.) Let the endpoint of this arc be at the point (x, y). Then the six circular functions of s are defined as follows.

CIRCULAR FUNCTIONS

$$\sin s = y \qquad \tan s = \frac{y}{x} \ (x \neq 0) \qquad \sec s = \frac{1}{x} \ (x \neq 0)$$

$$\cos s = x \qquad \cot s = \frac{x}{y} \ (y \neq 0) \qquad \csc s = \frac{1}{y} \ (y \neq 0)$$

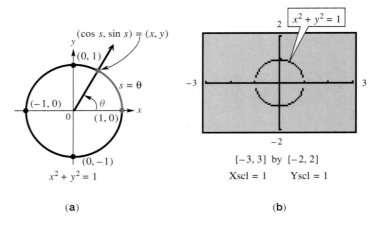

FIGURE 31

The circular functions (functions of real numbers) are closely related to the trigonometric functions of angles measured in radians. To see this, let us assume that angle θ is in standard position, superimposed on the unit circle as shown in Figure 31(a). Suppose further that θ is the *radian* measure of this angle. Using the arc length formula $s = r\theta$ with $r = 1$, we have $s = \theta$. Thus the length of the intercepted arc is the real number that corresponds to the radian measure of θ. Using the definitions of the trigonometric functions, we have

TECHNOLOGICAL NOTE
The material covered previously in this chapter involved both radian and degree measure, and as a result, you were probably often switching between degree and radian modes. The nature of the material concluding this chapter will necessitate using radian mode almost exclusively.

$$\sin \theta = \frac{y}{r} = \frac{y}{1} = y = \sin s,$$

$$\cos \theta = \frac{x}{r} = \frac{x}{1} = x = \cos s,$$

and so on. As shown here, the trigonometric functions and the circular functions lead to the same function values, provided we think of the angles in radian measure. This leads to the following important result concerning the evaluation of circular functions.

EVALUATING CIRCULAR FUNCTIONS

Circular function values of real numbers are obtained in the same manner as trigonometric function values of angles measured in radians. This applies to both methods of finding exact values (such as reference angle analysis) and calculator approximations. Calculators must be in radian mode when finding circular function values.

FOR **GROUP DISCUSSION**

With your calculator set for the window $[-3, 3]$ by $[-2, 2]$ and in radian mode, graph the unit circle $x^2 + y^2 = 1$ by graphing $y_1 = \sqrt{1 - x^2}$ and $y_2 = -\sqrt{1 - x^2}$. Now graph the line $y_3 = \frac{\sqrt{3}}{3}x$. You should get graphs similar to those seen in Figure 32.

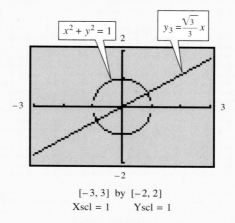

$[-3, 3]$ by $[-2, 2]$
Xscl = 1   Yscl = 1

FIGURE 32

1. Explain why the angle made by the line $y_3 = \frac{\sqrt{3}}{3}x$ makes an angle of $\frac{\pi}{6}$ radians with the positive x-axis.

2. Use the capabilities of your calculator to find the coordinates of the point of intersection of the line $y_3 = \frac{\sqrt{3}}{3}x$ and the portion of the unit circle in quadrant I.

3. What is the length of the arc from the point $(1, 0)$ to this point of intersection?

4. Find $\cos \frac{\pi}{6}$ using the cosine function of your calculator. How does this compare to the x-coordinate of the point of intersection found in Item 2? How does this support the discussion earlier in this section?

5. Find $\sin \frac{\pi}{6}$ using the sine function of your calculator. How does this compare to the y-coordinate of the point of intersection found in Item 2? How does this support the discussion earlier in this section?

6. If (x, y) denotes the point found in Item 2, find approximations for $\frac{y}{x}$ and for $\frac{\sqrt{3}}{3}$. How do they compare? How does the tangent of the angle relate to the slope of the line?

In Figure 33, you will find the graph of the unit circle $x^2 + y^2 = 1$ with a great deal of important information. This information is based on the development of the trigonometric functions earlier in this chapter along with the discussion of this section. For many special values, degree and radian measures are given for the first counterclockwise revolution, and the coordinates of the points on the circle are also given. This figure should prove invaluable in further work.

The graph in Figure 33 can be adapted to many coterminal arcs as well. For example, the circular function values of $-\frac{3\pi}{4}$ correspond to those of $\frac{5\pi}{4}$, those of $-\frac{3\pi}{2}$ to $\frac{\pi}{2}$, and so on.

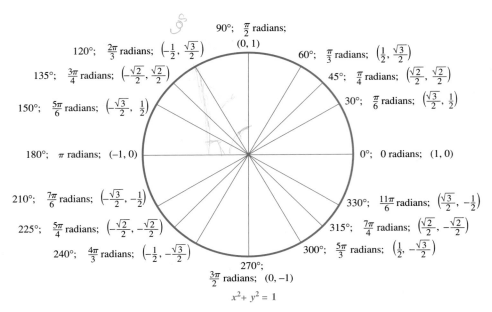

FIGURE 33

EXAMPLE 1

Finding Circular
Function Values

(a) Find the exact value of $\cos \frac{2\pi}{3}$ and of $\sin \frac{2\pi}{3}$.

SOLUTION An arc of length $\frac{2\pi}{3}$ has terminal point $\left(-\frac{1}{2}, \frac{\sqrt{3}}{2}\right)$. Therefore,

$$\cos \frac{2\pi}{3} = -\frac{1}{2} \quad \text{and} \quad \sin \frac{2\pi}{3} = \frac{\sqrt{3}}{2}.$$

(b) Find the exact value of $\tan \frac{\pi}{3}$.

SOLUTION By the definition of the circular tangent function, $\tan s = \frac{y}{x}$. When $s = \frac{\pi}{3}$, $(x, y) = \left(\frac{1}{2}, \frac{\sqrt{3}}{2}\right)$. Therefore

$$\tan \frac{\pi}{3} = \frac{\dfrac{\sqrt{3}}{2}}{\dfrac{1}{2}} = \sqrt{3}.$$

(c) Find a calculator approximation of $\cos 1.85$.

SOLUTION With the calculator in *radian* mode, we find

$$\cos 1.85 \approx -.2755902468.$$ ◘

A Brief Discussion of Parametric Equations

TECHNOLOGICAL NOTE
Refer to your owner's
manual to see how to put
your calculator in para-
metric mode.

Our treatment of graphs in Chapter 1 was based on functions defined in terms of the independent variable x (or in the case of the circular functions, s). In Chapter 6 we will investigate in more detail graphs defined *parametrically*. Because the circular functions can be illustrated beautifully using parametric equations, and because modern graphing calculators can graph curves parametrically, we present a brief discussion here.

> **PARAMETRIC EQUATIONS**
>
> Suppose that a set of points (x, y) is defined in such a way that $x = f(t)$ and $y = g(t)$. Then the set of points is said to be defined parametrically. The two equations are called **parametric equations,** and t is called the parameter. The **parameter** t is a real number in some interval I.

We will investigate the parametric equations

$$x = \cos t, \qquad y = \sin t$$

where the parameter t lies in the interval $[0, 2\pi)$. Based on our earlier discussion, the graph will be that of the unit circle. See Figure 34.

Modern graphing calculators have the capability of graphing curves defined parametrically. Figure 35 shows the unit circle $x = \cos t$, $y = \sin t$, for t in $[0, 2\pi)$ with t-step $= .1$ and in radian mode. You should read your owner's manual to see how your particular model can be used to graph parametric equations.

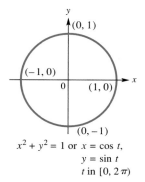

$x^2 + y^2 = 1$ or $x = \cos t,$
$y = \sin t$
t in $[0, 2\pi)$

FIGURE 34

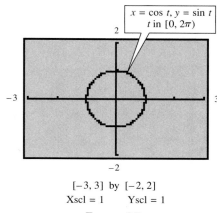

$[-3, 3]$ by $[-2, 2]$
Xscl $= 1$ Yscl $= 1$

FIGURE 35

EXAMPLE 2

Evaluating
Trigonometric
Functions Using the
Unit Circle Defined
Parametrically

(a)   Find values of $\cos \frac{2\pi}{3}$ and $\sin \frac{2\pi}{3}$ by determining the coordinates of the point (x, y) on the unit circle for which $t = \frac{2\pi}{3}$.

SOLUTION   As seen in Figure 36, when $t = \frac{2\pi}{3}$ (indicated by the approximation 2.0943951) $x = -.5$ and $y = .8660254$. Therefore, $\cos \frac{2\pi}{3} = -.5$ and $\sin \frac{2\pi}{3} \approx .8660254$. These results support the exact values found in part (a) of Example 1. (The displayed value of y, which is $\sin \frac{2\pi}{3}$, is a decimal approximation for the *exact* value, $\frac{\sqrt{3}}{2}$.)

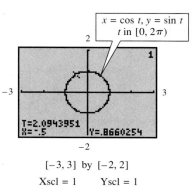

$[-3, 3]$ by $[-2, 2]$
Xscl $= 1$ Yscl $= 1$

FIGURE 36

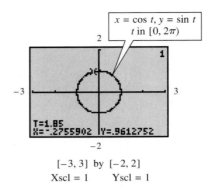

$$[-3, 3] \text{ by } [-2, 2]$$
$$\text{Xscl} = 1 \qquad \text{Yscl} = 1$$

FIGURE 37

(b) Find an approximation of cos 1.85 using the unit circle graphed parametrically.

SOLUTION Figure 37 shows the unit circle with $t = 1.85$. Since we want to approximate cos 1.85, we observe the *x-coordinate* of the terminal point of the arc with length 1.85. We find the approximation $-.2755902$, supporting our result in part (c) of Example 1. ∎

The Negative Number Identities

In Section 2.2 we introduced the reciprocal, Pythagorean, and quotient identities. There is another group of identities that will prove useful throughout our study. They are known as the negative number (or negative angle) identities. To illustrate these identities, see the unit circle in Figure 38. As suggested in the figure, an arc of length θ having the point (x, y) as its terminal point has a corresponding arc $-\theta$ with a point $(x, -y)$ as its terminal point. From the definition of sine,

$$\sin(-\theta) = -y \qquad \text{and} \qquad \sin \theta = y$$

so that $\sin(-\theta)$ and $\sin \theta$ are negatives of each other, or

$$\sin(-\theta) = -\sin \theta.$$

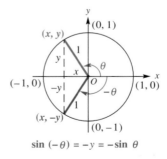

$$\sin(-\theta) = -y = -\sin \theta$$

FIGURE 38

The figure shows an arc θ in quadrant II, but the same result holds for θ in any quadrant. Also, by definition,

$$\cos(-\theta) = x \qquad \text{and} \qquad \cos \theta = x,$$

so that

$$\cos(-\theta) = \cos \theta.$$

These formulas for $\sin(-\theta)$ and $\cos(-\theta)$ can be used to find $\tan(-\theta)$ in terms of $\tan \theta$:

$$\tan(-\theta) = \frac{\sin(-\theta)}{\cos(-\theta)} = \frac{-\sin \theta}{\cos \theta} = -\frac{\sin \theta}{\cos \theta}$$

or

$$\tan(-\theta) = -\tan \theta.$$

Similar reasoning gives the remaining negative angle identities:

$$\csc(-\theta) = -\csc \theta, \qquad \sec(-\theta) = \sec \theta, \qquad \cot(-\theta) = -\cot \theta.$$

The negative number identities are summarized below. Again, we assume that the functions are defined.

NEGATIVE NUMBER IDENTITIES

$\sin(-\theta) = -\sin \theta$	$\csc(-\theta) = -\csc \theta$
$\cos(-\theta) = \cos \theta$	$\sec(-\theta) = \sec \theta$
$\tan(-\theta) = -\tan \theta$	$\cot(-\theta) = -\cot \theta$

FOR **GROUP DISCUSSION**

Put your calculator in radian mode. (This discussion is valid for degree-measured angles as well, but since we are studying circular functions, we will use radian measure here.) Let s be the number of letters in your last name.

1. Find $\sin s$ and then find $\sin(-s)$. How do they compare?
2. Find $\cos s$ and then find $\cos(-s)$. How do they compare?
3. Find $\tan s$ and then find $\tan(-s)$. How do they compare?

Discuss the results obtained and how they relate to the negative number identities.

2.4 EXERCISES

Use the unit circle shown in Figure 33, along with the definitions of the circular functions, to find the exact value for each of the following. Then support your answer by finding the circular function value on your calculator. Be sure the calculator is in radian mode.

1. $\sin \dfrac{7\pi}{6}$ **2.** $\cos \dfrac{5\pi}{3}$ **3.** $\tan \dfrac{3\pi}{4}$ **4.** $\sin \dfrac{5\pi}{3}$

5. $\cos \dfrac{7\pi}{6}$ **6.** $\tan \dfrac{4\pi}{3}$ **7.** $\sec \dfrac{2\pi}{3}$ **8.** $\csc \dfrac{11\pi}{6}$

9. $\cot \dfrac{5\pi}{6}$ **10.** $\cos\left(-\dfrac{4\pi}{3}\right)$ **11.** $\sin\left(-\dfrac{5\pi}{6}\right)$ **12.** $\tan \dfrac{17\pi}{3}$

13. $\sec \dfrac{23\pi}{6}$ **14.** $\csc \dfrac{13\pi}{3}$

Find a calculator approximation of each of the following. Be sure that your calculator is in radian mode.

15. sin .6109

16. sin .8203

17. cos(−1.1519)

18. cos(−5.2825)

19. tan 4.0203

20. tan 6.4752

21. csc(−9.4946)

22. csc 1.3875

23. sec 2.8440

24. sec(−8.3429)

25. cot 6.0301

26. cot 3.8426

In Exercises 27–30, a unit circle generated by the parametric equations $x = \cos t$, $y = \sin t$ is shown, along with a display for t, x, and y. Use this information to find cos t and sin t, and then verify by using the cosine and sine functions of your calculator.

27.

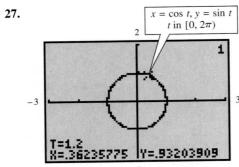

28.

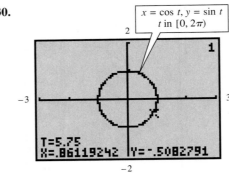

29.

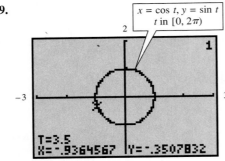

30.

Relating Concepts

With your calculator in radian mode, respond to Exercises 31–36 in order.

31. Find a calculator approximation for cos 4.9.

32. Use a negative number identity to give an approximation for cos(−4.9), referring only to your answer in Exercise 31.

33. Find a calculator approximation for sin 4.9.

34. Use a negative number identity to give an approximation for sin(−4.9), referring only to your answer in Exercise 33.

35. Use the results of Exercises 31 and 33 to fill in the blanks: Because cos 4.9 is _____ and sin 4.9 is _____ , the arc
     (positive/negative)   (positive/negative)
on the unit circle representing the real number 4.9 must terminate in quadrant

_____ .
    (I/II/III/IV)

36. Based on your results in Exercises 32 and 34, write a statement involving −4.9 analogous to the statement in Exercise 35.

Use the negative number identities to write each of the following as a circular function of a positive number. (For example, sin(−3.4) = −sin 3.4.)

37. $\cos(-4.38)$

38. $\cos(-5.46)$

39. $\sin(-.5)$

40. $\sin(-2.5)$

41. $\tan\left(-\dfrac{\pi}{7}\right)$

42. $\tan\left(-\dfrac{4\pi}{7}\right)$

43. $\sec(-8)$

44. $\sec(-.055)$

45. $\csc\left(-\dfrac{1}{4}\right)$

46. $\csc\left(-\dfrac{5}{9}\right)$

47. $\cot(-10^5)$

48. $\cot(-5^8)$

Further Explorations

While radian mode is very important for connecting angle measure and arc length, the graphing calculator cannot display values like π, $\frac{\pi}{2}$, and 2π in their exact forms. Instead the graphing calculator can only display and evaluate decimal approximations. In other words, it is unable to manipulate irrational numbers; it must first convert to a rational approximation. For this reason, it is sometimes preferable to use DEGREE mode when evaluating trigonometric functions. If your graphing calculator has a TABLE feature, it can be used to find patterns in trigonometric functions. The figure shows a TABLE where $Y_1 = \cos x$ and $Y_2 = \sin x$. The calculator is in DEGREE MODE and ΔTbl = 15°.

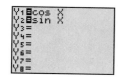  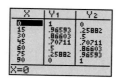

Note some patterns:

$$\cos 45° = \sin 45°$$

$\cos x$ decreases from 1 to 0 as x goes from 0° to 90°.

$\sin x$ increases from 0 to 1 as x goes from 0° to 90°.

1. Create the TABLE from the figure in your calculator. Scroll through the TABLE at least as far as $x = 360°$ in order to answer the following.
   (a) Find all values of x where $\cos x = \sin x$.
   (b) Describe all intervals where $\cos x$ is decreasing and all intervals where $\cos x$ is increasing.
   (c) Describe all intervals where $\sin x$ is decreasing and all intervals where $\sin x$ is increasing.
   (d) What are the maximum and minimum values for $\cos x$ and $\sin x$?
   (e) Find a ΔTbl and Tblmin where $\cos x$ always will equal 1. For what values of x does $\cos x = 1$?
   (f) Find a ΔTbl and Tblmin where $\cos x$ always will equal 0. For what values of x does $\cos x = 0$?
   (g) Find a ΔTbl and Tblmin where $\sin x$ always will equal 1. For what values of x does $\sin x = 1$?
   (h) Find a ΔTbl and Tblmin where $\sin x$ always will equal 0. For what values of x does $\sin x = 0$?

2.5 ANALYSIS OF THE SINE AND COSINE FUNCTIONS

Periodic Functions ▌ The Graph of the Sine Function ▌ The Graph of the Cosine Function ▌ Transformations of the Graphs of the Sine and Cosine Functions

Periodic Functions

FOR GROUP DISCUSSION

With your calculator in radian mode, do the following.

1. Let t represent the number of letters in your first name, and find a calculator approximation for $\cos t$. Store or write down your answer.
2. Let s represent the number of letters in your last name, find a calculator approximation for $\cos(t + s \cdot 2\pi)$. Compare your result to your answer in Item 1.

Everyone should get the same results for themselves in Items 1 and 2, although the answers will vary from student to student depending upon the values of t and s. Use the unit circle interpretation from the previous section to explain why the answers in Items 1 and 2 are the same. Will the same thing happen if you use the sine function rather than the cosine function?

The sine and cosine functions repeat their values over and over. They are examples of *periodic functions*.

PERIODIC FUNCTION

A **periodic function** is a function f such that

$$f(x) = f(x + np),$$

for every real number x in the domain of f, every integer n, and for some positive real number p. The smallest possible positive value of p is the **period** of the function.

The circumference of the unit circle is 2π, and therefore the smallest value of p for which the sine and cosine functions repeat is 2π. Therefore, the sine and cosine functions are periodic functions with period 2π.

The Graph of the Sine Function

For every real number x there is a real number y such that $y = \sin x$. This number y is a real number in the interval $[-1, 1]$, and can be determined by using a calculator in one of several ways. The value of $\sin x$ is the y-coordinate of the terminal point of an arc of length x that is represented on the unit circle; it may also be determined by using the sine function of the calculator, with the calculator in radian mode. The resulting points (x, y) form the graph of this function.

TECHNOLOGICAL NOTE
While we will refer to the window $[-2\pi, 2\pi]$ by $[-4, 4]$ as the *trig viewing window* in this text, your model may use a different "standard" viewing window for the graphs of circular functions.

Since the period of the sine function is 2π, the curve shown in Figure 39 repeats over and over. This curve is called a **sine wave,** or **sinusoid.** You should learn the shape of this graph and be able to sketch it quickly by hand. Graphing calculators often have a window designated for graphing circular functions. We will refer to the window $[-2\pi, 2\pi]$ by $[-4, 4]$ with Xscl $= \frac{\pi}{2}$ and Yscl $= 1$ as the *trig viewing window.*

SINE FUNCTION

$f(x) = \sin x$ (Figure 39)

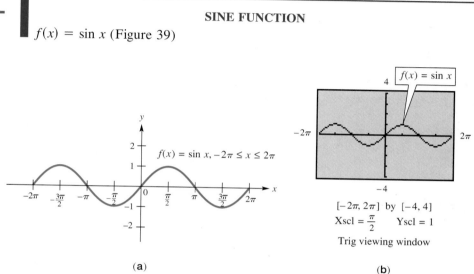

(a)   (b)

$[-2\pi, 2\pi]$ by $[-4, 4]$
Xscl $= \frac{\pi}{2}$ Yscl $= 1$
Trig viewing window

FIGURE 39

Domain: $(-\infty, \infty)$

Range: $[-1, 1]$

Over the interval $[0, 2\pi]$, the sine function exhibits the following behavior:

From 0 to $\frac{\pi}{2}$, $\sin x$ increases from 0 to 1.

From $\frac{\pi}{2}$ to π, $\sin x$ decreases from 1 to 0.

From π to $\frac{3\pi}{2}$, $\sin x$ decreases from 0 to -1.

From $\frac{3\pi}{2}$ to 2π, $\sin x$ increases from -1 to 0.

The graph is continuous over its entire domain. Its x-intercepts are of the form $n\pi$, where n is an integer. Its period is 2π. The graph is symmetric with respect to the origin. It is an odd function.

If a periodic function has maximum and minimum values, then its **amplitude** is defined to be half the difference between the maximum and minimum function values. Thus, for the sine function, the amplitude is $\frac{1}{2}[1 - (-1)] = \frac{1}{2}(2) = 1$.

AMPLITUDE OF THE SINE FUNCTION

The amplitude of the sine function is 1.

EXAMPLE **1**

Finding sin x Using
Several Different
Methods

Find an approximation of sin 1.25 using three different methods, including the graph of the sine function.

SOLUTION   We can find an approximation of sin 1.25 by using the sin key of a calculator set in radian mode. Verify that sin 1.25 ≈ .9489846194.

As shown in the previous section, we can graph the unit circle using the parametric equations $x = \cos t$, $y = \sin t$, and letting $t = 1.25$ we find the y-coordinate is approximately .94898462. See Figure 40.

We can also find sin 1.25 by graphing $f(x) = \sin x$, and using the capabilities of the calculator to locate the point with x-coordinate 1.25. As shown in Figure 41, this point has y-coordinate approximately .94898462, supporting our other two approximations.   ∎

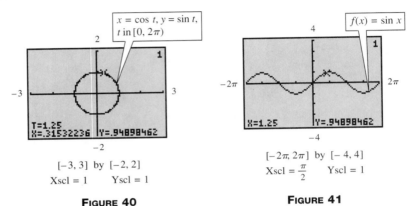

[−3, 3] by [−2, 2]
Xscl = 1   Yscl = 1

FIGURE 40

[−2π, 2π] by [−4, 4]
Xscl = $\frac{\pi}{2}$ Yscl = 1

FIGURE 41

As we mentioned above, the graph of the sine function is symmetric with respect to the origin. Recall from Chapter 1 that the graph of a function f is symmetric with respect to the origin if $f(-x) = -f(x)$ for all x in its domain. The negative number identity $\sin(-x) = -\sin x$ is an example of this general condition. The displays in Figure 42 support this fact for $x = 4.5$.

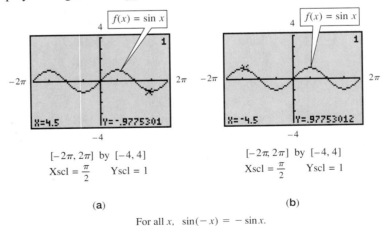

[−2π, 2π] by [−4, 4]
Xscl = $\frac{\pi}{2}$ Yscl = 1

(a)

[−2π, 2π] by [−4, 4]
Xscl = $\frac{\pi}{2}$ Yscl = 1

(b)

For all x, $\sin(-x) = -\sin x$.

FIGURE 42

The Graph of the Cosine Function

The same kind of analysis presented for the sine function can be applied to the cosine function, another function whose graph is a sinusoid.

COSINE FUNCTION

$f(x) = \cos x$ (Figure 43)

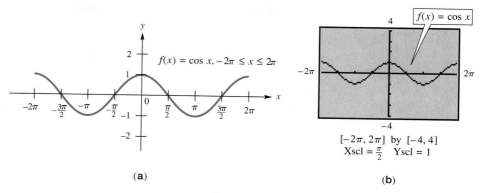

(a)   (b)

$[-2\pi, 2\pi]$ by $[-4, 4]$
Xscl $= \frac{\pi}{2}$ Yscl $= 1$

FIGURE 43

Domain: $(-\infty, \infty)$
Range: $[-1, 1]$

Over the interval $[0, 2\pi]$, the cosine function exhibits the following behavior:

From 0 to $\frac{\pi}{2}$, $\cos x$ decreases from 1 to 0.
From $\frac{\pi}{2}$ to π, $\cos x$ decreases from 0 to -1.
From π to $\frac{3\pi}{2}$, $\cos x$ increases from -1 to 0.
From $\frac{3\pi}{2}$ to 2π, $\cos x$ increases from 0 to 1.

The graph is continuous over its entire domain. Its x-intercepts are of the form $(2n + 1)\frac{\pi}{2}$, where n is an integer. Its period is 2π. The graph is symmetric with respect to the y-axis. It is an even function.

Verify that, like the sine function, the amplitude of the cosine function is 1.

AMPLITUDE OF THE COSINE FUNCTION

The amplitude of the cosine function is 1.

EXAMPLE 2

Interpreting the Coordinates of a Point on the Graph of $f(x) = \cos x$

Use the graph of $f(x) = \cos x$ and the display in Figure 44, along with the fact that $\pi \approx 3.1415927$, to reinforce earlier concepts involving the trigonometric and circular functions.

SOLUTION We see that $\cos \pi \approx \cos 3.1415927 = -1$. This supports the fact that the terminal point of an arc of length π on the unit circle has an x-coordinate of -1. Since the real number π corresponds to an angle of radian measure π (and thus degree measure $180°$), the terminal side of such an angle in standard position will contain the point $(-r, 0)$ for any $r > 0$. The definition of the trigonometric ratio for cosine tells us that $\cos \pi = \cos 180° = \frac{x}{r} = \frac{-r}{r} = -1$, supporting the result seen in Figure 44. ∎

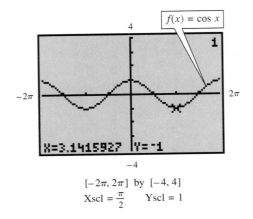

$$[-2\pi, 2\pi] \text{ by } [-4, 4]$$
$$\text{Xscl} = \frac{\pi}{2} \qquad \text{Yscl} = 1$$

FIGURE 44

FOR **GROUP DISCUSSION**

Graph the two functions $y_1 = \sin x$ and $y_2 = \cos(x - \frac{\pi}{2})$ in the trig viewing window of your calculator. Now answer the following items.

1. How do the two graphs compare?
2. Knowing that $\cos(-x) = \cos x$ for all x, how would the graph of $y_3 = \cos(\frac{\pi}{2} - x)$ compare to the graph of y_2? Verify this.
3. In Section 2.3 we learned that cofunctions of complementary angles are equal. How does your answer in Item 2 support this for the circular function *cosine*?

While the graph of the sine function is symmetric with respect to the origin, the graph of the cosine function is symmetric with respect to the y-axis. Recall that the graph of a function f is symmetric with respect to the y-axis if $f(-x) = f(x)$ for all x in its domain. The negative number identity $\cos(-x) = \cos x$ is an example of this general condition. The displays in Figure 45 indicate this fact for $x = 4.5$.

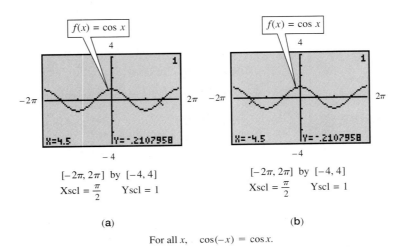

$$[-2\pi, 2\pi] \text{ by } [-4, 4] \qquad\qquad [-2\pi, 2\pi] \text{ by } [-4, 4]$$
$$\text{Xscl} = \frac{\pi}{2} \qquad \text{Yscl} = 1 \qquad\qquad \text{Xscl} = \frac{\pi}{2} \qquad \text{Yscl} = 1$$

(a) (b)

For all x, $\cos(-x) = \cos x$.

FIGURE 45

The most important points on the graph of the sine and cosine functions are the maximum and minimum points and the x-intercepts. The graph changes its concavity at these x-intercepts, and therefore the associated points are inflection points. We now give an interpretation of a "complete graph" of a sinusoid.

COMPLETE GRAPH OF A SINUSOID

A complete graph of a sinusoid will consist of at least one period of the graph. It will show the extreme points and the points of inflection in the interval.

Transformations of the Graphs of the Sine and Cosine Functions

In Chapter 1 we saw how graphs of functions may be transformed by stretching, shrinking, reflecting, and shifting. These transformations can be applied to the sine and cosine functions.

EXAMPLE 3

Analyzing the Graph of a Transformed Circular Function

The graphs of $y_1 = \sin x$ and $y_2 = 2 \sin x$ are shown in Figure 46. Notice that the graph of $y_2 = 2 \sin x$ can be obtained by stretching the graph of $y_1 = \sin x$ vertically by a factor of 2. The period of $y_2 = 2 \sin x$ is 2π, and the range is $[-2, 2]$, causing its amplitude to be 2. ◨

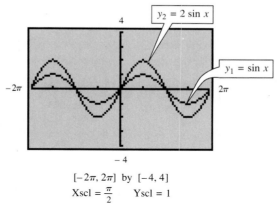

$[-2\pi, 2\pi]$ by $[-4, 4]$

$\text{Xscl} = \dfrac{\pi}{2}$ $\text{Yscl} = 1$

FIGURE 46

If you explore many problems similar to Example 3, you will discover the following (assume $a \neq 0$).

AMPLITUDE OF SINE AND COSINE

The graph of $y = a \sin x$ or $y = a \cos x$ will have the same basic shape as the graph of $y = \sin x$ or $y = \cos x$. The range of the function is $[-|a|, |a|]$ and the amplitude is $|a|$.

No matter what the value of the amplitude, the period of $y = a \sin x$ and $y = a \cos x$ is still 2π. However, the graph of a function of the form $y = \sin bx$ or $\cos bx$, for $b > 0$, $b \neq 1$, will have a period different from 2π. To see why this is so, remember that the values of $\sin bx$ or $\cos bx$ will take on all possible values as bx ranges from 0 to 2π. Therefore, to see what the period of either of these will be, we must solve the compound inequality

$$0 \le bx \le 2\pi.$$

Dividing by the positive number b gives

$$0 \le x \le \frac{2\pi}{b}.$$

Therefore, the period is $\frac{2\pi}{b}$.

PERIOD OF SINE AND COSINE

The graph of $y = \sin bx$ or $y = \cos bx$, $b > 0$, will have the same basic shape as the graph of $y = \sin x$ or $y = \cos x$. However, the period of the function is $\frac{2\pi}{b}$. The range is $[-1, 1]$ and the amplitude is 1.

By dividing the interval $[0, \frac{2\pi}{b}]$ into four equal parts, we obtain the values for which $\sin bx$ or $\cos bx$ is -1, 0, or 1. These will give minimum points, x-intercepts, and maximum points on the graph. For example, consider the function $y = \sin 2x$. To determine an interval that represents one period of the graph, we solve the inequality

$$0 \le 2x \le 2\pi.$$

We divide by 2 to get $0 \le x \le \pi$. The points $(0, 0)$, $(\frac{\pi}{4}, 1)$, $(\frac{\pi}{2}, 0)$, $(\frac{3\pi}{4}, -1)$, and $(\pi, 0)$ represent the important points for this period of the graph.

The following example provides a further analysis of the graph of $y = \sin 2x$.

EXAMPLE 4

Analyzing the Graph of a Transformed Circular Function

The graphs of $y_1 = \sin x$ and $y_2 = \sin 2x$ are shown in Figure 47, graphed over the interval $[-2\pi, 2\pi]$. Notice that while the amplitude of both functions is 1, the period of y_2 is π. This period is found by dividing the period of the sine function, 2π, by the coefficient of x in y_2, namely 2. That is,

$$\text{period of } y_2 = \frac{2\pi}{2} = \pi.$$

In both cases, the range is $[-1, 1]$. ◨

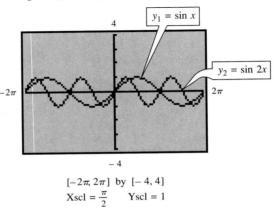

$[-2\pi, 2\pi]$ by $[-4, 4]$
$\text{Xscl} = \frac{\pi}{2}$ $\text{Yscl} = 1$

FIGURE 47

The next example shows how several transformations can be applied to the graph of $y = \cos x$.

<table>
<tr><td>

EXAMPLE 5

Analyzing the Graph of a Transformed Circular Function

</td><td>

Explain how the graph of $y_2 = -3 \cos \frac{1}{2}x$ can be obtained from the graph of $y_1 = \cos x$. Then graph both functions in the window $[-2\pi, 2\pi]$ by $[-4, 4]$.

SOLUTION   The graph of $y_2 = -3 \cos \frac{1}{2}x$ can be obtained from the graph of $y_1 = \cos x$ by changing the period from 2π to $\frac{2\pi}{1/2} = 4\pi$, stretching vertically by a factor of 3, and reflecting across the x-axis (because of the negative sign). See Figure 48.   ▌

</td></tr>
</table>

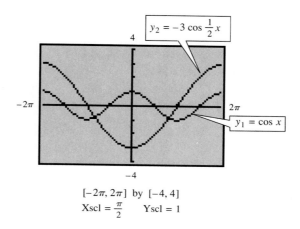

$$[-2\pi, 2\pi] \text{ by } [-4, 4]$$
$$\text{Xscl} = \frac{\pi}{2} \qquad \text{Yscl} = 1$$

FIGURE 48

The graphs of $y = \cos x$ and $y = \sin x$ can also be shifted horizontally and vertically. Recall that the graph of the function $y = f(x - d)$ is obtained by translating the graph of $y = f(x)$ d units to the right if $d > 0$ or $|d|$ units to the left if $d < 0$. In the case of a circular function, a horizontal translation is called a **phase shift.**

<table>
<tr><td>

EXAMPLE 6

Analyzing the Graph of a Circular Function with a Phase Shift

</td><td>

The graphs of $y_1 = \sin x$ and $y_2 = \sin(x - \frac{\pi}{3})$ are shown in Figure 49. The graph of y_2 can be obtained by shifting the graph of y_1 $\frac{\pi}{3}$ units to the right. The number $\frac{\pi}{3}$ is the phase shift for y_2. Notice that neither the period nor the amplitude is affected. For both functions the period is 2π and the amplitude is 1.   ▌

</td></tr>
</table>

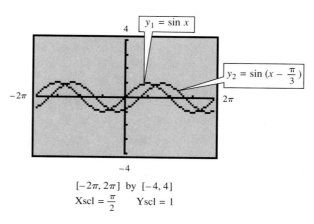

$$[-2\pi, 2\pi] \text{ by } [-4, 4]$$
$$\text{Xscl} = \frac{\pi}{2} \qquad \text{Yscl} = 1$$

FIGURE 49

EXAMPLE 7

Analyzing the Graph
of a Transformed
Circular Function

Explain how the graph of $y_2 = -2 \cos(3x + \pi)$ can be obtained from the graph of $y_1 = \cos x$. Then graph both functions in the window $[-2\pi, 2\pi]$ by $[-4, 4]$.

SOLUTION   The graph of $y_2 = -2 \cos(3x + \pi)$ can be obtained from the graph of $y_1 = \cos x$ by first noticing that the equation can be written $y_2 = -2 \cos 3(x + \frac{\pi}{3})$. The graph of $y_1 = \cos x$ must be shifted $\frac{\pi}{3}$ units to the left, the period must be changed to $\frac{2\pi}{3}$, the graph will be vertically stretched by a factor of 2, and will be reflected across the x-axis because of the negative sign on the coefficient of the cosine function -2. The amplitude of $y_2 = -2 \cos(3x + \pi)$ is 2, because $\frac{1}{2}[2 - (-2)] = 2$. See Figure 50.   ▉

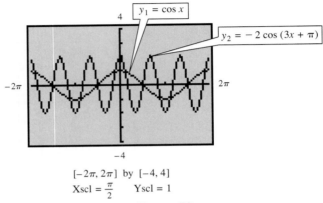

$[-2\pi, 2\pi]$ by $[-4, 4]$
$\text{Xscl} = \frac{\pi}{2}$ $\text{Yscl} = 1$

FIGURE 50

The graph of the function $y = f(x) + c$ can be obtained from the graph of $y = f(x)$ by a *vertical shift*. The shift is upward c units if $c > 0$, or downward $|c|$ units if $c < 0$.

EXAMPLE 8

Analyzing the Graph
of a Circular
Function with a
Vertical Shift

The graphs of $y_1 = \cos 2x$ and $y_2 = -2 + \cos 2x$ are shown in Figure 51. The graph of y_2 can be obtained by shifting the graph of y_1 2 units down, due to the term -2 which is added to $\cos 2x$. For both functions, the period is π and the amplitude is 1.   ▉

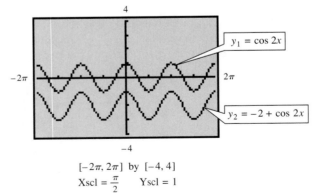

$[-2\pi, 2\pi]$ by $[-4, 4]$
$\text{Xscl} = \frac{\pi}{2}$ $\text{Yscl} = 1$

FIGURE 51

EXAMPLE 9

Analyzing the Graph
of a Transformed
Circular Function

Explain how the graph of $y_2 = -1 + 2 \sin 4(x + \frac{\pi}{4})$ can be obtained from the graph of $y_1 = \sin x$. Then graph both functions in the trig viewing window.

SOLUTION   The graph of $y_2 = -1 + 2 \sin 4(x + \frac{\pi}{4})$ can be obtained from the graph of $y_1 = \sin x$ by shifting $\frac{\pi}{4}$ units to the left, changing the period to $\frac{2\pi}{4} = \frac{\pi}{2}$, stretching vertically by a factor of 2, and shifting the graph down 1 unit. The amplitude of this function is 2. Notice also that its range is $[-3, 1]$. See Figure 52.

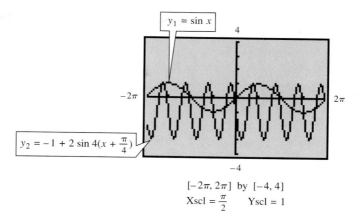

$$[-2\pi, 2\pi] \text{ by } [-4, 4]$$
$$\text{Xscl} = \frac{\pi}{2} \qquad \text{Yscl} = 1$$

FIGURE 52

In conclusion, we state the various characteristics of a function of the form $y = d + a \sin b(x + c)$ or $y = d + a \cos b(x + c)$, where $b > 0$.

For $y = d + a \sin b(x + c)$ or $y = d + a \cos b(x + c)$, where $b > 0$:

1. The period is $\dfrac{2\pi}{b}$.
2. The amplitude is $|a|$.
3. The phase shift is $-c$.
4. The domain is $(-\infty, \infty)$.
5. The range is the set of numbers between and inclusive of $d + |a|$ and $d - |a|$.

2.5 EXERCISES

In Exercises 1–6, you are given the graph of either $y = \sin x$ or $y = \cos x$ in the standard trig window, along with a display. Use the graph and the display to write an equation of the form $\sin x \approx k$, $\cos x \approx k$, $\sin x = k$, or $\cos x = k$, for specific values of x and k. Then verify your result using either the unit circle graphed parametrically or the appropriate function key on your calculator. Be sure your calculator is set to radian mode.

1.

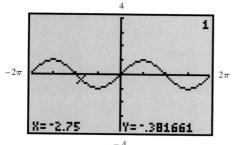

2.

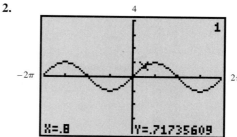

3.

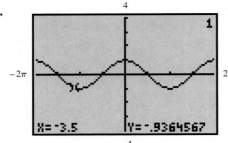

4.

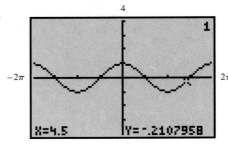

5. $\left(\dfrac{\pi}{2} \approx 1.5707963\right)$

6. $\left(\dfrac{3\pi}{2} \approx 4.712389\right)$

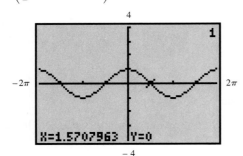

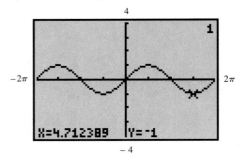

Without using a calculator to graph the functions, match each function in Exercises 7–14 with its graph.

7. $y = \sin x$

8. $y = \cos x$

9. $y = -\sin x$

10. $y = -\cos x$

11. $y = \sin 2x$

12. $y = \cos 2x$

13. $y = 2 \sin x$

14. $y = 2 \cos x$

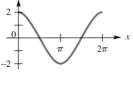

A

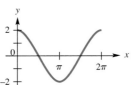

B

C

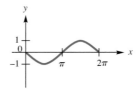

D

E

F

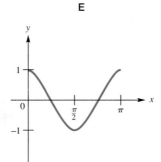

G

H

Without using a calculator to graph the functions, match each function in Exercises 15–22 with its graph.

15. $y = \sin(x - \frac{\pi}{4})$ D

16. $y = \sin(x + \frac{\pi}{4})$ G

17. $y = \cos(x - \frac{\pi}{4})$ H

18. $y = \cos(x + \frac{\pi}{4})$ A

19. $y = 1 + \sin x$ B

20. $y = -1 + \sin x$ E

21. $y = 1 + \cos x$ F

22. $y = -1 + \cos x$ C

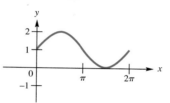

A   B

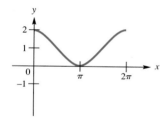

C   D   E

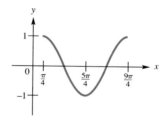

F   G   H

Suppose that a, b, c, and d are all positive numbers. Consider circular functions of the forms

$$y = d + a \sin b(x + c) \quad \text{and} \quad y = d + a \cos b(x + c)$$

and answer the questions in Exercises 23–26.

23. What is the amplitude of the function?

24. What is the period of the function?

25. What is the phase shift of the function?

26. Is there a vertical shift? If so, in which direction and by how many units?

For each of the following circular functions, find (a) the amplitude, (b) the period, (c) the phase shift (if any), (d) the vertical translation (if any), and (e) the range of the function.

27. $y = 2 \sin(x - \pi)$

28. $y = \frac{2}{3} \cos\left(x + \frac{\pi}{2}\right)$

29. $y = 4 \cos\left(\frac{1}{2}x + \frac{\pi}{2}\right)$

30. $y = -\cos\frac{2}{3}\left(x - \frac{\pi}{3}\right)$

31. $y = 2 - \sin\left(3x - \frac{\pi}{5}\right)$

32. $y = -1 + \frac{1}{2}\cos(2x - 3\pi)$

33. $y = 2 - 3 \cos \pi x$ $\quad 3, 2, 0, 2$

34. $y = 1 - \dfrac{2}{3} \sin \dfrac{3}{4}x$

Explain how the graph of the given function can be obtained from the graph of $y = \cos x$ or $y = \sin x$ by a transformation.

35. $y = -3 + 2 \sin\left(x + \dfrac{\pi}{2}\right)$

36. $y = 4 - 3 \cos(x - \pi)$

37. $y = -\dfrac{5}{2} + \cos 3\left(x - \dfrac{\pi}{6}\right)$

38. $y = \dfrac{1}{2} + \sin 2\left(x + \dfrac{\pi}{4}\right)$

Use a calculator to graph each function in the standard trig window.

39. $y = 3 \cos 2x$

40. $y = -2 \cos 4\left(x + \dfrac{\pi}{2}\right)$

41. $y = 1 - 2 \cos .5x$

42. $y = -1 + 3 \sin .5x$

R*elating Concepts*

Consider the function $f(x) = -5 + 3 \sin 2(x - \frac{\pi}{2})$. Answer the following exercises in order.

43. (a) Because the maximum value of the sine function is ——————— , the maximum value of $\sin 2(x - \frac{\pi}{2})$ is ——————— , the maximum value of $3 \sin 2(x - \frac{\pi}{2})$ is ——————— , and thus the maximum value of $-5 + 3 \sin 2(x - \frac{\pi}{2})$ is ——————— .

(b) Because the minimum value of the sine function is ——————— , the minimum value of $\sin 2(x - \frac{\pi}{2})$ is ——————— , the minimum value of $3 \sin 2(x - \frac{\pi}{2})$ is ——————— , and thus the minimum value of $-5 + 3 \sin 2(x - \frac{\pi}{2})$ is ——————— .

44. Based on the answers in Exercise 43, what is the range of f?

45. Why will the standard trig window as defined in the text not provide a complete graph of f?

46. In order to obtain a complete graph of f, Ymax must be at least ——————— and Ymin must be at most ——————— .

47. Explain why using Xmin $= -2\pi$ and Xmax $= 2\pi$ will show exactly 4 periods of the graph of f.

48. Use your calculator to graph f in the window $[-2\pi, 2\pi]$ by $[-10, 5]$.

49. Look at the calculator-generated graph from Exercise 48. A "border" of empty space appears above and below the graph. If you did not want such a border to appear, what Ymin and Ymax values should you use so that a complete graph would fit?

50. Evaluate the approximate values of the function for $x = -2$ and for $x = -2 + \pi$. What is the value in each case? Why is this so?

A function of the form $y = a \cos bx$ or $y = a \sin bx$, where $b < 0$, can be rewritten in an equivalent form so that the coefficient of x is positive. This is done using the appropriate negative number identity. For example,

$$y = 3 \cos(-2x) \text{ is equivalent to } y = 3 \cos 2x$$

and $\quad y = 3 \sin(-2x)$ is equivalent to $y = 3(-\sin 2x) = -3 \sin 2x$.

Write each of the following in an equivalent form so that the coefficient of x is positive.

51. $y = 4 \sin(-3x)$ **52.** $y = 5 \sin(-6x)$ **53.** $y = -2 \cos(-3x)$

54. $y = -\dfrac{1}{2} \cos(-\pi x)$ **55.** $y = -3 \sin(-6x)$ **56.** $y = -3 \cos(-4x)$

*F*urther Explorations

In each of the following TABLES, $Y_1 = \sin x$ and Y_2 is some transformation of Y_1. Use what you know about transformations to find the expression for Y_2. Confirm by duplicating each TABLE on your graphing calculator in RADIAN MODE. It is possible to find more than one solution for some of these exercises. In each exercise, $\Delta \text{Tbl} = \frac{\pi}{6}$. (*Note:* The calculator displays a decimal approximation of each.) The first TABLE in each exercise shows the interval $[0, \pi]$ and the second TABLE shows the interval $[\pi, 2\pi]$.

1.

X	Y1	Y2
0	0	0
.5236	.5	1
1.0472	.86603	1.7321
1.5708	1	2
2.0944	.86603	1.7321
2.618	.5	1
3.1416	0	0

X=0

X	Y1	Y2
3.1416	0	0
3.6652	-.5	-1
4.1888	-.866	-1.732
4.7124	-1	-2
5.236	-.866	-1.732
5.7596	-.5	-1
6.2832	0	0

X=3.14159265359

2.

X	Y1	Y2
0	0	0
.5236	.5	-1.5
1.0472	.86603	-2.598
1.5708	1	-3
2.0944	.86603	-2.598
2.618	.5	-1.5
3.1416	0	0

X=0

X	Y1	Y2
3.1416	0	0
3.6652	-.5	1.5
4.1888	-.866	2.5981
4.7124	-1	3
5.236	-.866	2.5981
5.7596	-.5	1.5
6.2832	0	0

X=3.14159265359

3.

X	Y1	Y2
0	0	0
.5236	.5	.86603
1.0472	.86603	.86603
1.5708	1	0
2.0944	.86603	-.866
2.618	.5	-.866
3.1416	0	0

X=0

X	Y1	Y2
3.1416	0	0
3.6652	-.5	.86603
4.1888	-.866	.86603
4.7124	-1	0
5.236	-.866	-.866
5.7596	-.5	-.866
6.2832	0	-2E-13

X=3.14159265359

4.

X	Y1	Y2
0	0	0
.5236	.5	1
1.0472	.86603	0
1.5708	1	-1
2.0944	.86603	0
2.618	.5	1
3.1416	0	0

X=0

X	Y1	Y2
3.1416	0	0
3.6652	-.5	-1
4.1888	-.866	-2E-13
4.7124	-1	1
5.236	-.866	0
5.7596	-.5	-1
6.2832	0	2E-13

X=3.14159265359

5.

X	Y1	Y2
0	0	-.5
.5236	.5	0
1.0472	.86603	.5
1.5708	1	.86603
2.0944	.86603	1
2.618	.5	.86603
3.1416	0	.5

X=0

X	Y1	Y2
3.1416	0	.5
3.6652	-.5	0
4.1888	-.866	-.5
4.7124	-1	-.866
5.236	-.866	-1
5.7596	-.5	-.866
6.2832	0	-.5

X=3.14159265359

6.

X	Y1	Y2
0	0	0
.5236	.5	-.5
1.0472	.86603	-.866
1.5708	1	-1
2.0944	.86603	-.866
2.618	.5	-.5
3.1416	0	0

X=0

X	Y1	Y2
3.1416	0	0
3.6652	-.5	.5
4.1888	-.866	.86603
4.7124	-1	1
5.236	-.866	.86603
5.7596	-.5	.5
6.2832	0	0

X=3.14159265359

7.

X	Y1	Y2
0	0	1
.5236	.5	1.5
1.0472	.86603	1.866
1.5708	1	2
2.0944	.86603	1.866
2.618	.5	1.5
3.1416	0	1

X=0

X	Y1	Y2
3.1416	0	1
3.6652	-.5	.5
4.1888	-.866	.13397
4.7124	-1	0
5.236	-.866	.13397
5.7596	-.5	.5
6.2832	0	1

X=3.14159265359

8.

X	Y1	Y2
0	0	-2
.5236	.5	-1.5
1.0472	.86603	-1.134
1.5708	1	-1
2.0944	.86603	-1.134
2.618	.5	-1.5
3.1416	0	-2

X=0

X	Y1	Y2
3.1416	0	-2
3.6652	-.5	-2.5
4.1888	-.866	-2.866
4.7124	-1	-3
5.236	-.866	-2.866
5.7596	-.5	-2.5
6.2832	0	-2

X=3.14159265359

2.6 ANALYSIS OF OTHER CIRCULAR FUNCTIONS

The Graphs of the Cosecant and Secant Functions ▮ The Graphs of the Tangent and Cotangent Functions ▮ Transformation of the Graphs of the Other Circular Functions

The Graphs of the Cosecant and Secant Functions

Since cosecant values are reciprocals of the corresponding sine values, the period of the function $y = \csc x$ is 2π, the same as for $y = \sin x$. When $\sin x = 1$, the value of $\csc x$ is also 1, and when $0 < \sin x < 1$, then $\csc x > 1$. Also, if $-1 < \sin x < 0$, then $\csc x < -1$. (Verify these statements with a calculator set in radian mode.) As $|x|$ approaches 0, $|\sin x|$ approaches 0, and $|\csc x|$ gets larger and larger. The graph of $\csc x$ approaches the vertical line $x = 0$ but never touches it. The line $x = 0$ is a *vertical asymptote*. In fact, the lines $x = n\pi$, where n is any integer, are all vertical asymptotes. Using this information and plotting a few points shows that the graph takes the shape of the solid curve shown in Figure 53. To show how the two graphs are related, the graph of $y = \sin x$ is also shown, as a dashed curve.

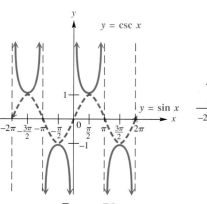

FIGURE 53

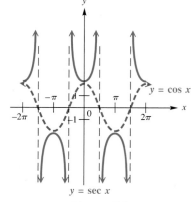

FIGURE 54

A similar analysis for the secant leads to the solid curve shown in Figure 54. The dashed curve, $y = \cos x$, is shown so that the relationship between these two reciprocal functions is seen.

When graphing these functions on a graphing calculator, we will use dot mode in order to get an accurate picture. (If connected mode is used, the calculator will attempt to connect points that are actually separated by vertical asymptotes.) We will enter csc x as $\frac{1}{\sin x}$, and sec x as $\frac{1}{\cos x}$.

COSECANT FUNCTION

$f(x) = \csc x$ (Figure 55)

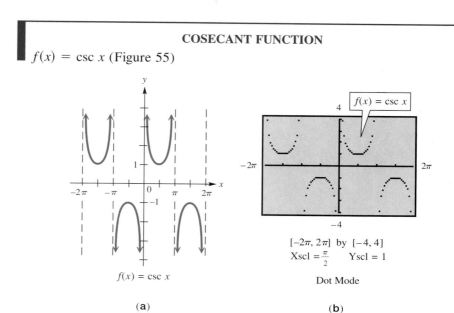

(a)   (b)

FIGURE 55

Domain: $\{x \mid x \neq n\pi, \text{ where } n \text{ is an integer}\}$

Range: $(-\infty, -1] \cup [1, \infty)$

Over the interval $(0, 2\pi)$, the cosecant function exhibits the following behavior:

From 0 to $\frac{\pi}{2}$, csc x decreases from ∞ to 1.

From $\frac{\pi}{2}$ to π, csc x increases from 1 to ∞.

From π to $\frac{3\pi}{2}$, csc x increases from $-\infty$ to -1.

From $\frac{3\pi}{2}$ to 2π, csc x decreases from -1 to $-\infty$.

The graph is discontinuous at values of x of the form $x = n\pi$, and has vertical asymptotes at these values. There are no x-intercepts. Its period is 2π. It has no amplitude, since there are no maximum and minimum values. The graph is symmetric with respect to the origin. It is an odd function.

SECANT FUNCTION

$f(x) = \sec x$ (Figure 56)

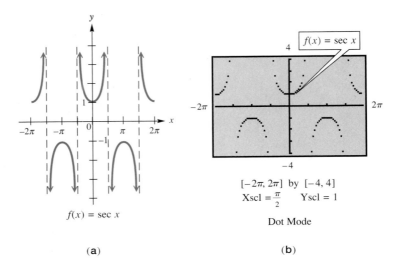

$[-2\pi, 2\pi]$ by $[-4, 4]$

$\text{Xscl} = \frac{\pi}{2}$ $\text{Yscl} = 1$

Dot Mode

(a)   (b)

FIGURE 56

Domain: $\{x \mid x \neq (2n + 1)\frac{\pi}{2}$, where n is an integer$\}$

Range: $(-\infty, -1] \cup [1, \infty)$

Over the interval $[0, 2\pi]$, the secant function exhibits the following behavior:

From 0 to $\frac{\pi}{2}$, $\sec x$ increases from 1 to ∞.

From $\frac{\pi}{2}$ to π, $\sec x$ increases from $-\infty$ to -1.

From π to $\frac{3\pi}{2}$, $\sec x$ decreases from -1 to $-\infty$.

From $\frac{3\pi}{2}$ to 2π, $\sec x$ decreases from ∞ to 1.

The graph is discontinuous at values of x of the form $x = (2n + 1)\frac{\pi}{2}$, and has vertical asymptotes at these values. There are no x-intercepts. Its period is 2π. It has no amplitude, since there are no maximum and minimum values. The graph is symmetric with respect to the y-axis. It is an even function.

The Graphs of the Tangent and Cotangent Functions

Because $\tan x = \frac{\sin x}{\cos x}$, the tangent function is undefined when $\cos x = 0$ and the graph of the tangent function has x-intercepts when $\sin x = 0$. Based on our discussion of the sine and cosine functions in the previous section, we may conclude that the tangent is undefined when $x = (2n + 1)\frac{\pi}{2}$, where n is an integer, and has x-intercepts when $x = n\pi$. Using a calculator in dot mode, we graph $y = \tan x$. See Figure 57.

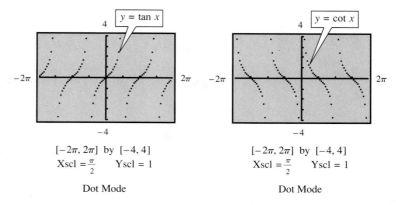

$[-2\pi, 2\pi]$ by $[-4, 4]$
Xscl $= \frac{\pi}{2}$ Yscl $= 1$

Dot Mode

FIGURE 57

$[-2\pi, 2\pi]$ by $[-4, 4]$
Xscl $= \frac{\pi}{2}$ Yscl $= 1$

Dot Mode

FIGURE 58

A similar analysis for the cotangent function leads to the graph in Figure 58. Using a graphing calculator, we may either enter cot x as $\frac{\cos x}{\sin x}$, or as $\frac{1}{\tan x}$. We see that the cotangent is undefined when $x = n\pi$, where n is an integer, and has x-intercepts when $x = (2n + 1)\frac{\pi}{2}$.

TANGENT FUNCTION

$f(x) = \tan x$ (Figure 59)

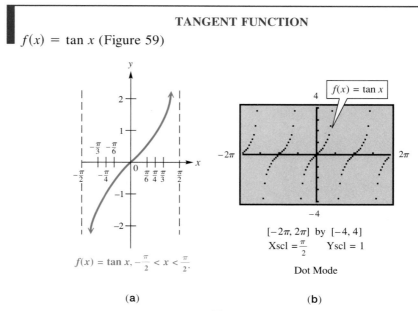

$f(x) = \tan x, -\frac{\pi}{2} < x < \frac{\pi}{2}$.

(a)

$[-2\pi, 2\pi]$ by $[-4, 4]$
Xscl $= \frac{\pi}{2}$ Yscl $= 1$

Dot Mode

(b)

FIGURE 59

Domain: $\{x \mid x \neq (2n + 1)\frac{\pi}{2}$, where n is an integer$\}$ Range: $(-\infty, \infty)$

Over the interval $[0, \pi]$, the tangent function exhibits the following behavior:

From 0 to $\frac{\pi}{2}$, tan x increases from 0 to ∞.

From $\frac{\pi}{2}$ to π, tan x increases from $-\infty$ to 0.

The graph is discontinuous at values of x of the form $x = (2n + 1)\frac{\pi}{2}$, and has vertical asymptotes at these values. The x-intercepts are of the form $x = n\pi$. Its period is π. It has no amplitude, since there are no minimum and maximum values. The graph is symmetric with respect to the origin. It is an odd function.

COTANGENT FUNCTION

$f(x) = \cot x$ (Figure 60)

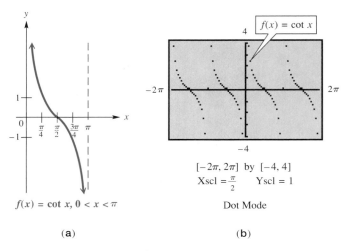

$f(x) = \cot x, \, 0 < x < \pi$

(a)

$[-2\pi, 2\pi]$ by $[-4, 4]$

$\text{Xscl} = \frac{\pi}{2}$ $\text{Yscl} = 1$

Dot Mode

(b)

FIGURE 60

Domain: $\{x \mid x \neq n\pi, \text{ where } n \text{ is an integer}\}$

Range: $(-\infty, \infty)$

Over the interval $(0, \pi)$, the cotangent function exhibits the following behavior:

From 0 to π, $\cot x$ decreases from ∞ to $-\infty$.

The graph is discontinuous at values of x of the form $x = n\pi$, and has vertical asymptotes at these values. The x-intercepts are of the form $x = (2n + 1)\frac{\pi}{2}$. Its period is π. It has no amplitude, since there are no minimum and maximum values. The graph is symmetric with respect to the origin. It is an odd function.

Transformation of the Graphs of the Other Circular Functions

In the following examples, we show how the graphs of the secant, cosecant, tangent, and cotangent functions can be transformed.

EXAMPLE 1

Analyzing a
Transformed Secant
Graph

Figure 61 shows the graph of $y = 2 \sec \frac{1}{2}x$ in both traditional and calculator-generated forms.

This graph is obtained by using the graph of $y = 2 \cos \frac{1}{2}x$ as a guide (shown as a dashed curve in Figure 61(a)). Its period is the same as that of this transformed cosine graph: $\frac{2\pi}{1/2} = 4\pi$. Notice that because the period is altered from the basic secant graph, the locations of the vertical asymptotes are also altered. They are of the form $x = (2n + 1)\pi$. The coefficient 2 affects the graph by stretching it vertically by a factor of 2. The y-intercept is now 2.   ∎

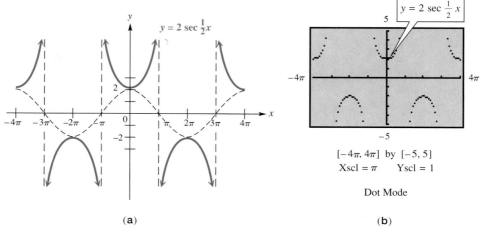

(a)   (b)

FIGURE 61

FOR **GROUP DISCUSSION**

If we know the location of the vertical asymptote $x = k$ with smallest positive value of k, along with the period of the function, we can determine the locations of all vertical asymptotes. Discuss how you might go about finding this smallest such positive value in Example 1.

EXAMPLE **2**

Analyzing a
Transformed
Cosecant Graph

Figure 62 shows the graph of $y = \frac{3}{2}\csc\left(x - \frac{\pi}{2}\right)$ in both traditional and calculator-generated forms.

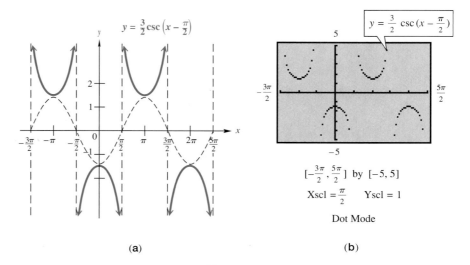

(a)   (b)

FIGURE 62

This graph is obtained by shifting the graph of $y = \csc x$ $\frac{\pi}{2}$ units to the right (the phase shift) and stretching vertically by a factor of $\frac{3}{2}$. The period is 2π, since this is the period of the associated reciprocal function $y = \frac{3}{2}\sin\left(x - \frac{\pi}{2}\right)$. Vertical asymptotes are of the form $x = (2n + 1)\frac{\pi}{2}$. (In Figure 62(a) we show the graph of the related reciprocal function as a dashed line.) ◼

FOR GROUP DISCUSSION

1. Use the graph in Figure 61 to determine the domain and the range of $y = 2 \sec \frac{1}{2}x$.
2. Use the graph in Figure 62 to determine the domain and the range of $y = \frac{3}{2} \csc(x - \frac{\pi}{2})$.
3. Use the capabilities of your calculator to support the following statement: The function $y = \frac{3}{2} \csc(x - \frac{\pi}{2})$ has a local minimum at $x = \pi$.
4. Repeat Item 3 for the following statement: The point $(\frac{\pi}{2}, 2\sqrt{2})$ lies on the graph of $y = 2 \sec \frac{1}{2}x$.

EXAMPLE 3

Analyzing a Transformed Tangent Graph

Figure 63 shows the graph of $y = -3 \tan \frac{1}{2}x$ in both traditional and calculator-generated forms.

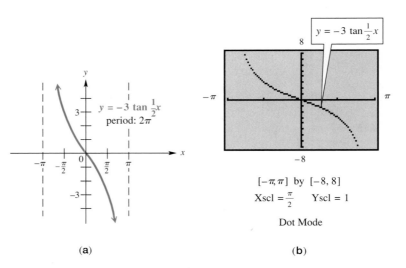

$[-\pi, \pi]$ by $[-8, 8]$

$\text{Xscl} = \frac{\pi}{2}$ $\text{Yscl} = 1$

Dot Mode

(a)   (b)

FIGURE 63

The graph of this function has period $\pi/(\frac{1}{2}) = 2\pi$. In addition to this, the fact that the coefficient is -3 causes a vertical stretch by a factor of 3 and a reflection across the x-axis. The vertical asymptotes are of the form $x = (2n + 1)\pi$, and the x-intercepts are of the form $2n\pi$. The domain is $\{x \mid x \neq (2n + 1)\pi\}$ and the range is $(-\infty, \infty)$. ◼

EXAMPLE **4**

Analyzing a
Transformed
Cotangent Graph

Figure 64 shows the graph of $y = -2 - \cot(x - \frac{\pi}{4})$ in both traditional and calculator-generated forms.

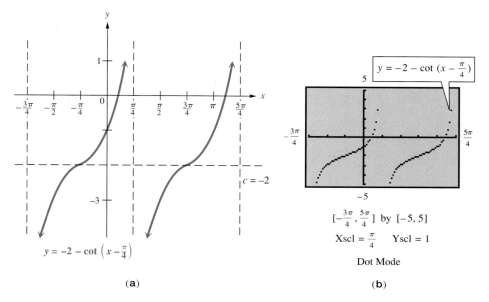

(a)

(b)

FIGURE 64

This graph is obtained from the graph of $y = \cot x$ by shifting $\frac{\pi}{4}$ units to the right (the phase shift), reflecting across the x-axis, and shifting 2 units downward. The period is π, there is no amplitude, and the range is $(-\infty, \infty)$. We will investigate the location of the vertical asymptotes and the x-intercepts of the function in Exercises 31–36. ∎

2.6 EXERCISES

In Exercises 1–8 you are given the graph of $y = f(x)$ in dot mode, where f is either tan, cot, sec, or csc in the standard trig window, along with a display. Use the graph and the display to write an equation of the form $f(x) \approx k$ or $f(x) = k$, for specific values of x and k. Then verify your result using the appropriate key or combination of keys on your calculator. Be sure your calculator is set to radian mode.

1.

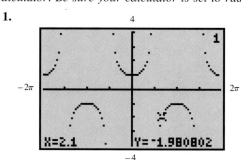

2.

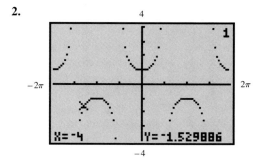

3.

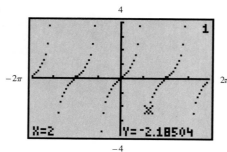

4.

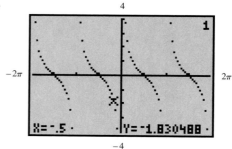

5.

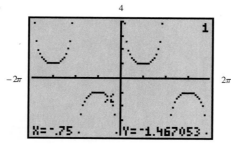

6.

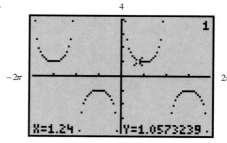

7. (*Hint:* $\dfrac{\pi}{4} \approx .78539816$)

8. (*Hint:* $-\dfrac{\pi}{4} \approx -.7853982$)

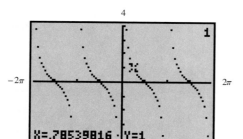

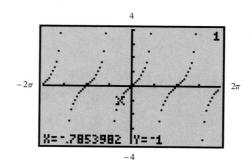

9. Between a pair of successive asymptotes, a portion of the graph of $y = \sec x$ or $y = \csc x$ resembles a parabola. However, these are not parabolas. Why not?

10. Use the graphs of $y = \sec x$ and $y = \csc x$ to respond to the following.

    (a) The function $y = \sec x$ has a local maximum for any x-value of the form _____, where n is an integer.

    (b) The function $y = \sec x$ has a local minimum for any x-value of the form _____, where n is an integer.

    (c) The function $y = \csc x$ has a local extremum for any x-value of the form _____, where n is an integer.

Without using a calculator to graph the functions, match each function in Exercises 11–16 with its graph.

11. $y = -\csc x$

12. $y = -\sec x$

13. $y = -\tan x$

14. $y = -\cot x$

15. $y = \tan(x - \frac{\pi}{4})$

16. $y = \cot(x - \frac{\pi}{4})$

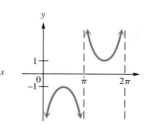

A

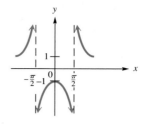

B

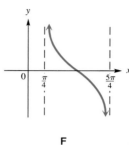

C

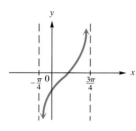

D

E

F

For each function in Exercises 17–26, do the following.
(a) *Graph the function in the trig viewing window. Use dot mode.*
(b) *Give the period of the function.*
(c) *Give the phase shift, if any.*
(d) *Give the range of the function.*
(e) *Give the smallest positive value for which the function is undefined.*

17. $y = 2 \csc \frac{1}{2}x$

18. $y = 3 \csc 2x$

19. $y = -2 \sec\left(x + \frac{\pi}{2}\right)$

20. $y = -\frac{3}{2} \sec(x - \pi)$

21. $y = \frac{5}{2} \cot \frac{1}{3}\left(x - \frac{\pi}{2}\right)$

22. $y = -3 \tan \frac{1}{2}\left(x + \frac{\pi}{4}\right)$

23. $f(x) = \frac{1}{2} \sec(2x + \pi)$

24. $f(x) = -\frac{1}{3} \csc\left(\frac{1}{2}x - \frac{\pi}{2}\right)$

25. $y = -1 - \tan\left(x + \frac{\pi}{4}\right)$

26. $y = 2 + \cot\left(2x - \frac{\pi}{3}\right)$

Explain how the graph of the given function can be obtained from the graph of one of the four circular functions analyzed in this section by a transformation.

27. $y = -2 + 3 \csc 2(x - \pi)$

28. $y = 5 + 2 \sec 3(x + \pi)$

29. $y = 4 - 2 \tan(2x - 2)$

30. $y = 5 - 3 \cot(3x + 3)$

Relating Concepts

Refer to Example 4 and Figure 64, which shows a portion of the graph of $y = -2 - \cot(x - \frac{\pi}{4})$.

31. What is the smallest positive number for which the graph of $y = \cot x$ is undefined?

32. Let k represent the number you found in Exercise 31. Set $x - \frac{\pi}{4}$ equal to k and solve to find the smallest positive number for which $\cot(x - \frac{\pi}{4})$ is undefined.

33. Based on your answer in Exercise 32 and the fact that the cotangent function has period π, give the general form of the equations of the asymptotes of the graph of $y = -2 - \cot(x - \frac{\pi}{4})$. Let n represent any integer.

34. Use the capabilities of your calculator to find the smallest positive x-intercept of the graph of this function.

35. Use the fact that the period of this function is π to find the next positive x-intercept.

36. Give the solution set of the equation $-2 - \cot(x - \frac{\pi}{4}) = 0$ over all real numbers. Let n represent any integer.

Observe the basic graphs of $y = \tan x$, $y = \cot x$, $y = \sec x$, and $y = \csc x$, and recall the ideas of symmetry with respect to the y-axis and symmetry with respect to the origin to answer Exercises 37–40.

37. The graph of $y = \tan x$ is symmetric with respect to the _____. It follows that $\tan(-x) =$ _____. (Notice that this supports a negative number identity introduced in Section 2.4.)

38. The graph of $y = \sec x$ is symmetric with respect to the _____. It follows that $\sec(-x) =$ _____.

39. For all x, $\cot(-x) = -\cot x$. Based on this fact, the graph of $y = \cot x$ is symmetric with respect to the _____. Furthermore, $\cot(-1.75) =$ _____, and this can be supported using calculator approximations. Since $\cot 1.75 \approx -.1811469526$, $\cot(-1.75) \approx$ _____.

40. When $\sin x = 0$, $\csc x$ is _____. As a result, the graph of $y = \csc x$ has a(n) _____ when $\sin x = 0$. Since $\cot x = \frac{\cos x}{\sin x}$, the cotangent is undefined when the _____ is undefined, and thus also has a(n) _____ when $\sin x = 0$.

Further Explorations

1. Recall that if θ is an angle in standard position and the point (a, b) lies on the terminal side of θ, then $\tan \theta = \frac{b}{a}$. The table shown in the figure was generated with the calculator in RADIAN MODE, with Tblmin $= 0$ and ΔTbl $= \frac{\pi}{4}$. Use the definition above to explain why error messages appear for two values of x in the table.

X	Y1
0	0
.7854	1
1.5708	ERROR
2.3562	-1
3.1416	0
3.927	1
4.7124	ERROR

X=0

Chapter 2   SUMMARY

An angle is formed by a rotation of a ray about its endpoint. Angles are measured in degrees or radians. Rotations smaller than a degree are measured in minutes and rotations smaller than one minute are measured in seconds. Sometimes decimal degrees are used instead of minutes and seconds. Radian measure is the ratio of the arc length cut off by a central angle of a circle to the radius of the circle. We use the relationship $180° = \pi$ radians to convert between degrees and radians. The definition of radian measure leads to this result: the length of an arc of a circle is given by $s = r\theta$, where θ is measured in radians and r is the radius of the circle.

The six trigonometric functions are defined as ratios of the quantities x, y, and r, where (x, y) is a point on the terminal side of an angle in standard position and r is the distance from the point to the origin. Since r is the square root of the sum of the squares of x and y, if the point (x, y) is known, r and the six trigonometric function values can be found. By knowing the signs of each of the trigonometric functions in each quadrant, we can find all trigonometric functions of an angle given a trigonometric function value and the quadrant. The terminal side of a quadrantal angle coincides with an axis. These angles have special trigonometric function values that may be undefined.

The definitions of the six trigonometric functions indicate that there are three pairs of reciprocal functions. These reciprocal relationships are called the reciprocal identities. Three Pythagorean identities are formed from the relationship $x^2 + y^2 = r^2$. The definitions of the tangent and cotangent functions lead to two quotient identities that express these functions as quotients of two trigonometric functions. These identities are useful for finding the trigonometric functions of an angle in standard position given one function value and the quadrant of the angle.

Certain function values, such as those for 30°, 45°, 60°, angles with these as reference angles, and quadrantal angles, can be determined *exactly* using analytic methods. Calculators can find approximations for function values of all angles.

The circular functions are defined in terms of real number arc lengths on a unit circle, a circle with center at the origin and radius 1. The circular functions are trigonometric functions with the argument in radians and $r = 1$ in the definitions. Thus, they have the same function values.

A set of points is defined parametrically by two equations $x = f(t)$ and $y = g(t)$, where t is the parameter. The unit circle can be graphed parametrically using $x = \cos t$ and $y = \sin t$.

The identities developed for the trigonometric functions also apply to the circular functions. The negative number identities give the function of the negative of a number in terms of the function of the number. One use of these identities, along with the reciprocal, quotient, and Pythagorean identities, is to express one circular (or trigonometric) function in terms of another. Later we will see that verifying identities helps to develop a skill which is useful in more advanced work in mathematics.

The circular (trigonometric) functions are periodic—that is, the function values repeat over and over. Sine and cosine and their reciprocals have a period of 2π; tangent and cotangent have a period of π. The sine and cosine functions have domain $(-\infty, \infty)$ and range $[-1, 1]$. For both functions the amplitude is half the length of the range, 1. The sine function is symmetric with respect to the origin; the cosine function is symmetric with respect to the y-axis. Both functions have the same wavy graph, called a sinusoid. Each one is a horizontal shift of the other. The graphs of the circular functions can be transformed by stretching, shrinking, reflecting, and shifting. Horizontal shifts of the graphs of circular functions are called phase shifts.

The graphs of cosecant and secant are derived from the fact that they are the reciprocal functions of sine and cosine. Both graphs have vertical asymptotes where the graphs of sine or cosine, respectively, are zero. Both have a range of $(-\infty, -1] \cup [1, \infty)$. Cosecant is symmetric with respect to the origin; secant is symmetric with respect to the y-axis.

Because $\tan x = \frac{\sin x}{\cos x}$, its graph has x-intercepts when $\sin x = 0$ and vertical asymptotes when $\cos x = 0$. The fact that $\cot x$ is the reciprocal of $\tan x$ means that its graph has x-intercepts when $\cos x = 0$ and vertical asymptotes when $\sin x = 0$. Both functions have graphs with a range $(-\infty, \infty)$. Both have graphs that are symmetric with respect to the origin. These graphs can be transformed in the same way as the graphs of other functions.

Key Terms

SECTION 2.1

line AB
segment AB
ray AB
angle
initial side
terminal side
vertex
positive angle
negative angle
degree
acute angle
right angle
obtuse angle
straight angle
complementary angles (complements)
supplementary angles (supplements)
minute
second
standard position
quadrantal angle
coterminal angles
central angle
radian
arc

SECTION 2.2

trigonometric functions
sine
cosine
tangent
cotangent
secant
cosecant

SECTION 2.3

cofunctions
reference angle

SECTION 2.4

circular functions
unit circle
parametric equations
parameter

SECTION 2.5

periodic function
period
sine wave or sinusoid
amplitude
phase shift

Chapter 2 REVIEW EXERCISES

Let θ represent a $-300°$ angle in Exercises 1–5.

1. Sketch θ in standard position.
2. Name the smallest positive angle coterminal with θ. Use degree measure.
3. Name a negative angle coterminal with θ. Use degree measure.
4. Give an expression that represents all angles coterminal with θ, using degree measure and letting n represent any integer.
5. Convert θ to radians. Leave π in your answer.

Use the formula $s = r\theta$ *to find the unknown value. Give the exact value.*

6. $r = 3$, $\theta = \frac{\pi}{3}$ **7.** $r = 5$, $s = 15$ **8.** $s = 12$, $\theta = 45°$

9. The radius of a circle is 15.2 centimeters. Find the length of an arc of the circle intercepted by a central angle of $\frac{3\pi}{4}$ radians.

10. What is the length of the arc intercepted by the hands of a clock at 5:00, if the radius is 12 inches?

11. Assuming that the radius of the earth is 6400 kilometers, what is the distance between cities on a north-south line that are on latitudes 28°N and 12°S, respectively?

Let θ *be an angle in standard position, with the point* $(-2, -7)$ *on its terminal side. Find the exact value of each of the following.*

12. $\sin \theta$ **13.** $\cos \theta$ **14.** $\tan \theta$

15. $\csc \theta$ **16.** $\sec \theta$ **17.** $\cot \theta$

18. For the angle θ described in the directions for Exercises 12–17 **(a)** give to the nearest hundredth the measure of θ if $0° \le \theta < 360°$, and **(b)** give the measure of the reference angle for θ to the nearest hundredth of a degree.

Consider an angle θ *in standard position whose terminal side has the equation* $y = -5x$, *with* $x \le 0$.

19. Sketch θ and use an arrow to show the rotation if $0° \le \theta < 360°$.

20. Find the exact values of $\sin \theta$ and $\cos \theta$.

21. Give the measure of θ to the nearest minute.

Find the quadrant in which the terminal side of angle θ *must lie given the conditions described.*

22. $\cos \theta < 0$ and $\tan \theta > 0$ **23.** $\sin \theta > 0$ and $\tan \theta < 0$

Suppose that $\sin \theta = \frac{\sqrt{3}}{5}$ *and* $\cos \theta < 0$. *Find each of the following exact function values.*

24. $\cos \theta$ **25.** $\csc \theta$ **26.** $\cot \theta$

Find the exact value for each function. If it is undefined, say so.

27. $\sin(-225°)$ **28.** $\cos(-3\pi)$ **29.** $\tan\left(-\frac{3\pi}{2}\right)$ **30.** $\csc \frac{5\pi}{3}$

31. $\sec 420°$ **32.** $\cot\left(-\frac{5\pi}{2}\right)$ **33.** $\sin\left(-\frac{15\pi}{4}\right)$ **34.** $\cos(-510°)$

35. $\tan \frac{13\pi}{3}$ **36.** $\csc(-13\pi)$ **37.** $\sec(180° + n \cdot 360°)$, where n is an integer   **38.** $\cot \frac{17\pi}{6}$

Use a calculator to find an approximation of each function value.

39. $\sin 146° \, 40'$ **40.** $\sec 5$ **41.** $\tan\left(-\frac{\pi}{5}\right)$ **42.** $\cot \frac{\pi}{7}$

Use Figure 33 to find the exact value of each of the following.

43. $\tan \frac{\pi}{3}$ **44.** $\cos \frac{2\pi}{3}$ **45.** $\sin\left(-\frac{5\pi}{6}\right)$ **46.** $\cot \frac{11\pi}{6}$

47. $\tan\left(-\frac{7\pi}{3}\right)$ **48.** $\sec \frac{\pi}{3}$ **49.** $\csc\left(-\frac{11\pi}{6}\right)$ **50.** $\cot\left(-\frac{17\pi}{3}\right)$

Find a calculator approximation of each of the following.

51. $\sin 1.0472$ **52.** $\tan 7.3159$ **53.** $\sec .4864$ **54.** $\csc(-.8385)$

Use the negative number identities to write each of the following as a function of a positive number.

55. $\cos(-3)$ **56.** $\sin(-3)$ **57.** $\tan(-3)$

58. $\sec(-3)$ **59.** $\csc(-3)$ **60.** $\cot(-3)$

Consider the function $y = -2 \sin(3x + \frac{\pi}{4}) + 5.$

61. What is the domain of this function?

62. What is its range?

63. Find the amplitude.

64. Find the period of the function.

65. What is the phase shift?

66. What is the vertical translation?

67. Use a calculator to graph the function in the window $[0, 2\pi]$ by $[0, 8]$.

Identify the one of the six circular functions that satisfies the description.

68. Period is π, x intercepts are of the form $n\pi$, where n is an integer

69. Period is 2π, passes through the origin

70. Period is 2π, passes through the point $(\frac{\pi}{2}, 0)$

71. Period is 2π, domain is $\{x \mid x \neq n\pi$, where n is an integer$\}$

72. Period is π, function is decreasing on the interval $0 < x < \pi$

73. Period is 2π, has vertical asymptotes of the form $x = (2n + 1)\frac{\pi}{2}$, where n is an integer

74. Explain why the domain of the tangent function is the same as that of the secant function.

For each of the graphs in Exercises 75–78 give the equation of a sine function having that graph.

75.

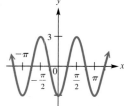

76.

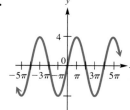

77.

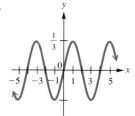

78.

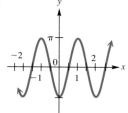

Give the equation of a cosine function having the graph of the figure.

79. Exercise 75   **80.** Exercise 76

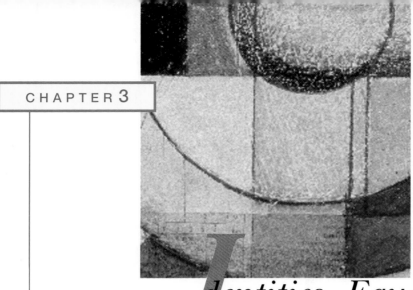

Identities, Equations, Inequalities, and Applications

3.1 IDENTITIES: ANALYTIC VERIFICATION AND GRAPHICAL SUPPORT

The Fundamental Identities Reviewed ▌ Expressing One Function in Terms of Another ▌ Verifying Identities Analytically and Supporting Graphically

The Fundamental Identities Reviewed

In Chapter 2 we introduced some fundamental identities: reciprocal, quotient, Pythagorean, and negative number (angle). We list them here for review. In all cases we assume that the functions are defined and that no denominators are 0.

FUNDAMENTAL IDENTITIES

Reciprocal Identities

$$\cot \theta = \frac{1}{\tan \theta} \qquad \sec \theta = \frac{1}{\cos \theta} \qquad \csc \theta = \frac{1}{\sin \theta}$$

Quotient Identities

$$\tan \theta = \frac{\sin \theta}{\cos \theta} \qquad \cot \theta = \frac{\cos \theta}{\sin \theta}$$

Pythagorean Identities

$$\sin^2 \theta + \cos^2 \theta = 1 \qquad \tan^2 \theta + 1 = \sec^2 \theta \qquad 1 + \cot^2 \theta = \csc^2 \theta$$

Negative Number Identities

$$\sin(-\theta) = -\sin \theta \qquad \csc(-\theta) = -\csc \theta$$
$$\cos(-\theta) = \cos \theta \qquad \sec(-\theta) = \sec \theta$$
$$\tan(-\theta) = -\tan \theta \qquad \cot(-\theta) = -\cot \theta$$

NOTE The forms of the identities given above are the most commonly recognized forms. Throughout this chapter it will be necessary to recognize alternate forms of these identities as well. For example, two other forms of $\sin^2 \theta + \cos^2 \theta = 1$ are

$$\sin^2 \theta = 1 - \cos^2 \theta$$

and

$$\cos^2 \theta = 1 - \sin^2 \theta.$$

You should be able to transform the basic identities using algebraic transformations.

Expressing One Function in Terms of Another

Any function of a number or angle can be expressed in terms of any other function, as shown in the following example.

EXAMPLE 1

Expressing One Function in Terms of Another

Express $\cos x$ in terms of $\tan x$.

SOLUTION Since $\sec x$ is related to both $\cos x$ and $\tan x$ by identities, start with $\tan^2 x + 1 = \sec^2 x$. Then take reciprocals to get

$$\frac{1}{\tan^2 x + 1} = \frac{1}{\sec^2 x}$$

or

$$\frac{1}{\tan^2 x + 1} = \cos^2 x$$

$$\pm\sqrt{\frac{1}{\tan^2 x + 1}} = \cos x \qquad \text{Take the square root of both sides.}$$

$$\cos x = \frac{\pm 1}{\sqrt{\tan^2 x + 1}}.$$

Rationalize the denominator to get

$$\cos x = \frac{\pm\sqrt{\tan^2 x + 1}}{\tan^2 x + 1}.$$

Choose the $+$ sign or the $-$ sign, depending on the quadrant of x. ∎

Each of $\tan \theta$, $\cot \theta$, $\sec \theta$, and $\csc \theta$ can easily be expressed in terms of $\sin \theta$ and/or $\cos \theta$. For this reason, we often make such substitutions in an expression so that the expression can be simplified. The next example shows such substitutions.

EXAMPLE 2

Rewriting an
Expression in Terms
of Sine and Cosine

Use the fundamental identities to write $\tan\theta + \cot\theta$ in terms of $\sin\theta$ and $\cos\theta$, and then simplify the expression.

SOLUTION From the fundamental identities,

$$\tan\theta + \cot\theta = \frac{\sin\theta}{\cos\theta} + \frac{\cos\theta}{\sin\theta}.$$

Simplify this expression by adding the two fractions on the right side, using the common denominator $\cos\theta\sin\theta$.

$$\tan\theta + \cot\theta = \frac{\sin^2\theta}{\cos\theta\sin\theta} + \frac{\cos^2\theta}{\cos\theta\sin\theta}$$

$$= \frac{\sin^2\theta + \cos^2\theta}{\cos\theta\sin\theta}$$

Now substitute **1** for $\sin^2\theta + \cos^2\theta$.

$$\tan\theta + \cot\theta = \frac{1}{\cos\theta\sin\theta} \qquad \blacksquare$$

CAUTION When working with circular and trigonometric expressions and identities, be sure to write the argument of the function. For example, we would *not* write $\sin^2 + \cos^2 = 1$; an argument such as θ is necessary in this identity.

Graphing calculators allow us to give graphical support to results like the one obtained in Example 2. If we graph $y_1 = \tan x + \cot x$ (entered as $\tan x + \frac{1}{\tan x}$) and $y_2 = \frac{1}{\cos x \sin x}$ in the trig viewing window of a calculator, we see that the graphs appear to coincide. See Figure 1. While this observation is not a *proof* that y_1 and y_2 are equivalent, it does provide excellent support. We will use this method of graphical support later in this section and in later sections of this chapter when we verify identities.

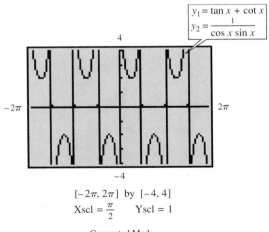

$[-2\pi, 2\pi]$ by $[-4, 4]$
$\text{Xscl} = \frac{\pi}{2}$ $\qquad \text{Yscl} = 1$

Connected Mode

FIGURE 1

Verifying Identities Analytically and Supporting Graphically

The result obtained in Example 2 allows us to write the following equation, which is an identity:

$$\tan \theta + \cot \theta = \frac{1}{\cos \theta \sin \theta}$$

One of the standard skills in studying the trigonometric and circular functions is the ability to show that an equation is an identity (for those values of the variable for which it is defined). To become proficient in verifying identities analytically, here are some hints that should prove helpful.

VERIFYING IDENTITIES

1. Learn the fundamental identities. Whenever you see either side of a fundamental identity, the other side should come to mind. Also, be aware of equivalent forms of the fundamental identities. For example $\sin^2 \theta = 1 - \cos^2 \theta$ is an alternate form of $\sin^2 \theta + \cos^2 \theta = 1$.

2. Try to rewrite the more complicated side of the equation so that it is identical to the simpler side.

3. It is often helpful to express all functions in the equation in terms of sine and cosine and then simplify the result.

4. Usually any factoring or indicated algebraic operations should be performed. For example, the expression $\sin^2 x + 2 \sin x + 1$ can be factored as $(\sin x + 1)^2$. The sum or difference of two expressions, such as

$$\frac{1}{\sin \theta} + \frac{1}{\cos \theta},$$

can be added or subtracted in the same way as any other rational expressions:

$$\frac{1}{\sin \theta} + \frac{1}{\cos \theta} = \frac{\cos \theta}{\sin \theta \cos \theta} + \frac{\sin \theta}{\sin \theta \cos \theta}$$

$$= \frac{\cos \theta + \sin \theta}{\sin \theta \cos \theta}.$$

5. As you select substitutions, keep in mind the side you are not changing, because it represents your goal. For example, to verify the identity

$$\tan^2 x + 1 = \frac{1}{\cos^2 x},$$

try to think of an identity that relates $\tan x$ to $\cos x$. Here, since $\sec x = \frac{1}{\cos x}$ and $\sec^2 x = \tan^2 x + 1$, the secant function is the best link between the two sides.

6. If an expression contains $1 + \sin x$, multiplying both numerator and denominator by $1 - \sin x$ would give $1 - \sin^2 x$, which could be replaced with $\cos^2 x$. Similar results for $1 - \sin x$, $1 + \cos x$, and $1 - \cos x$ are applicable.

EXAMPLE 3

Verifying an Identity
(Working with One
Side)

Verify that

$$\cot s + 1 = \csc s(\cos s + \sin s)$$

is an identity. Then support it graphically.

SOLUTION   Use the fundamental identities to rewrite one side of the equation so that it is identical to the other side. Since the right side is more complicated, it is probably a good idea to work with it. Here we use the method of changing all the functions to sine or cosine.

Steps   **Reasons**

$$\csc s(\cos s + \sin s) = \frac{1}{\sin s}(\cos s + \sin s) \qquad \csc s = \frac{1}{\sin s}$$

$$= \frac{\cos s}{\sin s} + \frac{\sin s}{\sin s} \qquad \text{Distributive property}$$

$$= \cot s + 1 \qquad \frac{\cos s}{\sin s} = \cot s;\ \frac{\sin s}{\sin s} = 1$$

The given equation is an identity since the right side equals the left side.

Now, we graph $y_1 = \cot x + 1$ and $y_2 = \csc x(\cos x + \sin x)$ in the trig viewing window. The graphs appear to coincide, supporting our analytic verification. See Figure 2.   ❚

TECHNOLOGICAL NOTE
To confirm the "coincidence" of two graphs like the ones indicated in Figure 2, trace to any x-value in the domain of y_1 and then move the tracing cursor to y_2. There should be no change in the value of y for that particular value of x.

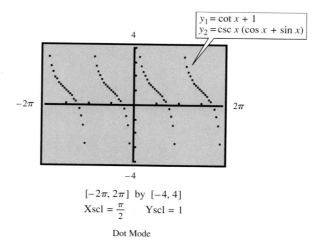

$y_1 = \cot x + 1$
$y_2 = \csc x\,(\cos x + \sin x)$

$[-2\pi, 2\pi]$ by $[-4, 4]$
$\text{Xscl} = \frac{\pi}{2} \qquad \text{Yscl} = 1$

Dot Mode

FIGURE 2

EXAMPLE 4

Verifying an Identity
(Working with One
Side)

Verify that

$$\frac{\tan t - \cot t}{\sin t \cos t} = \sec^2 t - \csc^2 t$$

is an identity. Then support it graphically.

SOLUTION Since the left side is the more complicated one, transform the left side to equal the right side.

$$\frac{\tan t - \cot t}{\sin t \cos t}$$

$$= \frac{\tan t}{\sin t \cos t} - \frac{\cot t}{\sin t \cos t} \qquad \frac{a-b}{c} = \frac{a}{c} - \frac{b}{c}$$

$$= \tan t \cdot \frac{1}{\sin t \cos t} - \cot t \cdot \frac{1}{\sin t \cos t} \qquad \frac{a}{b} = a \cdot \frac{1}{b}$$

$$= \frac{\sin t}{\cos t} \cdot \frac{1}{\sin t \cos t} - \frac{\cos t}{\sin t} \cdot \frac{1}{\sin t \cos t} \qquad \tan t = \frac{\sin t}{\cos t}; \cot t = \frac{\cos t}{\sin t}$$

$$= \frac{1}{\cos^2 t} - \frac{1}{\sin^2 t}$$

$$= \sec^2 t - \csc^2 t$$

Here, writing in terms of sine and cosine only was used in the third line.

Figure 3 shows the graphs of the two functions in the original identity. They appear to coincide, supporting our analytic work. ▯

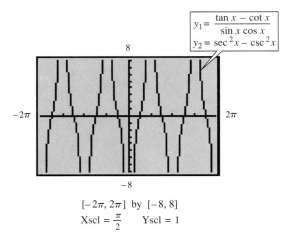

$$[-2\pi, 2\pi] \text{ by } [-8, 8]$$
$$\text{Xscl} = \frac{\pi}{2} \qquad \text{Yscl} = 1$$

FIGURE 3

If both sides of an identity appear to be equally complex, the identity can be verified by working independently on the left side and on the right side, until each side is changed into some common third result. *Each step, on each side, must be reversible.* With all steps reversible, the procedure is as follows.

left = right

common third

expression

The left side leads to the third expression, which leads back to the right side. This procedure is just a shortcut for the procedure used in the first examples of this section: the left side is changed into the right side, but by going through an intermediate step.

EXAMPLE **5**
Verifying an Identity (Working with Both Sides)

Verify that

$$\frac{\sec \alpha + \tan \alpha}{\sec \alpha - \tan \alpha} = \frac{1 + 2 \sin \alpha + \sin^2 \alpha}{\cos^2 \alpha}$$

is an identity. Support the result graphically.

SOLUTION   Both sides appear equally complex, so verify the identity by changing each side into a common third expression. Work first on the left, multiplying numerator and denominator by $\cos \alpha$.

$$\frac{\sec \alpha + \tan \alpha}{\sec \alpha - \tan \alpha} = \frac{(\sec \alpha + \tan \alpha)\cos \alpha}{(\sec \alpha - \tan \alpha)\cos \alpha} \qquad \frac{\cos \alpha}{\cos \alpha} = 1;$$
Multiplicative identity

$$= \frac{\sec \alpha \cos \alpha + \tan \alpha \cos \alpha}{\sec \alpha \cos \alpha - \tan \alpha \cos \alpha} \qquad \text{Distributive property}$$

$$= \frac{1 + \tan \alpha \cos \alpha}{1 - \tan \alpha \cos \alpha} \qquad \sec \alpha \cos \alpha = 1$$

$$= \frac{1 + \dfrac{\sin \alpha}{\cos \alpha} \cdot \cos \alpha}{1 - \dfrac{\sin \alpha}{\cos \alpha} \cdot \cos \alpha} \qquad \tan \alpha = \frac{\sin \alpha}{\cos \alpha}$$

$$= \frac{1 + \sin \alpha}{1 - \sin \alpha}$$

On the right side of the original statement, begin by factoring.

$$\frac{1 + 2 \sin \alpha + \sin^2 \alpha}{\cos^2 \alpha} = \frac{(1 + \sin \alpha)^2}{\cos^2 \alpha} \qquad a^2 + 2ab + b^2 = (a + b)^2$$

$$= \frac{(1 + \sin \alpha)^2}{1 - \sin^2 \alpha} \qquad \cos^2 \alpha = 1 - \sin^2 \alpha$$

$$= \frac{(1 + \sin \alpha)^2}{(1 + \sin \alpha)(1 - \sin \alpha)} \qquad \begin{array}{l}1 - \sin^2 \alpha = \\ (1 + \sin \alpha)(1 - \sin \alpha)\end{array}$$

$$= \frac{1 + \sin \alpha}{1 - \sin \alpha} \qquad \text{Reduce to lowest terms.}$$

We now have shown that

$$\frac{\sec \alpha + \tan \alpha}{\sec \alpha - \tan \alpha} = \frac{1 + \sin \alpha}{1 - \sin \alpha} = \frac{1 + 2 \sin \alpha + \sin^2 \alpha}{\cos^2 \alpha},$$

verifying that the original equation is an identity. In Figure 4 we graph the left side of the original equation as y_1 and the right side as y_2. They appear to coincide, supporting our work. ▯

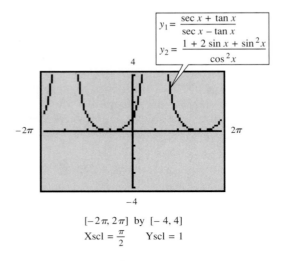

$$y_1 = \frac{\sec x + \tan x}{\sec x - \tan x}$$

$$y_2 = \frac{1 + 2 \sin x + \sin^2 x}{\cos^2 x}$$

$[-2\pi, 2\pi]$ by $[-4, 4]$

$\text{Xscl} = \frac{\pi}{2}$ $\quad$ $\text{Yscl} = 1$

FIGURE 4

EXAMPLE **6**

Using a Graph to Make Conjecture About an Identity

The expression

$$\frac{\sec x - \cos x}{\tan x}$$

simplifies to one of the six basic circular functions. Use a graph to make a conjecture as to which one it is, and then verify your conjecture analytically.

SOLUTION We begin by graphing $y = \frac{\sec x - \cos x}{\tan x}$ in the trig viewing window. As seen in Figure 5, the graph suggests that the expression might simplify to $\sin x$, since the graph there seems to be that of $y = \sin x$.

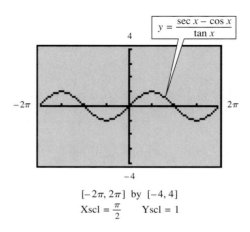

$$y = \frac{\sec x - \cos x}{\tan x}$$

$[-2\pi, 2\pi]$ by $[-4, 4]$

$\text{Xscl} = \frac{\pi}{2}$ $\quad$ $\text{Yscl} = 1$

FIGURE 5

Now we attempt to verify the identity

$$\frac{\sec x - \cos x}{\tan x} = \sin x.$$

$$\frac{\sec x - \cos x}{\tan x} = \frac{\dfrac{1}{\cos x} - \cos x}{\dfrac{\sin x}{\cos x}}$$

$$= \frac{\left(\dfrac{1}{\cos x} - \cos x\right)\cos x}{\left(\dfrac{\sin x}{\cos x}\right)\cos x}$$

$$= \frac{1 - \cos^2 x}{\sin x}$$

$$= \frac{\sin^2 x}{\sin x}$$

$$= \sin x$$

Our analytic verification supports our conjecture that the original expression is equivalent to $\sin x$. ∎

FOR **GROUP DISCUSSION**

Duplicate Figure 5 on your own calculator, letting $y_1 = \frac{\sec x - \cos x}{\tan x}$. Now use the capability of the calculator to attempt to locate the point for which $x = \frac{\pi}{2}$. What happens?

Now enter $y_2 = \sin x$ and repeat, attempting to locate the point for which $x = \frac{\pi}{2}$. What happens?

Discuss your results.

3.1 EXERCISES

For each expression in Column I, choose the expression from Column II that completes a fundamental identity.

Column I

1. $\dfrac{\cos x}{\sin x}$

2. $\tan x$

3. $\cos(-x)$

4. $\tan^2 x + 1$

5. 1

Column II

(a) $\sin^2 x + \cos^2 x$

(b) $\cot x$

(c) $\sec^2 x$

(d) $\dfrac{\sin x}{\cos x}$

(e) $\cos x$

For each expression in Column I, choose the expression from Column II that completes an identity. You may have to rewrite one or both expressions, using a fundamental identity, to recognize the matches.

Column I   Column II

6. $-\tan x \cos x$ _ (a) $\dfrac{\sin^2 x}{\cos^2 x}$

7. $\sec^2 x - 1$ (b) $\dfrac{1}{\sec^2 x}$

8. $\dfrac{\sec x}{\csc x}$ (c) $\sin(-x)$

9. $1 + \sin^2 x$ (d) $\csc^2 x - \cot^2 x + \sin^2 x$

10. $\cos^2 x$ (e) $\tan x$

11. A student writes "$1 + \cot^2 = \csc^2$." Comment on this student's work.

12. Another student makes the following claim: "Since $\sin^2 \theta + \cos^2 \theta = 1$, I should be able to also say $\sin \theta + \cos \theta = 1$ if I take the square root of both sides." Comment on this student's statement.

Complete this chart, so that each function in the column at the left is expressed in terms of the functions given across the top.

	sin θ	**cos θ**	**tan θ**	**cot θ**	**sec θ**	**csc θ**
13. sin θ	$\sin \theta$	$\pm\sqrt{1 - \cos^2 \theta}$	$\dfrac{\pm\tan \theta \sqrt{1 + \tan^2 \theta}}{1 + \tan^2 \theta}$			$\dfrac{1}{\csc \theta}$
14. cos θ		$\cos \theta$	$\dfrac{\pm\sqrt{\tan^2 \theta + 1}}{\tan^2 \theta + 1}$		$\dfrac{1}{\sec \theta}$	
15. tan θ			$\tan \theta$	$\dfrac{1}{\cot \theta}$		
16. cot θ			$\dfrac{1}{\tan \theta}$	$\cot \theta$	$\dfrac{\pm\sqrt{\sec^2 \theta - 1}}{\sec^2 \theta - 1}$	
17. sec θ		$\dfrac{1}{\cos \theta}$			$\sec \theta$	
18. csc θ	$\dfrac{1}{\sin \theta}$					$\csc \theta$

Each of the following expressions simplifies to a constant, a single circular function, or a power of a circular function. Use the fundamental identities to simplify each expression.

19. $\tan \theta \cos \theta$ **20.** $\cot \alpha \sin \alpha$ **21.** $\sec r \cos r$ **22.** $\cot t \tan t$

23. $\dfrac{\sin \beta \tan \beta}{\cos \beta}$ **24.** $\dfrac{\csc \theta \sec \theta}{\cot \theta}$ **25.** $\sec^2 x - 1$ **26.** $\csc^2 t - 1$

27. $\dfrac{\sin^2 x}{\cos^2 x} + \sin x \csc x$ **28.** $\dfrac{1}{\tan^2 \alpha} + \cot \alpha \tan \alpha$ **29.** $\dfrac{1 - \cos^2 x}{\sin x}$

30. $\dfrac{1 - \sin^2 x}{\cos x}$ **31.** $\cos^2 x (\tan^2 x + 1)$ **32.** $\dfrac{\cos^2 x}{\sin x} + \sin x$

Exercises 33–38 give identities involving the sine and/or cosine functions. For each identity (a) verify analytically and (b) support graphically using the trig viewing window, letting the left side be y_1 and the right side be y_2.

33. $(1 - \cos^2 x)(1 + \cos^2 x) = 2 \sin^2 x - \sin^4 x$

34. $2 \cos^2 x - \sin^2 x + 1 = 3 \cos^2 x$

35. $\cos^4 x + 2 \cos^2 x + 1 = (2 - \sin^2 x)^2$

36. $\sin^3 x + \cos^3 x = (\sin x + \cos x)(1 - \sin x \cos x)$

37. $(\sin x + 1)^2 - (\sin x - 1)^2 = 4 \sin x$

38. $(1 + \sin x)^2 + \cos^2 x = 2 + 2 \sin x$

Verify each of the following identities. You may wish to support your result graphically.

39. $\dfrac{\tan^2 \gamma + 1}{\sec \gamma} = \sec \gamma$

40. $\sin^2 \beta (1 + \cot^2 \beta) = 1$

41. $\sin^2 \alpha + \tan^2 \alpha + \cos^2 \alpha = \sec^2 \alpha$

42. $\cot s + \tan s = \sec s \csc s$

43. $\dfrac{\sin^2 \gamma}{\cos \gamma} = \sec \gamma - \cos \gamma$

44. $\dfrac{\cos \alpha}{\sec \alpha} + \dfrac{\sin \alpha}{\csc \alpha} = \sec^2 \alpha - \tan^2 \alpha$

45. $\dfrac{\cos \theta}{\sin \theta \cot \theta} = 1$

46. $\sin^4 \theta - \cos^4 \theta = 2 \sin^2 \theta - 1$

47. $\tan^2 \gamma \sin^2 \gamma = \tan^2 \gamma + \cos^2 \gamma - 1$

48. $(1 - \cos^2 \alpha)(1 + \cos^2 \alpha) = 2 \sin^2 \alpha - \sin^4 \alpha$

49. $\dfrac{(\sec \theta - \tan \theta)^2 + 1}{\sec \theta \csc \theta - \tan \theta \csc \theta} = 2 \tan \theta$

50. $\dfrac{\cos \theta + 1}{\tan^2 \theta} = \dfrac{\cos \theta}{\sec \theta - 1}$

51. $\dfrac{1}{1 - \sin \theta} + \dfrac{1}{1 + \sin \theta} = 2 \sec^2 \theta$

52. $\dfrac{1 - \cos x}{1 + \cos x} = (\cot x - \csc x)^2$

53. $\dfrac{1}{\tan \alpha - \sec \alpha} + \dfrac{1}{\tan \alpha + \sec \alpha} = -2 \tan \alpha$

54. $\dfrac{\csc \theta + \cot \theta}{\tan \theta + \sin \theta} = \cot \theta \csc \theta$

55. $\dfrac{\tan s}{1 + \cos s} + \dfrac{\sin s}{1 - \cos s} = \cot s + \sec s \csc s$

56. $\dfrac{\cot \alpha + 1}{\cot \alpha - 1} = \dfrac{1 + \tan \alpha}{1 - \tan \alpha}$

In Exercises 57–60, the expression on the left side of the incomplete equation is equivalent to a single circular function or a constant. Graph the expression as y_1 on your calculator and decide which function or what constant applies. Then fill in the blank on the right side, and verify the identity analytically.

57. $\dfrac{1 - \cos^2 x}{\sin x} = $ _____

58. $\dfrac{1 - \sin^2 x}{\cos x} = $ _____

59. $\cos^2 x (\tan^2 x + 1) = $ _____

60. $\csc^2 x (\cos^2 x - 1) = $ _____

61. A student claims that the equation

$$\cos \theta + \sin \theta = 1$$

is an identity, since by letting $\theta = \frac{\pi}{2}$, we get $0 + 1 = 1$, a true statement. Comment on this student's reasoning.

62. Explain why the method described in the text involving working on both sides of an identity to show that each side is equal to the same expression is a valid method of verifying an identity. When using this method, what must be true about each step taken? (*Hint:* See the discussion preceding Example 5.)

*R*elating *Concepts*

Consider the identity

$$\cos^4 x - \sin^4 x = 1 - 2 \sin^2 x.$$

Work Exercises 63–66 in order. You should see a relationship here regarding the material in this section and that of Section 2.4.

63. Verify the identity above. Use analytic methods.

64. The graphs of $y_1 = \cos^4 x - \sin^4 x$ and $y_2 = 1 - 2 \sin^2 x$ are shown in the screen on the next page.

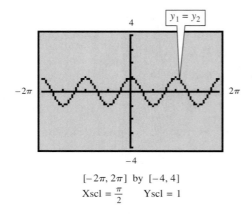

$$[-2\pi, 2\pi] \text{ by } [-4, 4]$$
$$\text{Xscl} = \frac{\pi}{2} \qquad \text{Yscl} = 1$$

Give the amplitude and the period of the function.

65. Based on your answer to Exercise 64, if you were to write the function in the form $y = a \cos bx$, for $a > 0$, $b > 0$, what would it be?

66. Complete this equation: $1 - 2\sin^2 x = \cos$ _____ . (This identity will be introduced later in the chapter.)

3.2 EQUATIONS AND INEQUALITIES I

Preliminary Considerations ▌ Equations and Inequalities Involving Circular Functions ▌ Solving Equations Using Strictly Graphical Methods

Preliminary Considerations

In Section 1.3 we introduced graphical methods of solving equations and inequalities. The two methods introduced were the *x-intercept method* and the *intersection-of-graphs method*. At this point you may wish to review these methods, as they will be used extensively in this section and in Section 3.4.

Equations and Inequalities Involving Circular Functions

We will now investigate methods of solving equations and associated inequalities involving circular functions. Because the circular functions are periodic, equations involving them will most often have infinitely many solutions unless we restrict the domain. Therefore, unless otherwise specified, we will solve equations and inequalities in this section over the interval $[0, 2\pi)$.

EXAMPLE 1

Solve $2 \sin x - 1 = 0$ over the interval $[0, 2\pi)$ and support the solution set graphically.

Solving an Equation Involving a Circular Function (Linear Method)

SOLUTION   Since this equation involves the first power of $\sin x$, it is linear in $\sin x$ and we will solve it using the usual method of solving a linear equation.

$$2 \sin x - 1 = 0$$
$$2 \sin x = 1$$
$$\sin x = \frac{1}{2}$$

The two values of x in the interval $[0, 2\pi)$ that have a sine value of $\frac{1}{2}$ are $\frac{\pi}{6}$ and $\frac{5\pi}{6}$. This can be determined by using the unit circle (see Figure 33 in Chapter 2) or reference angle analysis (see Section 2.3). Therefore, the solution set in the specified interval is $\{\frac{\pi}{6}, \frac{5\pi}{6}\}$.

To support this result graphically, we graph $y = 2 \sin x - 1$ over the interval $[0, 2\pi]$ and use the capabilities of the calculator to verify that the x-intercepts have the same decimal approximations as $\frac{\pi}{6}$ and $\frac{5\pi}{6}$. See Figure 6.   ◘

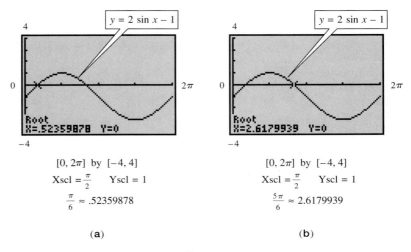

$[0, 2\pi]$ by $[-4, 4]$	$[0, 2\pi]$ by $[-4, 4]$
$\text{Xscl} = \frac{\pi}{2}$ $\text{Yscl} = 1$	$\text{Xscl} = \frac{\pi}{2}$ $\text{Yscl} = 1$
$\frac{\pi}{6} \approx .52359878$	$\frac{5\pi}{6} \approx 2.6179939$
(a)	(b)

FIGURE 6

FOR GROUP DISCUSSION

Refer to the equation in Example 1 as you discuss these items.

1. What is the period of the function $y = 2 \sin x - 1$?
2. Whenever an integer multiple of the period is added to a solution of the equation, another solution is found (if we allow a larger interval over which we are solving). Let k be your favorite integer between 1 and 3 inclusive. With Xmin = 0 and Xmax = 10π, graph $y_1 = 2 \sin x - 1$ and show that $\frac{\pi}{6} + 2k\pi$ and $\frac{5\pi}{6} + 2k\pi$ are also solutions of $2 \sin x - 1 = 0$.
3. Discuss why the following statement is true: The solution set of $2 \sin x - 1 = 0$ over all real numbers is $\{x \mid x = \frac{\pi}{6} + 2n\pi, x = \frac{5\pi}{6} + 2n\pi$, where n is an integer$\}$.

EXAMPLE 2

Solving Inequalities Associated with an Equation Involving a Circular Function

Use the result of Example 1 and the graph in Figure 6 to solve (a) $2 \sin x - 1 > 0$ and (b) $2 \sin x - 1 < 0$ over the interval $[0, 2\pi)$.

SOLUTION   To solve the inequality in (a), we must identify the x-values in the interval $[0, 2\pi)$ for which the graph is *above* the x-axis. Similarly, to solve the inequality in (b) we must identify the x-values for which the graph is *below* the *x*-axis. The graph in Figure 6 indicates the following:

$$\text{The solution set of } 2\sin x - 1 > 0 \text{ is } \left(\frac{\pi}{6}, \frac{5\pi}{6}\right).$$

$$\text{The solution set of } 2\sin x - 1 < 0 \text{ is } \left[0, \frac{\pi}{6}\right) \cup \left(\frac{5\pi}{6}, 2\pi\right). \quad \blacksquare$$

The equation and inequalities in Examples 1 and 2 were linear in the expression $\sin x$. The next example shows how methods of solving quadratic equations and inequalities, discussed in Chapter 1, can be extended to equations and inequalities involving circular functions.

EXAMPLE 3

Solving an Equation and Associated Inequalities Involving a Circular Function (Quadratic Method)

(a) Solve $\tan^2 x + \tan x - 2 = 0$ over the interval $[0, 2\pi)$ by using a method of solving quadratic equations analytically.

SOLUTION We could use either the quadratic formula or the zero-factor property to solve this equation. We will use the latter.

$$\tan^2 x + \tan x - 2 = 0$$
$$(\tan x - 1)(\tan x + 2) = 0$$

Set each factor equal to 0.

$$\tan x - 1 = 0 \quad \text{or} \quad \tan x + 2 = 0$$
$$\tan x = 1 \quad \text{or} \quad \tan x = -2$$

The solutions for $\tan x = 1$ in the interval $[0, 2\pi)$ are $x = \frac{\pi}{4}$ or $\frac{5\pi}{4}$. To solve $\tan x = -2$ in the interval, use a calculator set in the radian mode. We find that $\tan^{-1}(-2) \approx -1.107148718$. However, due to the method in which the calculator determines this value (justified in Section 3.5), this number is not in the desired interval. Because the period of the tangent function is π, we will add π and then add 2π to $\tan^{-1}(-2)$ to obtain the solutions in the desired interval:

$$x = \tan^{-1}(-2) + \pi \approx 2.034443936$$
$$x = \tan^{-1}(-2) + 2\pi \approx 5.176036589.$$

The solution set is

$$\left\{ \underbrace{\frac{\pi}{4}, \frac{5\pi}{4}}_{\substack{\text{Exact} \\ \text{values}}}, \quad \underbrace{2.03, 5.18}_{\substack{\text{Approximate} \\ \text{values to the} \\ \text{nearest} \\ \text{hundredth}}} \right\}.$$

(b) Support the solution in part (a) by graphing $y = \tan^2 x + \tan x - 2$ in the window $[0, 2\pi]$ by $[-4, 4]$, and finding the *x*-intercepts.

SOLUTION As seen in Figure 7, there are indeed four *x*-intercepts in the interval. Using the capabilities of the calculator, we can support the solutions in part (a). The figure shows the support for the solution $\frac{5\pi}{4}$, since a decimal approximation of this number is 3.9269908. The other three solutions can be supported similarly.

(c) Use the result of part (a) and Figure 7 to find the solution set of $\tan^2 x + \tan x - 2 > 0$ over the interval $[0, 2\pi)$.

SOLUTION   The graph of $y = \tan^2 x + \tan x - 2$ lies *above* the x-axis for the following subset of the interval $[0, 2\pi)$:

$$\left(\frac{\pi}{4}, \frac{\pi}{2}\right) \cup \left(\frac{\pi}{2}, 2.03\right) \cup \left(\frac{5\pi}{4}, \frac{3\pi}{2}\right) \cup \left(\frac{3\pi}{2}, 5.18\right).$$

This is the solution set of the inequality. Notice that we used the fact that the tangent is not defined when $x = \frac{\pi}{2}$ and when $x = \frac{3\pi}{2}$. These values must be excluded from the solution set.   ∎

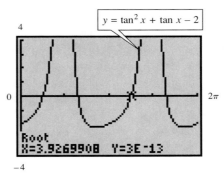

$[0, 2\pi]$ by $[-4, 4]$
$$\text{Xscl} = \frac{\pi}{2} \qquad \text{Yscl} = 1$$

FIGURE 7

FOR GROUP DISCUSSION

What is the solution set of $\tan^2 x + \tan x - 2 < 0$ over the interval $[0, 2\pi)$? (*Hint:* Refer to Figure 7.)

The next example illustrates how the quadratic formula can be used to solve an equation involving a circular function.

EXAMPLE 4

Solving an Equation Involving a Circular Function Using the Quadratic Formula

Solve the equation

$$\cot x(\cot x + 3) = 1$$

over the interval $[0, 2\pi)$. Use the quadratic formula, and support your solution graphically.

SOLUTION   We begin by multiplying the factors on the left and subtracting 1 to get the equation in the standard form of a quadratic equation.

$$\cot^2 x + 3 \cot x - 1 = 0$$

Since this equation cannot be solved by factoring, use the quadratic formula, with $a = 1$, $b = 3$, $c = -1$, and $\cot x$ as the variable.

$$\cot x = \frac{-3 \pm \sqrt{9 + 4}}{2} = \frac{-3 \pm \sqrt{13}}{2}$$

Using a calculator, we find

$$\cot x \approx -3.302775638 \qquad \text{or} \qquad \cot x \approx .3027756377.$$

Since we cannot find inverse cotangent values directly on a calculator, we use the fact that $\cot x = \frac{1}{\tan x}$, and take reciprocals to get

$$\tan x \approx -.3027756377 \qquad \text{or} \qquad \tan x \approx 3.302775638$$
$$x \approx -.2940013018 \qquad\qquad\qquad x \approx 1.276795025$$

(These x-values were found by using the inverse tangent function key.)

The first of these, $-.2940013018$, is not in the desired interval. Since the period of the cotangent function is π, we add π and then add 2π to $-.2940013018$ to get 2.847591352 and 5.989184005.

The second value, 1.276795025, is in the desired interval. We add π to it to get another solution in the interval: $1.276795025 + \pi$. Rounding to the nearest hundredth, the four solutions in the interval are 1.28, 2.85, 4.42, and 5.99, and the solution set is $\{1.28, 2.85, 4.42, 5.99\}$.

The graph of $y = \cot^2 x + 3 \cot x - 1$ is shown in Figure 8. Using the capabilities of the calculator, we can support the solutions found above. The figure shows the support for the solution 1.276795025. The others can be supported similarly. ◻

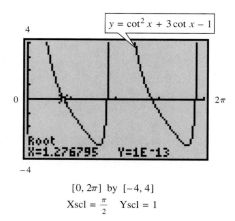

$[0, 2\pi]$ by $[-4, 4]$

$\text{Xscl} = \frac{\pi}{2} \qquad \text{Yscl} = 1$

FIGURE 8

Solving Equations Using Strictly Graphical Methods

Equations such as

$$\cos x = 2x, \qquad e^x = \sin x + 3, \qquad \text{and} \qquad \ln x = \cos x$$

are not solvable by analytic methods. However, they can be solved for approximate solutions graphically, using either the x-intercept method or the intersection-of-graphs method. The final example shows how the equation $\cos x = 2x$ can be solved graphically.

EXAMPLE 5

Solving an Equation
Using a Graph

Consider the equation $\cos x = 2x$.

(a) Use a graph to determine the number of solutions this equation has over the set of real numbers.

(b) Use the intersection-of-graphs method to solve it over the interval $[0, 2\pi)$. Give the solution to the nearest hundredth.

(c) Repeat part (b), but use the x-intercept method.

SOLUTION

(a)   Graphing $y_1 = \cos x$ and $y_2 = 2x$ in the trig viewing window, it is apparent that over the set of real numbers, this equation has only one solution, as indicated by the single point of intersection of the two graphs. See Figure 9.

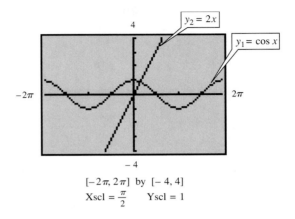

$[-2\pi, 2\pi]$ by $[-4, 4]$

$\text{Xscl} = \dfrac{\pi}{2}$ $\text{Yscl} = 1$

FIGURE 9

(b)   Using the intersection feature of the calculator, we determine that the x-coordinate of the point of intersection is approximately .45018361. See Figure 10. Rounding this solution to the nearest hundredth, we obtain the solution set $\{.45\}$.

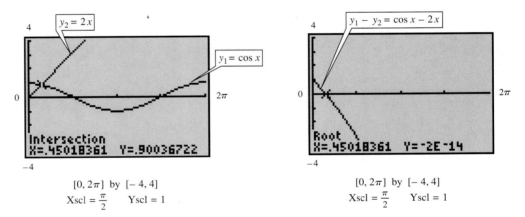

$[0, 2\pi]$ by $[-4, 4]$
$\text{Xscl} = \dfrac{\pi}{2}$ $\text{Yscl} = 1$

FIGURE 10

$[0, 2\pi]$ by $[-4, 4]$
$\text{Xscl} = \dfrac{\pi}{2}$ $\text{Yscl} = 1$

FIGURE 11

(c)   First graph $y_1 - y_2 = \cos x - 2x$ in the window $[0, 2\pi]$ by $[-4, 4]$. See Figure 11. Then use the calculator to find the x-intercept of the graph. As seen in Figure 11, it is approximately .45018361, corresponding to the value found in part (b).   ◼

FOR **GROUP DISCUSSION**

Repeat part (c) of Example 5, but enter $y_2 - y_1 = 2x - \cos x$. Do you get the same result? Why or why not?

3.2 EXERCISES

The graph of y = −2 cos x + 1 is shown in both screens below. It is graphed in the window [0, 2π] by [−4, 4]. Use the screens, along with the facts that

$$\frac{\pi}{3} \approx 1.0471976 \qquad and \qquad \frac{5\pi}{3} \approx 5.2359878$$

to respond to Exercises 1–6.

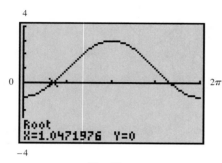

First View   Second View

1. Find the solution set of

$$-2 \cos x + 1 = 0$$

over the interval [0, 2π). Give exact values.

2. Find the solution set of

$$-2 \cos x + 1 > 0$$

over the interval [0, 2π). Give exact values for endpoints.

3. Find the solution set of

$$-2 \cos x + 1 \leq 0$$

over the interval [0, 2π). Give exact values for endpoints.

4. Over the interval [0, 2π), how many solutions does the equation

$$-2 \cos x + 1 = 2$$

have? (*Hint:* Imagine that the graph of y = 2 also appears on the screen.)

5. Over the interval [0, 2π), how many solutions does the equation

$$-2 \cos x + 1 = -2$$

have? (*Hint:* Imagine that the graph of y = −2 also appears on the screen.)

6. Find the solution set of −2 cos x + 1 = 0 over all real numbers.

For the function f in Exercises 7–16,
(a) *Solve f(x) = 0 analytically over the interval [0, 2π), and give solutions in exact forms.*
(b) *Graph y = f(x) in the window [0, 2π] by [−4, 4].*
(c) *Use the results of (a) and the graph in (b) to give the exact solution set of f(x) > 0 over [0, 2π).*
(d) *Use the results of (a) and the graph in (b) to give the exact solution set of f(x) < 0 over [0, 2π).*

7. $f(x) = 2 \cos x + 1$

8. $f(x) = 2 \sin x + 1$

9. $f(x) = \tan^2 x - 3$

10. $f(x) = \sec^2 x - 1$

11. $f(x) = 2 \cos^2 x - \sqrt{3} \cos x$

12. $f(x) = 2 \sin^2 x + 3 \sin x + 1$

13. $f(x) = \cos^2 x - \sin^2 x$
(*Hint:* Use $\cos^2 x = 1 - \sin^2 x$)

14. $f(x) = \cos^2 x - \sin^2 x - 1$

15. $f(x) = \csc^2 x - 2 \cot x$
(*Hint:* Use $\csc^2 x = 1 + \cot^2 x$)

16. $f(x) = \sin^2 x \cos x - \cos x$

In Exercises 17–22, repeat the directions given for Exercises 7–16, except give solutions and endpoints as approximations to the nearest hundredth when exact values cannot be determined. You may need to use the quadratic formula.

17. $f(x) = 3 \sin^2 x - \sin x - 2$

18. $f(x) = 9 \sin^2 x - 6 \sin x - 1$

19. $f(x) = \tan^2 x + 4 \tan x + 2$

20. $f(x) = 3 \cot^2 x - 3 \cot x - 1$

21. $f(x) = 2 \cos^2 x + 2 \cos x - 1$

22. $f(x) = \sin^2 x - 2 \sin x + 3$

Use a strictly graphical approach to solve the equation over the interval $[0, 2\pi)$. Express solutions to the nearest hundredth.

23. $\cot x + 2 \csc x = 3$

24. $2 \sin x = 1 - 2 \cos x$

25. $\sin^3 x + \sin x = 1$

26. $2 \cos^3 x + \sin x = -1$

27. $e^x = \sin x + 3$

28. $\ln x = \cos x$

3.3 FURTHER IDENTITIES

Sum and Difference Identities ▌ Double Number Identities ▌ Half Number Identities

Sum and Difference Identities

FOR GROUP DISCUSSION

Use different values for A and B to investigate whether the given statement is true for all values of A and B.

1. $\cos(A - B) = \cos A - \cos B$

2. $\cos(A + B) = \cos A + \cos B$

The results of the preceding group discussion should convince you that for the cosine function, the cosine of the difference (or sum) of two numbers is not, in general, equal to the difference (or sum) of the cosines of the numbers.

NOTE While we will discuss the identities in this section with respect to real number domains, they apply also to degree-measured angles.

In order to derive the identity for $\cos(A - B)$ in terms of functions of A and B, consider A and B to be real numbers that correspond to radian-measured angles A and B. Start by locating angles A and B in standard position on a unit circle, with $B < A$. Let S and Q be the points where angles A and B, respectively, intersect the circle. Locate point R on the unit circle so that angle POR equals the difference $A - B$. See Figure 12.

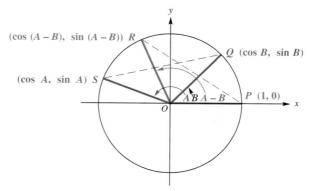

FIGURE 12

Point Q is on the unit circle, so by the work with circular functions, the x-coordinate of Q is given by the cosine of angle B, while the y-coordinate of Q is given by the sine of angle B:

Q has coordinates $(\cos B, \sin B)$.

In the same way,

S has coordinates $(\cos A, \sin A)$,

and

R has coordinates $(\cos(A - B), \sin(A - B))$.

Angle SOQ also equals $A - B$. Since the central angles SOQ and POR are equal, chords PR and SQ are equal. By the distance formula, since $PR = SQ$,

$$\sqrt{[\cos(A - B) - 1]^2 + [\sin(A - B) - 0]^2}$$
$$= \sqrt{(\cos A - \cos B)^2 + (\sin A - \sin B)^2}.$$

Squaring both sides and clearing parentheses gives

$$\cos^2(A - B) - 2\cos(A - B) + 1 + \sin^2(A - B)$$
$$= \cos^2 A - 2\cos A \cos B + \cos^2 B + \sin^2 A - 2\sin A \sin B + \sin^2 B.$$

Since $\sin^2 x + \cos^2 x = 1$ for any value of x, rewrite the equation as

$$2 - 2\cos(A - B) = 2 - 2\cos A \cos B - 2\sin A \sin B$$
$$\cos(A - B) = \cos A \cos B + \sin A \sin B.$$

This is the identity for $\cos(A - B)$. Although Figure 12 shows angles A and B in the second and first quadrants, respectively, it can be shown that this result is the same for any values of these angles.

To find a similar expression for $\cos(A + B)$, rewrite $A + B$ as $A - (-B)$ and use the identity for $\cos(A - B)$ found above, along with the fact that $\cos(-B) = \cos B$ and $\sin(-B) = -\sin B$.

$$\cos(A + B) = \cos[A - (-B)]$$
$$= \cos A \cos(-B) + \sin A \sin(-B)$$
$$= \cos A \cos B + \sin A(-\sin B)$$
$$\cos(A + B) = \cos A \cos B - \sin A \sin B$$

The two formulas we have just derived are summarized as follows.

COSINE OF A SUM OR DIFFERENCE

$$\cos(A - B) = \cos A \cos B + \sin A \sin B$$
$$\cos(A + B) = \cos A \cos B - \sin A \sin B$$

These identities can be used to derive similar identities for the sine and tangent of the difference and the sum of two numbers. Because of the cofunction relationship, we have

$$\sin \theta = \cos\left(\frac{\pi}{2} - \theta\right).$$

Now replace θ with $A + B$.

$$\sin(A + B) = \cos\left[\frac{\pi}{2} - (A + B)\right]$$
$$= \cos\left[\left(\frac{\pi}{2} - A\right) - B\right]$$

Using the formula for $\cos(A - B)$ from the previous discussion gives

$$\sin(A + B) = \cos\left(\frac{\pi}{2} - A\right)\cos B + \sin\left(\frac{\pi}{2} - A\right)\sin B$$

or $\quad \sin(A + B) = \sin A \cos B + \cos A \sin B.$

(The cofunction relationships were used in the last step.)

Now we write $\sin(A - B)$ as $\sin[A + (-B)]$ and use the identity for $\sin(A + B)$ to get

$$\sin(A - B) = \sin[A + (-B)]$$
$$= \sin A \cos(-B) + \cos A \sin(-B)$$
$$= \sin A \cos B - \cos A \sin B,$$

since $\cos(-B) = \cos B$ and $\sin(-B) = -\sin B$. In summary,

$$\sin(A - B) = \sin A \cos B - \cos A \sin B.$$

SINE OF A SUM OR DIFFERENCE

$$\sin(A + B) = \sin A \cos B + \cos A \sin B$$
$$\sin(A - B) = \sin A \cos B - \cos A \sin B$$

Using the identities for $\sin(A + B)$, $\cos(A + B)$, and the identity $\tan A = \frac{\sin A}{\cos A}$, we can derive the identity for $\tan(A + B)$. Start with

$$\tan(A + B) = \frac{\sin(A + B)}{\cos(A + B)}$$
$$= \frac{\sin A \cos B + \cos A \sin B}{\cos A \cos B - \sin A \sin B}.$$

To express this result in terms of the tangent function, multiply both numerator and denominator by $\frac{1}{\cos A \cos B}$.

$$\tan(A + B) = \frac{\dfrac{\sin A \cos B + \cos A \sin B}{1}}{\dfrac{\cos A \cos B - \sin A \sin B}{1}} \cdot \frac{\dfrac{1}{\cos A \cos B}}{\dfrac{1}{\cos A \cos B}}$$

$$= \frac{\dfrac{\sin A \cos B}{\cos A \cos B} + \dfrac{\cos A \sin B}{\cos A \cos B}}{\dfrac{\cos A \cos B}{\cos A \cos B} - \dfrac{\sin A \sin B}{\cos A \cos B}}$$

$$= \frac{\dfrac{\sin A}{\cos A} + \dfrac{\sin B}{\cos B}}{1 - \dfrac{\sin A}{\cos A} \cdot \dfrac{\sin B}{\cos B}}$$

Using the identity $\tan \theta = \frac{\sin \theta}{\cos \theta}$, we finally obtain

$$\tan(A + B) = \frac{\tan A + \tan B}{1 - \tan A \tan B}.$$

By replacing B with $-B$ and using the fact that $\tan(-B) = -\tan B$, we are able to find the identity for the tangent of the difference of two numbers. The two tangent identities follow.

TANGENT OF A SUM OR DIFFERENCE

$$\tan(A + B) = \frac{\tan A + \tan B}{1 - \tan A \tan B}$$

$$\tan(A - B) = \frac{\tan A - \tan B}{1 + \tan A \tan B}$$

NOTE The list that follows will help you in understanding the examples and exercises that follow in this section.

$$\frac{\pi}{3} = \frac{4\pi}{12}$$

$$\frac{\pi}{4} = \frac{3\pi}{12}$$

$$\frac{\pi}{6} = \frac{2\pi}{12}$$

Using this list, for example, we see that $\frac{\pi}{12} = \frac{\pi}{3} - \frac{\pi}{4}$ (or $\frac{\pi}{4} - \frac{\pi}{6}$).

EXAMPLE **1**
Finding an Exact Function Value and Supporting Graphically

Consider the number $\cos \frac{\pi}{12}$.

(a) Use the identity for $\cos(A - B)$ to find its exact value.

(b) Show that a calculator approximation for $\cos \frac{\pi}{12}$ corresponds to an approximation for the exact value found in part (a).

(c) Support the result of part (a) by locating the appropriate point on the graph of $y = \cos x$.

SOLUTION

(a) Because $\frac{\pi}{12} = \frac{\pi}{3} - \frac{\pi}{4}$, we can find the *exact* value of $\cos \frac{\pi}{12}$ as follows.

$$\cos \frac{\pi}{12} = \cos\left(\frac{\pi}{3} - \frac{\pi}{4}\right)$$

$$= \cos \frac{\pi}{3} \cos \frac{\pi}{4} + \sin \frac{\pi}{3} \sin \frac{\pi}{4} \qquad \text{Use the cosine difference identity.}$$

$$= \frac{\sqrt{3}}{2} \cdot \frac{\sqrt{2}}{2} + \frac{1}{2} \cdot \frac{\sqrt{2}}{2} \qquad \text{Use exact values.}$$

$$= \frac{\sqrt{6} + \sqrt{2}}{4} \qquad \text{Exact value}$$

(b) Using a calculator in *radian* mode, we can find that an approximation for $\cos \frac{\pi}{12}$ is .9659258263. Also, if we use the square root function, along with the addition and division operations, we find that an approximation for $\frac{\sqrt{6} + \sqrt{2}}{4}$ is also .9659258263. This supports our answer in part (a).

(c) Figure 13 shows the graph of $y = \cos x$, and the display at the bottom indicates that when $x = \frac{\pi}{12}$ (as indicated by the approximation .26179939), $y \approx .96592583$, supporting our result in part (a). ∎

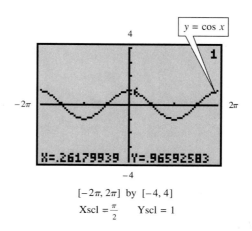

$[-2\pi, 2\pi]$ by $[-4, 4]$

$\text{Xscl} = \frac{\pi}{2}$ $\text{Yscl} = 1$

FIGURE 13

FOR **GROUP DISCUSSION**

Use the graph in Figure 13 to explain why the exact value of $\cos\left(-\frac{\pi}{12}\right)$ is also $\frac{\sqrt{6} + \sqrt{2}}{4}$. Support this using your own calculator.

EXAMPLE 2 Finding an Exact Function Value and Supporting Graphically	Repeat Example 1 for $\tan \frac{7\pi}{12}$.

SOLUTION   To find the exact value of $\tan \frac{7\pi}{12}$, we recognize that $\frac{7\pi}{12} = \frac{\pi}{3} + \frac{\pi}{4}$. Now use the identity for $\tan(A + B)$.

$$\tan \frac{7\pi}{12} = \tan\left(\frac{\pi}{3} + \frac{\pi}{4}\right)$$

$$= \frac{\tan \frac{\pi}{3} + \tan \frac{\pi}{4}}{1 - \tan \frac{\pi}{3} \tan \frac{\pi}{4}}$$

$$= \frac{\sqrt{3} + 1}{1 - \sqrt{3} \cdot 1} \qquad \text{Use exact values.}$$

$$= \frac{\sqrt{3} + 1}{1 - \sqrt{3}} \cdot \frac{1 + \sqrt{3}}{1 + \sqrt{3}} \qquad \text{Multiply numerator and denominator by the conjugate of the denominator.}$$

$$= \frac{\sqrt{3} + 3 + 1 + \sqrt{3}}{1 - 3} \qquad \text{Multiply binomials.}$$

$$= \frac{4 + 2\sqrt{3}}{-2}$$

$$= -2 - \sqrt{3} \qquad \text{Reduce to lowest terms.}$$

Using a calculator in radian mode, we can show that $\tan \frac{7\pi}{12} \approx -3.732050808$, which is the same approximation found for the exact value, $-2 - \sqrt{3}$. Figure 14 shows the graph of $y = \tan x$ in the window $\left[\frac{\pi}{2}, \frac{3\pi}{2}\right]$ by $[-5, 5]$. As indicated in the display at the bottom of the screen, $\tan \frac{7\pi}{12} \approx \tan 1.8325957 \approx -3.732051$, supporting our earlier result.   ◨

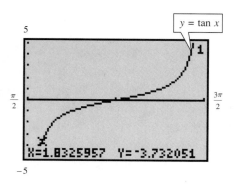

$\left[\frac{\pi}{2}, \frac{3\pi}{2}\right]$ by $[-5, 5]$

$\text{Xscl} = \frac{\pi}{2} \qquad \text{Yscl} = 1$

FIGURE 14

EXAMPLE **3**

Finding Function
Values and the
Quadrant of $A + B$

Suppose that A and B are angles in standard position, with $\sin A = \frac{4}{5}, \frac{\pi}{2} < A < \pi$, $\cos B = -\frac{5}{13}$, and $\pi < B < \frac{3\pi}{2}$. Find (a) $\sin(A + B)$, (b) $\tan(A + B)$, and (c) the quadrant of $A + B$.

SOLUTION

(a) The identity for $\sin(A + B)$ requires $\sin A$, $\cos A$, $\sin B$, and $\cos B$. Two of these values are given. The two missing values, $\cos A$ and $\sin B$, must be found first. These values can be found with the identity $\sin^2 x + \cos^2 x = 1$. To find $\cos A$, use

$$\sin^2 A + \cos^2 A = 1$$

$$\frac{16}{25} + \cos^2 A = 1 \qquad \sin A = \frac{4}{5}$$

$$\cos^2 A = \frac{9}{25}$$

$$\cos A = -\frac{3}{5}. \qquad \text{Since } \frac{\pi}{2} < A < \pi, \cos A < 0.$$

In the same way, $\sin B = -\frac{12}{13}$. Now use the formula for $\sin(A + B)$.

$$\sin(A + B) = \frac{4}{5}\left(-\frac{5}{13}\right) + \left(-\frac{3}{5}\right)\left(-\frac{12}{13}\right)$$

$$= -\frac{20}{65} + \frac{36}{65} = \frac{16}{65}$$

(b) Use the values of sine and cosine from part (a) to get $\tan A = -\frac{4}{3}$ and $\tan B = \frac{12}{5}$. Then

$$\tan(A + B) = \frac{-\dfrac{4}{3} + \dfrac{12}{5}}{1 - \left(-\dfrac{4}{3}\right)\left(\dfrac{12}{5}\right)} = \frac{\dfrac{16}{15}}{1 + \dfrac{48}{15}} = \frac{\dfrac{16}{15}}{\dfrac{63}{15}} = \frac{16}{63}.$$

(c) From the results of parts (a) and (b), we find that $\sin(A + B)$ is positive and $\tan(A + B)$ is also positive. Therefore, $A + B$ must be in quadrant I, since it is the only quadrant in which both sine and tangent are positive. ∎

Double Number Identities

Some special cases of the identities for the sum of two numbers are used often enough to be expressed as separate identities. These are the identities that result from the addition identities when $A = B$, so that $A + B = 2A$. These identities, called the **double number identities,** are now derived.

In the identity $\cos(A + B) = \cos A \cos B - \sin A \sin B$, let $B = A$ to derive an expression for $\cos 2A$.

$$\cos 2A = \cos(A + A)$$

$$= \cos A \cos A - \sin A \sin A$$

$$\cos 2A = \cos^2 A - \sin^2 A$$

Two other useful forms of this identity can be obtained by substituting either $\cos^2 A = 1 - \sin^2 A$ or $\sin^2 A = 1 - \cos^2 A$. Replace $\cos^2 A$ with $1 - \sin^2 A$ to get

$$\cos 2A = \cos^2 A - \sin^2 A$$
$$= (1 - \sin^2 A) - \sin^2 A$$
$$\cos 2A = 1 - 2 \sin^2 A,$$

and replace $\sin^2 A$ with $1 - \cos^2 A$ to get

$$\cos 2A = \cos^2 A - (1 - \cos^2 A)$$
$$= \cos^2 A - 1 + \cos^2 A$$
$$\cos 2A = 2 \cos^2 A - 1.$$

These double number identities for the cosine can be supported graphically by graphing the functions $y_1 = \cos 2x$, $y_2 = \cos^2 x - \sin^2 x$, $y_3 = 1 - 2 \sin^2 x$, and $y_4 = 2 \cos^2 x - 1$ all in the same viewing window. The graphs will all coincide, supporting our analytic work above. Figure 15 shows this single graph in the trig viewing window.

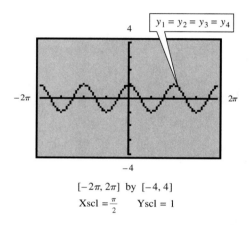

$$[-2\pi, 2\pi] \text{ by } [-4, 4]$$
$$\text{Xscl} = \frac{\pi}{2} \qquad \text{Yscl} = 1$$

FIGURE 15

We can find $\sin 2A$ with the identity $\sin(A + B) = \sin A \cos B + \cos A \sin B$, by letting $B = A$.

$$\sin 2A = \sin(A + A)$$
$$= \sin A \cos A + \cos A \sin A$$
$$\sin 2A = 2 \sin A \cos A$$

Similarly, the identity for $\tan(A + B)$ is used to find $\tan 2A$.

$$\tan 2A = \tan(A + A)$$
$$= \frac{\tan A + \tan A}{1 - \tan A \tan A}$$
$$\tan 2A = \frac{2 \tan A}{1 - \tan^2 A}$$

FOR **GROUP DISCUSSION**

1. Give graphical support for the identity for sin 2A by graphing both $y_1 = \sin 2x$ and $y_2 = 2 \sin x \cos x$ in the same viewing window. The graphs should coincide.
2. Repeat Item 1 for tan 2A, graphing $y_1 = \tan 2x$ and $y_2 = \frac{2 \tan x}{1 - \tan^2 x}$.

A summary of the double number identities follows.

DOUBLE NUMBER IDENTITIES

$$\cos 2A = \cos^2 A - \sin^2 A \qquad \cos 2A = 1 - 2 \sin^2 A$$

$$\cos 2A = 2 \cos^2 A - 1 \qquad \sin 2A = 2 \sin A \cos A$$

$$\tan 2A = \frac{2 \tan A}{1 - \tan^2 A}$$

The next example shows how we can find function values of twice a number θ if we know the cosine of θ and the sign of $\sin \theta$.

EXAMPLE 4

Finding Function Values of 2θ Given Information About θ

Given $\cos \theta = \frac{3}{5}$ and $\sin \theta < 0$, find $\sin 2\theta$, $\cos 2\theta$, and $\tan 2\theta$.

SOLUTION   In order to find $\sin 2\theta$, we must first find the value of $\sin \theta$. From the identity $\sin^2 \theta + \cos^2 \theta = 1$, we obtain

$$\sin^2 \theta + \left(\frac{3}{5}\right)^2 = 1 \qquad \cos \theta = \frac{3}{5}$$

$$\sin^2 \theta = \frac{16}{25}$$

$$\sin \theta = -\frac{4}{5}. \qquad \text{Choose the negative square root, since } \sin \theta < 0.$$

Using the double number identity for sine, we get

$$\sin 2\theta = 2 \sin \theta \cos \theta = 2\left(-\frac{4}{5}\right)\left(\frac{3}{5}\right) = -\frac{24}{25}.$$

Now find $\cos 2\theta$, using the first form of the identity. (Any form may be used.)

$$\cos 2\theta = \cos^2 \theta - \sin^2 \theta = \frac{9}{25} - \frac{16}{25} = -\frac{7}{25}$$

The value of $\tan 2\theta$ can be found in either of two ways. We can use the double number identity, and the fact that $\tan \theta = \frac{\sin \theta}{\cos \theta} = \frac{-\frac{4}{5}}{\frac{3}{5}} = -\frac{4}{3}$.

$$\tan 2\theta = \frac{2 \tan \theta}{1 - \tan^2 \theta} = \frac{2\left(-\frac{4}{3}\right)}{1 - \frac{16}{9}} = \frac{-\frac{8}{3}}{-\frac{7}{9}} = \frac{24}{7}$$

As an alternative method, we can find $\tan 2\theta$ by finding the quotient of $\sin 2\theta$ and $\cos 2\theta$.

$$\tan 2\theta = \frac{\sin 2\theta}{\cos 2\theta} = \frac{-\dfrac{24}{5}}{-\dfrac{7}{25}} = \frac{24}{7} \qquad ▯$$

Other identities involving double number identities can be verified analytically and supported graphically like those first seen in Sections 3.1 and 3.2.

EXAMPLE 5

Verifying an Identity Analytically and Supporting Graphically

---

Verify the identity

$$\cot x \sin 2x = 1 + \cos 2x$$

analytically, and support graphically.

SOLUTION   Let us start by working on the left side.

$$\cot x \sin 2x = \frac{\cos x}{\sin x} \cdot \sin 2x \qquad \cot x = \tfrac{\cos x}{\sin x}$$

$$= \frac{\cos x}{\sin x}(2 \sin x \cos x) \qquad \sin 2x = 2 \sin x \cos x$$

$$= 2 \cos^2 x$$

$$= 1 + \cos 2x \qquad 2 \cos^2 x - 1 = \cos 2x$$

To support our work graphically, we show that the graphs of $y_1 = \cot x \sin 2x$ and $y_2 = 1 + \cos 2x$ coincide. Notice that by observing the form of y_2, we see that the graph is obtained from that of $y = \cos x$ with period changed to $\frac{2\pi}{2} = \pi$, shifted 1 unit upward. See Figure 16.   ▯

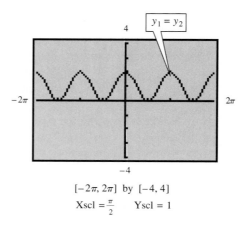

$[-2\pi, 2\pi]$ by $[-4, 4]$
$\text{Xscl} = \frac{\pi}{2} \qquad \text{Yscl} = 1$

FIGURE 16

Identities involving larger multiples of the variable can be derived by repeated use of the double number identities and the Pythagorean identities. In the following example, we find an expression for $\sin 3x$ in terms of $\sin x$.

EXAMPLE **6**

Deriving and
Supporting an
Identity for sin 3x

Write $\sin 3x$ in terms of $\sin x$, and then support the result graphically.

SOLUTION   Write $3x$ as $2x + x$, and then apply identities.

$$
\begin{aligned}
\sin 3x &= \sin(2x + x) \\
&= \sin 2x \cos x + \cos 2x \sin x \\
&= (2 \sin x \cos x) \cos x + (\cos^2 x - \sin^2 x) \sin x \\
&= 2 \sin x \cos^2 x + \cos^2 x \sin x - \sin^3 x \\
&= 2 \sin x(1 - \sin^2 x) + (1 - \sin^2 x) \sin x - \sin^3 x \\
&= 2 \sin x - 2 \sin^3 x + \sin x - \sin^3 x - \sin^3 x \\
&= -4 \sin^3 x + 3 \sin x
\end{aligned}
$$

By graphing $y_1 = \sin 3x$ and $y_2 = -4 \sin^3 x + 3 \sin x$ in the same viewing window, we find that the graphs coincide, supporting our analytic result. See Figure 17.   ❑

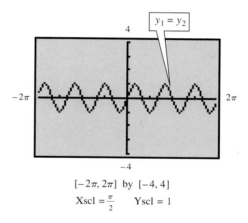

$[-2\pi, 2\pi]$ by $[-4, 4]$

$\mathrm{Xscl} = \frac{\pi}{2}$ $\mathrm{Yscl} = 1$

FIGURE 17

Half Number Identities

From the alternative forms of the identity for $\cos 2A$, we can derive three additional identities for $\sin \frac{A}{2}$, $\cos \frac{A}{2}$, and $\tan \frac{A}{2}$. These are known as **half number identities.**

To derive the identity for $\sin \frac{A}{2}$, start with the following double number identity for cosine.

$$\cos 2x = 1 - 2 \sin^2 x$$

Then solve for $\sin x$.

$$
\begin{aligned}
2 \sin^2 x &= 1 - \cos 2x \\
\sin x &= \pm \sqrt{\frac{1 - \cos 2x}{2}}
\end{aligned}
$$

Now let $2x = A$, so that $x = \frac{A}{2}$, and substitute into this last expression.

$$\sin \frac{A}{2} = \pm \sqrt{\frac{1 - \cos A}{2}}$$

The $\pm$ sign in the identity above indicates that, in practice, the appropriate sign is chosen depending upon the quadrant of $\frac{A}{2}$. For example, if $\frac{A}{2}$ is a third quadrant number on the unit circle, we choose the negative sign since the sine function is negative there.

The identity for $\cos \frac{A}{2}$ is derived in a very similar way, starting with the double number identity $\cos 2x = 2 \cos^2 x - 1$. Solve for $\cos x$.

$$\cos 2x + 1 = 2 \cos^2 x$$

$$\cos x = \pm \sqrt{\frac{1 + \cos 2x}{2}}$$

Replacing x with $\frac{A}{2}$ gives

$$\cos \frac{A}{2} = \pm \sqrt{\frac{1 + \cos A}{2}}.$$

The $\pm$ sign is used as described earlier.

Finally, an identity for $\tan \frac{A}{2}$ comes from the half number identities for sine and cosine.

$$\tan \frac{A}{2} = \frac{\pm \sqrt{\dfrac{1 - \cos A}{2}}}{\pm \sqrt{\dfrac{1 + \cos A}{2}}}$$

or $$\tan \frac{A}{2} = \pm \sqrt{\frac{1 - \cos A}{1 + \cos A}} \qquad \text{$\pm$ chosen depending upon quadrant of $\frac{A}{2}$}$$

An alternative identity for $\tan \frac{A}{2}$ can be derived using the fact that $\tan \frac{A}{2} = \frac{\sin \frac{A}{2}}{\cos \frac{A}{2}}$.

$$\tan \frac{A}{2} = \frac{\sin \dfrac{A}{2}}{\cos \dfrac{A}{2}}$$

$$= \frac{2 \sin \dfrac{A}{2} \cos \dfrac{A}{2}}{2 \cos^2 \dfrac{A}{2}} \qquad \text{Multiply by 2 $\cos \frac{A}{2}$ in numerator and denominator.}$$

$$= \frac{\sin 2\left(\dfrac{A}{2}\right)}{1 + \cos 2\left(\dfrac{A}{2}\right)} \qquad \text{Use double number identities.}$$

$$\tan \frac{A}{2} = \frac{\sin A}{1 + \cos A}$$

From this identity for $\tan \frac{A}{2}$, we can also derive

$$\tan \frac{A}{2} = \frac{1 - \cos A}{\sin A}.$$

See Exercise 61. These last two identities for $\tan \frac{A}{2}$ do not require a sign choice, as the first one does.

HALF NUMBER IDENTITIES

$$\cos\frac{A}{2} = \pm\sqrt{\frac{1 + \cos A}{2}} \qquad \sin\frac{A}{2} = \pm\sqrt{\frac{1 - \cos A}{2}}$$

$$\tan\frac{A}{2} = \pm\sqrt{\frac{1 - \cos A}{1 + \cos A}} \qquad \tan\frac{A}{2} = \frac{\sin A}{1 + \cos A}$$

$$\tan\frac{A}{2} = \frac{1 - \cos A}{\sin A}$$

In Example 1, we showed that the exact value of $\cos\frac{\pi}{12}$ is $\frac{\sqrt{6} + \sqrt{2}}{4}$. This was accomplished by using the identity for $\cos(A - B)$. Another form of the exact value of $\cos\frac{\pi}{12}$ can be found by using the identity for $\cos\frac{A}{2}$.

$$\cos\frac{\pi}{12} = \cos\frac{\frac{\pi}{6}}{2} \qquad\qquad \frac{\pi}{12} = \frac{\frac{\pi}{6}}{2}$$

$$= \sqrt{\frac{1 + \cos\frac{\pi}{6}}{2}} \qquad\qquad \text{Use the identity.}$$

$$= \sqrt{\frac{1 + \frac{\sqrt{3}}{2}}{2}} \qquad\qquad \cos\frac{\pi}{6} = \frac{\sqrt{3}}{2}$$

$$= \sqrt{\frac{\left(1 + \frac{\sqrt{3}}{2}\right) \cdot 2}{2 \cdot 2}} \qquad\qquad \text{Multiply by } \frac{2}{2} \text{ under the radical.}$$

$$= \frac{\sqrt{2 + \sqrt{3}}}{2}$$

This final expression, $\frac{\sqrt{2 + \sqrt{3}}}{2}$, has a calculator approximation of .9659258263, the same one found for $\frac{\sqrt{6} + \sqrt{2}}{4}$. Furthermore, if we graph $y = \sqrt{\frac{1 + \cos x}{2}}$ and let $x = \frac{\pi}{6} \approx .52359878$, we see that the corresponding y-value is .96592583, once again supporting our result. See Figure 18.

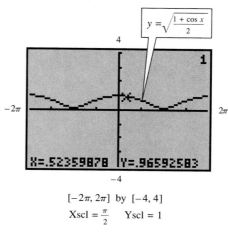

$[-2\pi, 2\pi]$ by $[-4, 4]$

$\text{Xscl} = \frac{\pi}{2} \qquad \text{Yscl} = 1$

FIGURE 18

The final example shows that if the cosine of a number is known and an appropriate interval of the number is given, we can find the cosine, sine, and tangent of half the number.

EXAMPLE 7
Finding Function Values of $\frac{x}{2}$ Given Information About x

Given $\cos x = \frac{2}{3}$, with $\frac{3\pi}{2} < x < 2\pi$, find $\cos \frac{x}{2}$, $\sin \frac{x}{2}$, and $\tan \frac{x}{2}$.

SOLUTION   Since

$$\frac{3\pi}{2} < x < 2\pi,$$

dividing through by 2 gives

$$\frac{3\pi}{4} < \frac{x}{2} < \pi,$$

showing that $\frac{x}{2}$ terminates in quadrant II on the unit circle. In this quadrant the value of $\cos \frac{x}{2}$ is negative and the value of $\sin \frac{x}{2}$ is positive. Use the appropriate half number identities to get

$$\sin \frac{x}{2} = \sqrt{\frac{1 - \frac{2}{3}}{2}} = \sqrt{\frac{1}{6}} = \frac{\sqrt{6}}{6};$$

and

$$\cos \frac{x}{2} = -\sqrt{\frac{1 + \frac{2}{3}}{2}} = -\sqrt{\frac{5}{6}} = -\frac{\sqrt{30}}{6}.$$

Also,

$$\tan \frac{x}{2} = \frac{\sin \frac{x}{2}}{\cos \frac{x}{2}} = \frac{\frac{\sqrt{6}}{6}}{-\frac{\sqrt{30}}{6}} = -\frac{\sqrt{5}}{5}.$$

Notice that it is not necessary to use a half number identity for $\tan \frac{x}{2}$ once we find $\sin \frac{x}{2}$ and $\cos \frac{x}{2}$. However, using this identity would provide an excellent check.

3.3 EXERCISES

*For each of the following circular function values (**a**) find the exact value using the appropriate sum or difference identity, (**b**) find an approximation in two ways: by using the appropriate circular function of your calculator, making sure it is in radian mode, and then by finding a calculator approximation for the exact value found in part (a), and (**c**) using the appropriate circular function graph, support the fact that your answer in part (b) is the approximation for y when the domain value is entered.*

1. $\sin \dfrac{\pi}{12}$

2. $\tan \dfrac{\pi}{12}$

3. $\cos \dfrac{7\pi}{12}$

4. $\sin \dfrac{7\pi}{12}$

5. $\sin \dfrac{5\pi}{12}$

6. $\tan \dfrac{5\pi}{12}$

7. $\cos \dfrac{5\pi}{12}$

8. $\cos\left(\dfrac{-5\pi}{12}\right)$

9. $\sin\left(\dfrac{-5\pi}{12}\right)$

10. $\tan\left(\dfrac{-5\pi}{12}\right)$

11. $\sin\left(\dfrac{13\pi}{12}\right)$

12. $\cos\left(\dfrac{13\pi}{12}\right)$

The exact value of a trigonometric function can often be found by using a sum or differ-ence identity, along with the fact that exact function values of the special angles 30°, 45°, and 60° are known. For each of the following trigonometric function values (**a**) find the exact value using one of these identities, and (**b**) find an approximation in two ways: by using the appropriate trigonometric function of your calculator, making sure it is in degree mode, and then by finding a calculator approximation for the exact value found in part (a).

13. $\sin 15°$ **14.** $\cos 15°$ **15.** $\tan 15°$ **16.** $\cos 75°$

17. $\sin 75°$ **18.** $\tan 75°$ **19.** $\sin 105°$ **20.** $\cos 105°$

21. $\tan 105°$ **22.** $\sin(-15°)$ **23.** $\cos(-15°)$ **24.** $\tan(-75°)$

Use the appropriate sum or difference identity to write the given expression as a func-tion of x alone. (For example, using the identity for $\sin(A - B)$, it can be shown that $\sin(\frac{3\pi}{2} - x) = -\cos x$.)

25. $\cos\left(\dfrac{\pi}{2} - x\right)$ **26.** $\cos(\pi - x)$ **27.** $\cos\left(\dfrac{3\pi}{2} + x\right)$

28. $\sin(\pi - x)$ **29.** $\sin(\pi + x)$ **30.** $\tan(2\pi - x)$

31. $\tan(\pi + x)$ **32.** $\tan(\pi - x)$ **33.** $\sin\left(\dfrac{3\pi}{2} - x\right)$

34. Explain how the identities for $\sec(A + B)$, $\csc(A + B)$, and $\cot(A + B)$ can be easily found by using the sum identities given in this section.

*R*elating Concepts

35. Use a calculator in radian mode to graph the function $y_1 = \cos(x + \frac{\pi}{2})$ in the trig viewing window.

36. Explain how the graph in Exercise 35 can be obtained by a translation of the graph of $y = \cos x$.

37. Let $x = 1$ in y_1 from Exercise 35, and find an approximation for $\cos(1 + \frac{\pi}{2})$ from the graph.

38. Use the identity for $\cos(A + B)$ to show analytically that $\cos(x + \frac{\pi}{2}) = -\sin x$.

39. Graph $y_2 = -\sin x$ in the trig viewing window. How does this graph support your work in Exercise 38?

40. Let $x = 1$ in y_2 from Exercise 39, and find an approximation for $-\sin 1$. How does it compare to the result in Exercise 37?

Suppose that A and B are angles in standard position. Use the given information to find (**a**) $\sin(A + B)$, (**b**) $\sin(A - B)$, (**c**) $\tan(A + B)$, (**d**) $\tan(A - B)$, (**e**) the quadrant of $(A + B)$, and (**f**) the quadrant of $(A - B)$.

41. $\cos A = \dfrac{3}{5}$, $\sin B = \dfrac{5}{13}$, $0 < A < \dfrac{\pi}{2}$, $0 < B < \dfrac{\pi}{2}$

42. $\sin A = \dfrac{3}{5}$, $\sin B = -\dfrac{12}{13}$, $0 < A < \dfrac{\pi}{2}$, $\pi < B < \dfrac{3\pi}{2}$

43. $\cos A = -\dfrac{8}{17}$, $\cos B = -\dfrac{3}{5}$, $\pi < A < \dfrac{3\pi}{2}$, $\pi < B < \dfrac{3\pi}{2}$

44. $\cos A = -\dfrac{15}{17}$, $\sin B = \dfrac{4}{5}$, $\dfrac{\pi}{2} < A < \pi$, $0 < B < \dfrac{\pi}{2}$

Suppose that x is a real number. Use the given information and the double number and reciprocal identities to find all six circular function values of 2x.

45. $\cos x = -\dfrac{12}{13},\ \sin x > 0$ **46.** $\tan x = 2,\ \cos x > 0$

47. $\tan x = \dfrac{5}{3},\ \sin x < 0$ **48.** $\sin x = \dfrac{2}{5},\ \cos x < 0$

*For each of the following, (**a**) use the appropriate half number identity to find the exact value and (**b**) use the appropriate circular or trigonometric function key on your calculator to find an approximation. Then show that this approximation is the same as the one found for your result in part (a).*

49. $\sin \dfrac{\pi}{12}$ **50.** $\cos \dfrac{\pi}{8}$ **51.** $\tan\left(-\dfrac{\pi}{8}\right)$

52. $\cos 67.5°$ **53.** $\sin 67.5°$ **54.** $\tan 195°$

Use a half number identity to find an exact value of each of the following, given the information about x.

55. $\cos \dfrac{x}{2}$, given $\cos x = \dfrac{1}{4}$, with $0 < x < \dfrac{\pi}{2}$

56. $\sin \dfrac{x}{2}$, given $\cos x = -\dfrac{5}{8}$, with $\dfrac{\pi}{2} < x < \pi$

57. $\tan \dfrac{x}{2}$, given $\sin x = \dfrac{3}{5}$, with $\dfrac{\pi}{2} < x < \pi$

58. $\cos \dfrac{x}{2}$, given $\sin x = -\dfrac{4}{5}$, with $\dfrac{3\pi}{2} < x < 2\pi$

59. $\tan \dfrac{x}{2}$, given $\tan x = \dfrac{\sqrt{7}}{3}$, with $\pi < x < \dfrac{3\pi}{2}$

60. $\tan \dfrac{x}{2}$, given $\tan x = -\dfrac{\sqrt{5}}{2}$, with $\dfrac{\pi}{2} < x < \pi$

61. Use the identity $\tan \dfrac{A}{2} = \dfrac{\sin A}{1 + \cos A}$ to derive the equivalent identity $\tan \dfrac{A}{2} = \dfrac{1 - \cos A}{\sin A}$ by multiplying both the numerator and denominator by $1 - \cos A$.

62. Consider the expression $\tan\left(\dfrac{\pi}{2} + x\right)$.
   (a) Why can't we use the identity for $\tan(A + B)$ to express it as a function of x alone?
   (b) Use the identity $\tan \theta = \dfrac{\sin \theta}{\cos \theta}$ to rewrite the expression in terms of sine and cosine.
   (c) Use the result of part (b) to show that $\tan\left(\dfrac{\pi}{2} + x\right) = -\cot x$.

63. The identity

$$\tan \frac{A}{2} = \pm\sqrt{\frac{1 - \cos A}{1 + \cos A}}$$

can be used to find $\tan 22.5° = \sqrt{3 - 2\sqrt{2}}$, and the identity

$$\tan \frac{A}{2} = \frac{\sin A}{1 + \cos A}$$

can be used to get $\tan 22.5° = \sqrt{2} - 1$. Show that these answers are the same, without using a calculator. (*Hint:* If $a > 0$ and $b > 0$ and $a^2 = b^2$, then $a = b$.)

64. Explain how you could use an identity of this section to find the exact value of $\sin 7.5°$.

Use the method of Example 6 to do each of the following. Then support your result graphically, using the trig viewing window of your calculator.

65. Express $\cos 3x$ in terms of $\cos x$.

66. Express $\tan 3x$ in terms of $\tan x$.

67. Express $\tan 4x$ in terms of $\tan x$.

68. Express $\cos 4x$ in terms of $\cos x$.

Verify the identity analytically.

69. $\dfrac{\cos(A - B)}{\cos A \sin B} = \tan A + \cot B$

70. $\dfrac{\sin(A + B)}{\cos A \cos B} = \tan A + \tan B$

71. $\dfrac{\sin(A - B)}{\sin(A + B)} = \dfrac{\tan A - \tan B}{\tan A + \tan B}$

72. $\dfrac{\sin(A + B)}{\cos(A - B)} = \dfrac{\cot A + \cot B}{1 + \cot A \cot B}$

73. $\dfrac{\sin(A - B)}{\sin B} + \dfrac{\cos(A - B)}{\cos B} = \dfrac{\sin A}{\sin B \cos B}$

74. $\dfrac{\tan(A + B) - \tan B}{1 + \tan(A + B) \tan B} = \tan A$

Verify the identity analytically. Then graph both sides of the equation as separate functions in the trig viewing window of your calculator, entering them as y_1 and y_2, to support your work graphically.

75. $\sin 2x = \dfrac{2 \tan x}{1 + \tan^2 x}$

76. $\cos 2x = \dfrac{2 - \sec^2 x}{\sec^2 x}$

77. $\dfrac{2 \cos 2x}{\sin 2x} = \cot x - \tan x$

78. $\cos 2x = \dfrac{1 - \tan^2 x}{1 + \tan^2 x}$

79. $\cot x = \dfrac{1 + \cos 2x}{\sin 2x}$

80. $\dfrac{2}{1 + \cos x} - \tan^2 \dfrac{x}{2} = 1$

81. $1 - \tan^2 \dfrac{x}{2} = \dfrac{2 \cos x}{1 + \cos x}$

82. $\tan \dfrac{x}{2} = \csc x - \cot x$

*F*urther Explorations

Identities that can be supported graphically can also be supported numerically with the TABLE. If $Y_1 = Y_2$, then the values under Y_1 correspond to the values under Y_2. However, false conclusions could occur with certain values of ΔTbl. If ΔTbl is chosen carefully so that only points of intersection for the two graphs are evaluated in the TABLE, and Tblmin is one of the points of intersection, then the TABLE could lead to the false conclusion that the expressions are equivalent. For example:

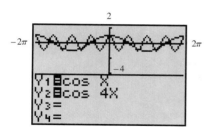

X	Y₁	Y₂
0	1	1
2.0944	-.5	-.5
4.1888	-.5	-.5
6.2832	1	1
8.3776	-.5	-.5
10.472	-.5	-.5
12.566	1	1

$X=0$

Clearly, these equations
are not equivalent.

ΔTbl $= \dfrac{2\pi}{3} \approx 2.0943951$

Tblmin $= 0$

Careful selection of ΔTbl and
Tblmin can give misleading results.

If your graphing calculator has a TABLE feature, find a ΔTbl and Tblmin for each of the following pairs of nonequivalent expressions so that the TABLE will evaluate only values for x where $Y_1 = Y_2$. (Hint: Graph each pair of expressions and find the frequency of the points where they intersect.) Use radian mode.

1. $Y_1 = \cos x$
   $Y_2 = \sin x$

2. $Y_1 = \sin(x + \pi)$
   $Y_2 = .5 \sin x$

3. $Y_1 = \cos^2 x + \sin^2 x$
   $Y_2 = \cos 2x + 1$

4. $Y_1 = \cos x$
   $Y_2 = 1 + \sin x$

5. $Y_1 = 4 \sin x \cos x$
   $Y_2 = 2 \cos 2x$

3.4 EQUATIONS AND INEQUALITIES II

Equations and Inequalities Involving Double Number Identities ▮ Equations and Inequalities Involving Half Number Identities

In this section we will study methods of solution of equations and inequalities involving circular functions of double a number (for example, $\cos 2x$) and half a number (for example, $\sin \frac{x}{2}$). As usual, we will present analytic solutions and graphical support. Because the circular functions are periodic, we will restrict our domain in most cases to the interval $[0, 2\pi)$, in order to make our work less cumbersome. If solutions over larger intervals were required, then we could use the periodic nature of these functions to write expressions for the solutions in those intervals.

Equations and Inequalities Involving Double Number Identities

> **EXAMPLE 1**
>
> Solving an Equation Involving a Function of 2x

Solve the equation $\cos 2x = \cos x$ over the interval $[0, 2\pi)$ using analytic methods. Then support the solutions with a graph.

SOLUTION First change $\cos 2x$ to a circular function of x. Use the identity $\cos 2x = 2 \cos^2 x - 1$ so that the equation involves only the cosine of x.

$$\cos 2x = \cos x$$

$$2 \cos^2 x - 1 = \cos x$$

$$2 \cos^2 x - \cos x - 1 = 0 \qquad \text{Standard quadratic form}$$

$$(2 \cos x + 1)(\cos x - 1) = 0 \qquad \text{Factor.}$$

$$2 \cos x + 1 = 0 \qquad \text{or} \qquad \cos x - 1 = 0 \qquad \text{Solve.}$$

$$\cos x = -\frac{1}{2} \qquad \text{or} \qquad \cos x = 1$$

In the required interval,

$$x = \underbrace{\frac{2\pi}{3} \quad \text{or} \quad \frac{4\pi}{3}}_{\text{for } \cos x = -\frac{1}{2}} \qquad \text{or} \qquad \underbrace{x = 0.}_{\text{for } \cos x = 1}$$

TECHNOLOGICAL NOTE
If we let Xscl = $\frac{\pi}{3}$, we can see that the graph of $y = \cos 2x - \cos x$ intersects the x-axis at the *second* tick mark, further supporting our result that $\frac{2\pi}{3}$ is a solution in Example 1.

The solution set of exact values is $\{0, \frac{2\pi}{3}, \frac{4\pi}{3}\}$. If we graph the function $y = \cos 2x - \cos x$ (obtained by subtracting the right side of the original equation from the left side) and use the x-intercept method of solution, we can support our analytic solution. Notice that in Figure 19, there are three x-intercepts over the interval $[0, 2\pi)$. As shown on the screen, one solution is approximated by 2.0943951, which corresponds to an approximation of $\frac{2\pi}{3}$. The other solutions can be supported similarly. ∎

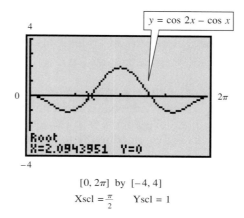

$[0, 2\pi]$ by $[-4, 4]$

Xscl = $\frac{\pi}{2}$ Yscl = 1

FIGURE 19

NOTE In this section, we will round solutions of equations and endpoints of intervals of solutions of inequalities to the nearest thousandth, for convenience. Using this agreement, the solution set in Example 1 would be expressed as $\{0, 2.094, 4.189\}$.

CAUTION In Example 1 it is important to notice that $\cos 2x$ cannot be changed to $\cos x$ by dividing by 2, since 2 is not a factor of the numerator.

$$\frac{\cos 2x}{2} \neq \cos x$$

The only way to change $\cos 2x$ to a circular function of x is by using one of the identities for $\cos 2x$.

EXAMPLE 2

Solving an Inequality Involving a Function of 2x

Refer to Example 1 and Figure 19 to do each of the following.

(a) Find the solution set of $\cos 2x < \cos x$ over the interval $[0, 2\pi)$, giving exact values.

(b) Find the solution set of $\cos 2x > \cos x$ over the interval $[0, 2\pi)$, giving approximations when appropriate.

SOLUTION

(a) The given inequality is equivalent to $\cos 2x - \cos x < 0$, and the solution set of this inequality can be determined by finding x-values for which the graph of $y = \cos 2x - \cos x$ is *below* the x-axis. Referring to Figure 19 and knowing that the exact solutions of the equation $\cos 2x = \cos x$ in the interval $[0, 2\pi)$ are $0, \frac{2\pi}{3}$, and $\frac{4\pi}{3}$, we determine the solution set to be

$$\left(0, \frac{2\pi}{3}\right) \cup \left(\frac{4\pi}{3}, 2\pi\right).$$

(b)   We must find the x-values for which the graph of $y = \cos 2x - \cos x$ is *above* the x-axis. Using approximations as directed, we find that the solution set is the open interval $(2.094, 4.189)$.   ∎

FOR **GROUP DISCUSSION**

Refer to Example 2 and discuss how the solution sets would be affected with the following minor changes.

1.   Solve $\cos 2x \leq \cos x$.
2.   Solve $\cos 2x \geq \cos x$ over the interval $[0, 4\pi)$.

Conditional equations in which a double number or half number function is involved often require an additional step to solve. This step involves adjusting the interval of solution to fit the requirements of the double number or half number. This is illustrated in the following examples.

EXAMPLE 3

Solving an Equation Using a Double Number Identity

Solve $4 \sin x \cos x = \sqrt{3}$ over the interval $[0, 2\pi)$.

SOLUTION   The identity $2 \sin x \cos x = \sin 2x$ is useful here.

$$4 \sin x \cos x = \sqrt{3}$$

$$2(2 \sin x \cos x) = \sqrt{3} \qquad 4 = 2 \cdot 2$$

$$2 \sin 2x = \sqrt{3} \qquad 2 \sin x \cos x = \sin 2x$$

$$\sin 2x = \frac{\sqrt{3}}{2} \qquad \text{Divide by 2.}$$

Because the equation now involves a function of $2x$, we must be careful to observe that since $0 \leq x < 2\pi$, it follows that $0 \leq 2x < 4\pi$. Now we list all solutions in the interval $[0, 4\pi)$:

$$2x = \frac{\pi}{3}, \frac{2\pi}{3}, \frac{7\pi}{3}, \frac{8\pi}{3}$$

or, dividing through by 2,

$$x = \frac{\pi}{6}, \frac{\pi}{3}, \frac{7\pi}{6}, \frac{4\pi}{3},$$

giving the solution set $\left\{\frac{\pi}{6}, \frac{\pi}{3}, \frac{7\pi}{6}, \frac{4\pi}{3}\right\}$ (exact values).

In Example 1, we supported our solution by using the x-intercept method. For variety, we can support this solution by graphing $y_1 = 4 \sin x \cos x$ and $y_2 = \sqrt{3}$, and locating the points of intersection over the interval $[0, 2\pi)$. Figure 20 indicates that one of these points has x-coordinate $.52359878$, which is an approximation for the exact value $\frac{\pi}{6}$. The other three solutions over this interval can be supported similarly.   ∎

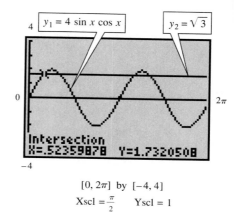

$[0, 2\pi]$ by $[-4, 4]$

$\text{Xscl} = \dfrac{\pi}{2}$ $\quad \text{Yscl} = 1$

FIGURE 20

EXAMPLE 4

Solving an Equation
That Involves
Squaring Both Sides

Solve $\tan 3x + \sec 3x = 2$ over the interval $[0, 2\pi)$.

SOLUTION Since the tangent and secant functions are related by the identity $1 + \tan^2 \theta = \sec^2 \theta$, one way to begin is to express everything in terms of the secant. This may be done by subtracting $\sec 3x$ from both sides and then squaring.

$$\tan 3x + \sec 3x = 2$$

$$\tan 3x = 2 - \sec 3x \qquad \text{Subtract } \sec 3x.$$

$$\tan^2 3x = 4 - 4 \sec 3x + \sec^2 3x \qquad \begin{array}{l}\text{Square both sides;} \\ (a - b)^2 = a^2 - 2ab + b^2\end{array}$$

$$\sec^2 3x - 1 = 4 - 4 \sec 3x + \sec^2 3x \qquad \text{Replace } \tan^2 3x \text{ with } \sec^2 3x - 1.$$

$$0 = 5 - 4 \sec 3x$$

$$4 \sec 3x = 5$$

$$\sec 3x = \frac{5}{4}$$

$$\frac{1}{\cos 3x} = \frac{5}{4} \qquad \sec \theta = \frac{1}{\cos \theta}$$

$$\cos 3x = \frac{4}{5} \qquad \text{Use reciprocals.}$$

Multiply the inequality $0 \le x < 2\pi$ by 3 to find the interval for $3x$: $[0, 6\pi)$. Using a calculator and knowing that cosine is positive in quadrants I and IV, we get

$$3x \approx .64350111, 5.6396842, 6.9266864, 11.922870, 13.209872, 18.206055.$$

Dividing by 3 gives

$$x \approx .21450037, 1.8798947, 2.3088955, 3.9742898, 4.4032906, 6.0686849.$$

Recall from algebra that when both sides of an equation are squared, there is a possibility of introducing extraneous solutions. Observing the graph of $y = \tan 3x + \sec 3x - 2$ in Figure 21, we see that in the interval $[0, 2\pi)$ there are only three x-intercepts. One of these is approximately 4.4032906, which is one of the six possible solutions shown above (see the display in the figure). The other two can be verified by the calculator in a similar manner. They are .21450037 and 2.3088955. Expressing these to the nearest thousandth, the solution set in the interval $[0, 2\pi)$ is $\{.215, 2.309, 4.403\}$. ◪

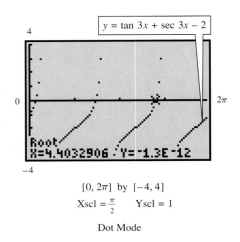

$[0, 2\pi]$ by $[-4, 4]$

$\text{Xscl} = \frac{\pi}{2}$ $\text{Yscl} = 1$

Dot Mode

FIGURE 21

Equations and Inequalities Involving Half Number Identities

EXAMPLE 5

Solving an Equation Involving a Function of $\frac{x}{2}$

Solve $2 \sin \frac{x}{2} = 1$ over the interval $[0, 2\pi)$ and support graphically.

SOLUTION   The interval $[0, 2\pi)$ may be written $0 \le x < 2\pi$. Dividing the three expressions by 2 gives

$$0 \le \frac{x}{2} < \pi.$$

To find all values for $\frac{x}{2}$ satisfying the equation, we begin by dividing both sides by 2.

$$\sin \frac{x}{2} = \frac{1}{2} \qquad \text{Divide by 2.}$$

In the interval $[0, \pi)$, the two numbers that have sine value of $\frac{1}{2}$ are $\frac{\pi}{6}$ and $\frac{5\pi}{6}$. To solve for the corresponding values of x, we multiply by 2.

$$\frac{x}{2} = \frac{\pi}{6} \qquad \text{or} \qquad \frac{x}{2} = \frac{5\pi}{6}$$

$$x = \frac{\pi}{3} \qquad\qquad x = \frac{5\pi}{3}$$

The solutions in the given interval are $\frac{\pi}{3}$ and $\frac{5\pi}{3}$.

Figure 22 shows the graph of $y = 2 \sin \frac{x}{2} - 1$ over the interval $[0, 2\pi)$. The solution $\frac{\pi}{3}$ is approximated by 1.0471976, as shown in the figure. The other solution, $\frac{5\pi}{3}$, can be supported similarly. The solution set of exact values is $\{\frac{\pi}{3}, \frac{5\pi}{3}\}$. ∎

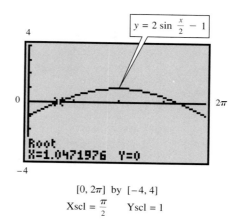

$[0, 2\pi]$ by $[-4, 4]$
$$\text{Xscl} = \frac{\pi}{2} \qquad \text{Yscl} = 1$$

FIGURE 22

EXAMPLE 6

Solving an Inequality Involving a Function of $\frac{x}{2}$

Refer to Example 5 and Figure 22 to do each of the following.

(a) Find the solution set of $2 \sin \frac{x}{2} > 1$ over the interval $[0, 2\pi)$, giving exact values.

(b) Find the solution set of $2 \sin \frac{x}{2} < 1$ over the interval $[0, 2\pi)$, giving approximations when appropriate.

SOLUTION

(a) Referring to Figure 22 and using the same procedure described in Example 2, we find that the graph of $y = 2 \sin \frac{x}{2} - 1$ lies *above* the x-axis between $\frac{\pi}{3}$ and $\frac{5\pi}{3}$. Therefore, the solution set of $2 \sin \frac{x}{2} > 1$ is the open interval $(\frac{\pi}{3}, \frac{5\pi}{3})$.

(b) Using approximations as directed and approximating the other x-intercept, we determine the solution set of $2 \sin \frac{x}{2} < 1$ to be $[0, 1.047) \cup (5.236, 2\pi)$.

∎

If an equation or inequality involving circular functions cannot be solved analytically, a purely graphical approach can be used. The final example of this section illustrates such an equation.

EXAMPLE 7

Solving an Equation Using a Purely Graphical Approach

Solve $\sin 3x + \cos 2x = \cos \frac{x}{2}$ over the interval $[0, 2\pi)$, using a graph. Give approximate solutions when appropriate.

SOLUTION We will solve this equation using the x-intercept method. The graph of $y = \sin 3x + \cos 2x - \cos \frac{x}{2}$ is shown in Figure 23. The least positive solution, .72739787, is indicated on the screen. The solution 0 can be verified by direct substitution. The solution set for the interval $[0, 2\pi)$ is

$$\{0, .727, 2.288, 3.524, 4.189\}. \quad ∎$$

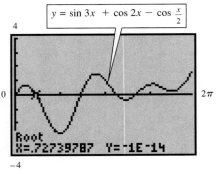

$$[0, 2\pi] \text{ by } [-4, 4]$$
$$\text{Xscl} = \tfrac{\pi}{2} \qquad \text{Yscl} = 1$$

FIGURE 23

FOR GROUP DISCUSSION

Use Figure 23 and the solution set given in Example 7 to solve the following inequalities over the interval $[0, 2\pi)$.

1. $\sin 3x + \cos 2x < \cos \dfrac{x}{2}$ **2.** $\sin 3x + \cos 2x > \cos \dfrac{x}{2}$

3.4 EXERCISES

*Solve the equation in part (**a**) analytically. Then use a graph to solve the inequalities in parts (**b**) and (**c**). In all cases, solve over the interval $[0, 2\pi)$, giving exact values for solutions and endpoints (that is, rational multiples of π).*

1. (a) $\cos 2x = \dfrac{\sqrt{3}}{2}$

　(b) $\cos 2x > \dfrac{\sqrt{3}}{2}$

　(c) $\cos 2x < \dfrac{\sqrt{3}}{2}$

2. (a) $\cos 2x = -\dfrac{1}{2}$

　(b) $\cos 2x > -\dfrac{1}{2}$

　(c) $\cos 2x < -\dfrac{1}{2}$

3. (a) $\sin 3x = -1$
　(b) $\sin 3x > -1$
　(c) $\sin 3x < -1$

4. (a) $\sin 3x = 0$
　(b) $\sin 3x > 0$
　(c) $\sin 3x < 0$

5. (a) $\sqrt{2} \cos 2x = -1$
　(b) $\sqrt{2} \cos 2x \geq -1$
　(c) $\sqrt{2} \cos 2x \leq -1$

6. (a) $2\sqrt{3} \sin 2x = \sqrt{3}$
　(b) $2\sqrt{3} \sin 2x \geq \sqrt{3}$
　(c) $2\sqrt{3} \sin 2x \leq \sqrt{3}$

7. (a) $\sin \dfrac{x}{2} = \sqrt{2} - \sin \dfrac{x}{2}$

　(b) $\sin \dfrac{x}{2} > \sqrt{2} - \sin \dfrac{x}{2}$

　(c) $\sin \dfrac{x}{2} < \sqrt{2} - \sin \dfrac{x}{2}$

8. (a) $\sin x = \sin 2x$

　(b) $\sin x > \sin 2x$

　(c) $\sin x < \sin 2x$

9. (a) $\cos 2x - \cos x = 0$

(b) $\cos 2x - \cos x \leq 0$

(c) $\cos 2x - \cos x \geq 0$

10. (a) $\sin^2 \dfrac{x}{2} - 1 = 0$

(b) $\sin^2 \dfrac{x}{2} - 1 \leq 0$

(c) $\sin^2 \dfrac{x}{2} - 1 \geq 0$

11. (a) $\sin \dfrac{x}{2} = \cos \dfrac{x}{2}$

(b) $\sin \dfrac{x}{2} > \cos \dfrac{x}{2}$

(c) $\sin \dfrac{x}{2} < \cos \dfrac{x}{2}$

12. (a) $\sec \dfrac{x}{2} = \cos \dfrac{x}{2}$

(b) $\sec \dfrac{x}{2} > \cos \dfrac{x}{2}$

(c) $\sec \dfrac{x}{2} < \cos \dfrac{x}{2}$

13. (a) $\cos 2x + \cos x = 0$

(b) $\cos 2x + \cos x \geq 0$

(c) $\cos 2x + \cos x \leq 0$

14. (a) $\sin x \cos x = \dfrac{1}{4}$

(b) $\sin x \cos x \geq \dfrac{1}{4}$

(c) $\sin x \cos x \leq \dfrac{1}{4}$

15. (a) $\tan 3x + \sec 3x = 1$

(b) $\tan 3x + \sec 3x \geq 1$

(c) $\tan 3x + \sec 3x \leq 1$

16. (a) $\csc^2 \dfrac{x}{2} = 2 \sec x$

(b) $\csc^2 \dfrac{x}{2} > 2 \sec x$

(c) $\csc^2 \dfrac{x}{2} < 2 \sec x$

17. (a) $\cos x - 1 = \cos 2x$
(b) $\cos x - 1 \leq \cos 2x$
(c) $\cos x - 1 \geq \cos 2x$

18. (a) $1 - \sin x = \cos 2x$
(b) $1 - \sin x \leq \cos 2x$
(c) $1 - \sin x \geq \cos 2x$

Use a purely graphical approach to solve each equation over the interval $[0, 2\pi)$.

19. $\sin x + \sin 3x = \cos x$

20. $\sin 3x - \sin x = 0$

21. $\cos 2x + \cos x = 0$

22. $\sin 4x + \sin 2x = 2 \cos x$

23. $\cos \frac{x}{2} = 2 \sin 2x$

24. $\sin \frac{x}{2} + \cos 3x = 0$

25. What is wrong with the following solution? Solve $\tan 2\theta = 2$ over the interval $[0, 2\pi)$.

$$\tan 2\theta = 2$$

$$\frac{\tan 2\theta}{2} = \frac{2}{2}$$

$$\tan \theta = 1$$

$$\theta = \frac{\pi}{4} \quad \text{or} \quad \theta = \frac{5\pi}{4}$$

The solutions are $\frac{\pi}{4}$ and $\frac{5\pi}{4}$.

26. The equation

$$\cot \frac{x}{2} - \csc \frac{x}{2} - 1 = 0$$

has no solution over the interval $[0, 2\pi)$. Using this information, what can be said about the graph of $y = \cot \frac{x}{2} - \csc \frac{x}{2} - 1$ over this interval? Confirm your answer by actually graphing the function over the interval.

R*elating Concepts*

It can be shown using the methods of calculus that the x-coordinates of the local extrema of the function $y_1 = \sin^2 2x + \cos \frac{1}{2}x$ *are the same as the zeros of the function* $y_2 = 4 \sin 2x \cos 2x - \frac{1}{2} \sin \frac{1}{2}x.$ *Work Exercises 27 and 28 in order.*

27. Graph both y_1 and y_2 with a calculator over the interval $[0, 2\pi)$. Use the same screen for both.

28. Verify that the least positive zero of y_2 corresponds to the x-coordinate of the first local extreme point of y_1 in the interval.

3.5 THE INVERSE CIRCULAR FUNCTIONS

Preliminary Considerations ▮ The Inverse Sine Function ▮ The Inverse
Cosine Function ▮ The Inverse Tangent Function ▮ Miscellaneous
Problems Involving Inverse Functions

Preliminary Considerations

In our work with angles in Chapter 2, we learned how to use the inverse trigonometric functions to find the measure of an angle θ if we knew one of the values $\sin \theta$, $\cos \theta$, or $\tan \theta$. Now we will investigate the inverse circular functions $y = \sin^{-1} x$, $y = \cos^{-1} x$, and $y = \tan^{-1} x$. Recall from college algebra that a function must be one-to-one* for an inverse (function) to exist. Since the circular functions $y = \sin x$, $y = \cos x$, and $y = \tan x$ are not one-to-one if their natural domains are chosen, in order to define their inverses we must restrict their domains so that the ranges are unchanged and each y-value corresponds to one and only one x-value.

The Inverse Sine Function

From Figure 24 and the horizontal line test, it is clear that $y = \sin x$ is not a one-to-one function. By suitably restricting the domain of the sine function, however, a one-to-one function can be defined. It is generally agreed upon by mathematicians that the interval $\left[-\frac{\pi}{2}, \frac{\pi}{2}\right]$ be chosen for this restriction. This gives the portion of the graph shown as a solid curve in Figure 24.

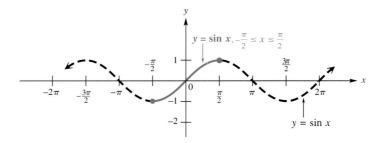

FIGURE 24

---

*A function is *one-to-one* if no two different domain values yield the same range value. For example, $f(x) = x$ is one-to-one, while $f(x) = x^2$ is not. A horizontal line will intersect the graph of a one-to-one function in at most one point.

In algebra it is shown that the graph of f^{-1} can be obtained by reflecting the graph of f across the line $y = x$. If we reflect the graph of this restricted portion of $y = \sin x$ across the line $y = x$, we obtain the graph of the inverse of the function. It is symbolized $\sin^{-1} x$. The graph of $y = \sin^{-1} x$ is shown in two forms in Figure 25; part (a) shows a traditional graph, with selected points labeled, while part (b) shows a graphing calculator-generated graph in the window $[-1, 1]$ by $\left[-\frac{\pi}{2}, \frac{\pi}{2}\right]$. (The alternate notation *arcsin x* is sometimes used to denote $\sin^{-1} x$.)

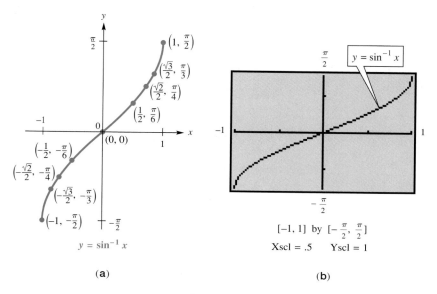

(a)

(b)

FIGURE 25

THE INVERSE SINE FUNCTION

$y = \sin^{-1} x$ or $y = \arcsin x$ means that $x = \sin y$, for $-\frac{\pi}{2} \le y \le \frac{\pi}{2}$. The domain of $y = \sin^{-1} x$ is $[-1, 1]$ and the range is $\left[-\frac{\pi}{2}, \frac{\pi}{2}\right]$.

We may think of $y = \sin^{-1} x$ or $y = \arcsin x$ as "y is the number in the interval $\left[-\frac{\pi}{2}, \frac{\pi}{2}\right]$ whose sine is x." Both notations will be used in this book.

EXAMPLE 1
Finding Inverse Sine Values

(a) Use the graph in Figure 25(a) to find $y = \sin^{-1} \frac{1}{2}$.

SOLUTION The figure shows that the point $\left(\frac{1}{2}, \frac{\pi}{6}\right)$ lies on the graph of $y = \sin^{-1} x$. Therefore, $\sin^{-1} \frac{1}{2} = \frac{\pi}{6}$.

(b) Use a graphing calculator to support the result in part (a).

SOLUTION We use the capability of the calculator to locate the point on the graph with x-coordinate $\frac{1}{2} = .5$. We find that $y \approx .52359878$, which is an approximation for $\frac{\pi}{6}$. See Figure 26.

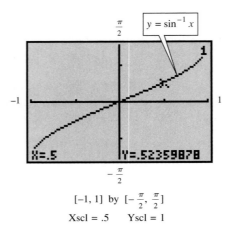

$$[-1, 1] \text{ by } [-\tfrac{\pi}{2}, \tfrac{\pi}{2}]$$
$$\text{Xscl} = .5 \qquad \text{Yscl} = 1$$

FIGURE 26

(c) Find an approximation for arcsin(−.36) in two ways, using the capabilities of a graphing calculator.

SOLUTION One way to approximate arcsin(−.36) is to put the calculator in radian mode and allow the calculator to compute it and show it in its display, as in Figure 27(a). Another way is to locate the point on the graph of $y = \sin^{-1} x$ with x-coordinate −.36. As seen in Figure 27(b), the y-value is approximately, −.3682679, which corresponds to the approximation shown in part (a) of the figure.

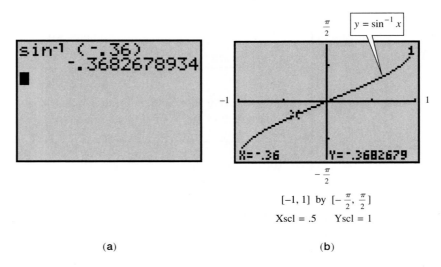

$$[-1, 1] \text{ by } [-\tfrac{\pi}{2}, \tfrac{\pi}{2}]$$
$$\text{Xscl} = .5 \qquad \text{Yscl} = 1$$

(a)   (b)

FIGURE 27

(d) Find the exact value of arcsin(−1).

SOLUTION Because the point $(-1, -\tfrac{\pi}{2})$ lies on the graph of $y = \sin^{-1} x$, arcsin $(-1) = -\tfrac{\pi}{2}$.

(e) Explain why arcsin 2 does not exist.

TECHNOLOGICAL NOTE
Some of the latest models of graphing calculators *will* give a complex number display for arcsin 2. The interpretation of this result is beyond the scope of this text, and students at this level of mathematics should realize that the development of the circular and inverse circular functions as presented here is restricted to domains and ranges consisting of real numbers only.

SOLUTION   Because 2 is not in the domain of the inverse sine function, $\sin^{-1} 2$ does not exist. If we try to find a calculator approximation for $\sin^{-1} 2$, an error message will appear.   ▯

CAUTION   In Example 1(d), it is tempting to give the value of $\arcsin(-1)$ as $\frac{3\pi}{2}$, since $\sin\frac{3\pi}{2} = -1$. Notice, however, that $\frac{3\pi}{2}$ is not in the range of the inverse sine function. Be certain, in dealing with *all* inverse circular functions, that the number given for the function value is in the range of the particular inverse function being considered.

Important information about the inverse sine function is summarized in the box that follows.

INVERSE SINE FUNCTION

$y = \sin^{-1} x$ or $y = \arcsin x$
(Figure 28)
Domain: $[-1, 1]$
Range: $\left[-\frac{\pi}{2}, \frac{\pi}{2}\right]$

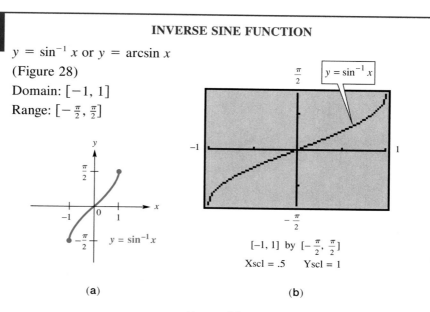

$[-1, 1]$ by $\left[-\frac{\pi}{2}, \frac{\pi}{2}\right]$
Xscl = .5   Yscl = 1

(a)   (b)

FIGURE 28

Over the interval $(-1, 1)$, the inverse sine function is increasing. Its x-intercept is 0 and its y-intercept is 0. Its graph is symmetric with respect to the origin. In general, $\sin^{-1}(-x) = -\sin^{-1} x$. It is an odd function.

The Inverse Cosine Function

The function $y = \cos^{-1} x$ (or $y = \arccos x$) is defined by restricting the domain of the function $y = \cos x$ to the interval $[0, \pi]$, and then reversing the roles of x and y. The graph of $y = \cos^{-1} x$ is shown in Figure 29, in both traditional and calculator-generated forms. Again, some key points are shown on the traditional graph.

THE INVERSE COSINE FUNCTION

$y = \cos^{-1} x$ or $y = \arccos x$ means that $x = \cos y$, for $0 \leq y \leq \pi$. The domain of $y = \cos^{-1} x$ is $[-1, 1]$ and the range is $[0, \pi]$.

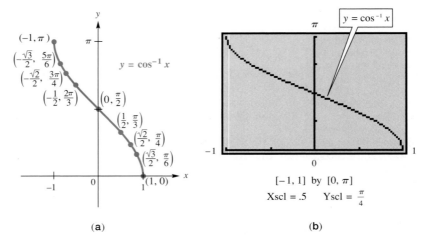

(a)

(b)

FIGURE 29

FOR **GROUP DISCUSSION**

Earlier in this section we stated that $y = \sin^{-1} x$ means "y is the number in the interval $\left[-\frac{\pi}{2}, \frac{\pi}{2}\right]$ whose sine is x." Make a similar statement for

$$y = \cos^{-1} x.$$

EXAMPLE 2

Finding Inverse Cosine Values

(a) Use the graph in Figure 29(a) to find y, if $y = \cos^{-1}\left(-\frac{1}{2}\right)$.

SOLUTION Since the point $\left(-\frac{1}{2}, \frac{2\pi}{3}\right)$ lies on the graph of $y = \cos^{-1} x$, $\cos^{-1}\left(-\frac{1}{2}\right) = \frac{2\pi}{3}$.

(b) Use a graphing calculator to support the result in part (a).

SOLUTION We graph $y = \cos^{-1} x$ in the window $[-1, 1]$ by $[0, \pi]$ and determine the y-value when $x = -\frac{1}{2} = -.5$. The display in Figure 30 shows that this y-value is 2.0943951, which is a decimal approximation for $\frac{2\pi}{3}$. (Verify this on your own calculator.)

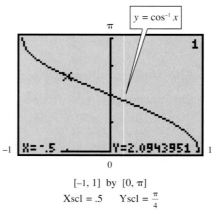

$[-1, 1]$ by $[0, \pi]$
Xscl = .5   Yscl = $\frac{\pi}{4}$

FIGURE 30

(c) Find an approximation for $\cos^{-1}(-.75)$ in two ways, using the capabilities of a graphing calculator.

SOLUTION   The display in Figure 31(a) and the graph in Figure 31(b) both indicate that $\cos^{-1}(-.75) \approx 2.4188584$. (The calculator must be in radian mode.)

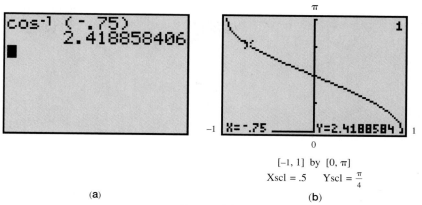

(a)   (b)

FIGURE 31

TECHNOLOGICAL NOTE
Again some current models of graphing calculators will give a complex number display for $\cos^{-1} 3$. See the technological note accompanying Example 1 in this section.

(d)   Find the exact value of $\cos^{-1} \frac{\sqrt{2}}{2}$.

SOLUTION   The point $(\frac{\sqrt{2}}{2}, \frac{\pi}{4})$ lies on the graph of $y = \cos^{-1} x$. Therefore, $\cos^{-1} \frac{\sqrt{2}}{2} = \frac{\pi}{4}$.

(e)   Why does a calculator give an error message for $\cos^{-1} 3$?

SOLUTION   Because 3 is not in the domain of the inverse cosine function, the expression $\cos^{-1} 3$ is not defined. (Another way to think of this is: there is no number whose cosine is 3, because $\cos x$ must be in the interval $[-1, 1]$ for all numbers x.)   ◼

A summary of the important information about the inverse cosine function follows.

INVERSE COSINE FUNCTION

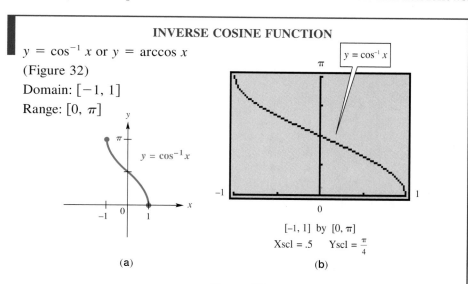

$y = \cos^{-1} x$ or $y = \arccos x$
(Figure 32)
Domain: $[-1, 1]$
Range: $[0, \pi]$

(a)   (b)

FIGURE 32

Over the interval $(-1, 1)$, the inverse cosine function is decreasing. Its x-intercept is 1 and its y-intercept is $\frac{\pi}{2}$. The graph is neither symmetric with respect to the y-axis nor symmetric with respect to the origin.

The Inverse Tangent Function

Restricting the domain of the function $y = \tan x$ to the open interval $\left(-\frac{\pi}{2}, \frac{\pi}{2}\right)$ yields a one-to-one function. By interchanging the roles of x and y, we obtain the inverse tangent function $y = \tan^{-1} x$ or $y = \arctan x$. Both traditional and calculator-generated graphs of this function are given in Figure 33.

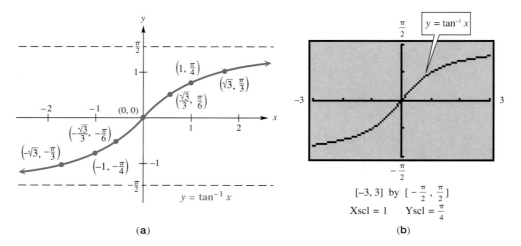

(a)   (b)

FIGURE 33

THE INVERSE TANGENT FUNCTION

$y = \tan^{-1} x$ or $y = \arctan x$ means that $x = \tan y$, for $-\frac{\pi}{2} < y < \frac{\pi}{2}$. The domain of $y = \tan^{-1} x$ is $(-\infty, \infty)$, and the range is $\left(-\frac{\pi}{2}, \frac{\pi}{2}\right)$. The lines $y = \frac{\pi}{2}$ and $y = -\frac{\pi}{2}$ are horizontal asymptotes of the graph.

FOR GROUP DISCUSSION

With calculators in hand, discuss the following.

1. Refer to Figure 33(a) to explain why $\tan^{-1} \sqrt{3} = \frac{\pi}{3}$.
2. Use your calculator to support the equality in Item 1.
3. Find an approximation for $\tan^{-1} 2$ in two ways, using your calculator.
4. Find the exact value of $\tan^{-1}(-1)$.
5. Discuss the symmetry of the graph of $y = \tan^{-1} x$.
6. Discuss the following: We will never get an error message for $\tan^{-1} x$, no matter what value of x we enter into our calculators.

The following box summarizes important information about the inverse tangent function.

INVERSE TANGENT FUNCTION

$y = \tan^{-1} x$ or $y = \arctan x$

(Figure 34)

Domain: $(-\infty, \infty)$

Range: $\left(-\frac{\pi}{2}, \frac{\pi}{2}\right)$

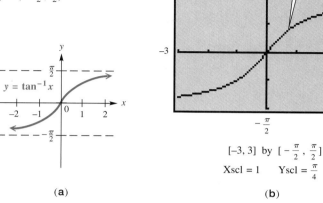

$[-3, 3]$ by $\left[-\frac{\pi}{2}, \frac{\pi}{2}\right]$

$Xscl = 1 \qquad Yscl = \frac{\pi}{4}$

(a)   (b)

FIGURE 34

Over $(-\infty, \infty)$, the inverse tangent function is increasing. Its x-intercept is 0 and its y-intercept is 0. The graph is symmetric with respect to the origin. In general, $\tan^{-1}(-x) = -\tan^{-1} x$. As $x \to \infty$, $y \to \frac{\pi}{2}$ from below, meaning that the line $y = \frac{\pi}{2}$ is a horizontal asymptote. Similarly, the line $y = -\frac{\pi}{2}$ is also a horizontal asymptote, since as $x \to -\infty$, $y \to -\frac{\pi}{2}$ from above. It is an odd function.

We have defined the inverse sine, cosine, and tangent functions with suitable restrictions on the domains. The other three inverse trigonometric functions are similarly defined. The six inverse trigonometric functions with their domains and ranges are given in the table.* This information, particularly the range for each function, should be learned. (The graphs of the last three inverse trigonometric functions are left for the exercises.)

Function	Domain	Range
$y = \sin^{-1} x$	$[-1, 1]$	$\left[-\frac{\pi}{2}, \frac{\pi}{2}\right]$
$y = \cos^{-1} x$	$[-1, 1]$	$[0, \pi]$
$y = \tan^{-1} x$	$(-\infty, \infty)$	$\left(-\frac{\pi}{2}, \frac{\pi}{2}\right)$
$y = \cot^{-1} x$	$(-\infty, \infty)$	$(0, \pi)$
$y = \sec^{-1} x$	$(-\infty, -1] \cup [1, \infty)$	$[0, \pi], y \neq \frac{\pi}{2}$
$y = \csc^{-1} x$	$(-\infty, -1] \cup [1, \infty)$	$\left[-\frac{\pi}{2}, \frac{\pi}{2}\right], y \neq 0$

* The inverse secant and inverse cosecant functions are sometimes defined differently.

Miscellaneous Problems Involving Inverse Functions

We will now illustrate how problems such as those in the remaining examples are solved using identities and inverse function concepts.

EXAMPLE 3

Finding Function Values Using Inverse Circular Functions

Evaluate each of the following without a calculator.

(a) $\sin\left(\tan^{-1}\dfrac{3}{2}\right)$ **(b)** $\tan\left(\cos^{-1}\left(-\dfrac{5}{13}\right)\right)$

SOLUTION

(a) To find the value of $\sin(\tan^{-1}\frac{3}{2})$, let $\theta = \tan^{-1}\frac{3}{2}$. We must find $\sin\theta$. It follows that $\tan\theta = \frac{3}{2}$ and $0 < \theta < \frac{\pi}{2}$, based on the range of the inverse tangent function.

$$1 + \tan^2\theta = \sec^2\theta \qquad \text{Pythagorean identity}$$

$$1 + \left(\frac{3}{2}\right)^2 = \sec^2\theta \qquad \text{Replace } \tan\theta \text{ with } \tfrac{3}{2}.$$

$$1 + \frac{9}{4} = \sec^2\theta$$

$$\sec^2\theta = \frac{13}{4}$$

$$\sec\theta = \frac{\sqrt{13}}{2} \qquad \begin{array}{l}\text{Choose the positive square}\\ \text{root, since } \sec\theta > 0 \text{ when}\\ 0 < \theta < \tfrac{\pi}{2}.\end{array}$$

Since $\sec\theta = \frac{\sqrt{13}}{2}$, $\cos\theta = \frac{2}{\sqrt{13}}$ by a reciprocal identity. Now use $\sin\theta = \sqrt{1 - \cos^2\theta}$ to find $\sin(\tan^{-1}\frac{3}{2})$.

$$\sin\theta = \sqrt{1 - \cos^2\theta} \qquad \begin{array}{l}\text{Use the positive square}\\ \text{root, since } 0 < \theta < \tfrac{\pi}{2}.\end{array}$$

$$= \sqrt{1 - \left(\frac{2}{\sqrt{13}}\right)^2}$$

$$= \sqrt{1 - \frac{4}{13}}$$

$$= \sqrt{\frac{9}{13}}$$

$$= \frac{3}{\sqrt{13}} = \frac{3\sqrt{13}}{13}.$$

Therefore, $\sin(\tan^{-1}\frac{3}{2}) = \frac{3\sqrt{13}}{13}$.

(b) To find $\tan(\cos^{-1}(-\frac{5}{13}))$, start by letting $\theta = \cos^{-1}(-\frac{5}{13})$. Then $\cos\theta = -\frac{5}{13}$ and $\frac{\pi}{2} \le \theta \le \pi$, based on the range of the inverse cosine function. We must find $\tan\theta$. To do this we will find $\sin\theta$ first.

$$\sin^2 \theta = 1 - \cos^2 \theta$$

$$= 1 - \left(-\frac{5}{13}\right)^2$$

$$= 1 - \frac{25}{169}$$

$$= \frac{144}{169}$$

$$\sin \theta = \frac{12}{13}$$ Choose the positive square root, since $\frac{\pi}{2} \le \theta \le \pi$.

Now use the quotient identity $\tan \theta = \frac{\sin \theta}{\cos \theta}$ to find $\tan(\cos^{-1}(-\frac{5}{13}))$.

$$\tan \theta = \frac{\dfrac{12}{13}}{-\dfrac{5}{13}} = -\frac{12}{5}.$$

Therefore, $\tan(\cos^{-1}(-\frac{5}{13})) = -\frac{12}{5}.$ ∎

FOR **GROUP DISCUSSION**

Discuss how the screen in Figure 35 supports the analytic work in Example 3(b). Does it matter whether the calculator is in radian mode or degree mode?

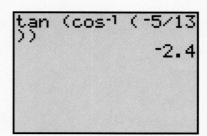

FIGURE 35

EXAMPLE **4**
Using a Double Number Identity to Evaluate an Expression

Evaluate $\tan(2 \arcsin \frac{2}{5})$ without using a calculator.

SOLUTION

Let $\arcsin \frac{2}{5} = \theta$. Then from the double number tangent identity,

$$\tan\left(2 \arcsin \frac{2}{5}\right) = \tan 2\theta$$

$$= \frac{2 \tan \theta}{1 - \tan^2 \theta}.$$

Since $\arcsin \frac{2}{5} = \theta$, $\sin \theta = \frac{2}{5}$ and $0 \le \theta \le \frac{\pi}{2}$. We will first find $\cos \theta$, then $\tan \theta$, and finally $\tan 2\theta$, as required.

$$\cos \theta = \sqrt{1 - \sin^2 \theta}$$

$$= \sqrt{1 - \left(\frac{2}{5}\right)^2} \qquad \sin \theta = \frac{2}{5}$$

$$= \sqrt{1 - \frac{4}{25}}$$

$$= \sqrt{\frac{21}{25}} = \frac{\sqrt{21}}{5}.$$

Since $\tan \theta = \dfrac{\sin \theta}{\cos \theta}$, $\tan \theta = \dfrac{\dfrac{2}{5}}{\dfrac{\sqrt{21}}{5}} = \dfrac{2}{\sqrt{21}}$.

Therefore, $\tan \left(2 \arcsin \dfrac{2}{5} \right) = \tan 2\theta$

$$= \frac{2\left(\dfrac{2}{\sqrt{21}}\right)}{1 - \left(\dfrac{2}{\sqrt{21}}\right)^2}$$

$$= \frac{\dfrac{4}{\sqrt{21}}}{1 - \dfrac{4}{21}}$$

$$= \frac{\dfrac{4}{\sqrt{21}}}{\dfrac{17}{21}} = \frac{4\sqrt{21}}{17}. \qquad ▯$$

FOR **GROUP DISCUSSION**

Discuss how the screen in Figure 36 supports the analytic work in Example 4. Does it matter whether the calculator is in radian mode or degree mode?

```
tan (2sin-1 (2/5)
)
            1.078253105
4√21/17
            1.078253105
```

FIGURE 36

EXAMPLE 5

Writing a Function Value in Terms of u

Write $\sin(\tan^{-1} u)$ as an algebraic expression in u, $u > 0$.

SOLUTION Let $\theta = \tan^{-1} u$. Then $\tan \theta = u$, and $0 < \theta < \frac{\pi}{2}$.

$$1 + \tan^2 \theta = \sec^2 \theta$$
$$1 + u^2 = \sec^2 \theta$$
$$\sec \theta = \sqrt{1 + u^2} \qquad \text{sec } \theta > 0 \text{ since } 0 < \theta < \frac{\pi}{2}.$$

$$\cos \theta = \frac{1}{\sqrt{1 + u^2}} \qquad \cos \theta = \frac{1}{\sec \theta}$$

Now use $\sin^2 \theta = 1 - \cos^2 \theta$ to find $\sin \theta$.

$$\sin^2 \theta = 1 - \cos^2 \theta$$
$$= 1 - \left(\frac{1}{\sqrt{1 + u^2}}\right)^2$$
$$= 1 - \frac{1}{1 + u^2}$$
$$= \frac{1 + u^2}{1 + u^2} - \frac{1}{1 + u^2} \qquad \text{Get a common denominator.}$$
$$= \frac{u^2}{1 + u^2}$$
$$\sin \theta = \frac{u}{\sqrt{1 + u^2}} \qquad \text{Find the square root.}$$
$$= \frac{u\sqrt{1 + u^2}}{1 + u^2} \qquad \text{Rationalize the denominator.}$$

Therefore, $\sin(\tan^{-1} u) = \frac{u\sqrt{1 + u^2}}{1 + u^2}$ for $u > 0$. ∎

3.5 EXERCISES

Use the graph of the appropriate inverse circular function, found in Figure 25(a), 29(a), or 33(a), to find the exact value of each of the following.

1. $\arcsin\left(-\frac{1}{2}\right)$

2. $\arccos \frac{\sqrt{3}}{2}$

3. $\tan^{-1} 1$

4. $\sin^{-1} 0$

5. $\cos^{-1}(-1)$

6. $\cos^{-1} \frac{1}{2}$

7. $\sin^{-1}\left(-\frac{\sqrt{3}}{2}\right)$

8. $\cos^{-1} 0$

9. $\arctan(-1)$

10. $\arccos\left(-\frac{1}{2}\right)$

11. $\arcsin \frac{\sqrt{2}}{2}$

12. $\arcsin\left(-\frac{\sqrt{2}}{2}\right)$

13. $\arccos\left(-\frac{\sqrt{3}}{2}\right)$

14. $\arcsin\left(-\frac{\sqrt{3}}{2}\right)$

15. $\tan^{-1} 0$

16. $\tan^{-1}(-\sqrt{3})$

For each of the following, (a) find an approximation of the expression using a calculator-generated graph of the appropriate inverse circular function, and (b) support your answer in part (a) by using the inverse function key on your calculator. Be sure the calculator is in radian mode.

17. $\sin^{-1} .35$ **18.** $\cos^{-1} .35$ **19.** $\cos^{-1}(-.6)$ **20.** $\sin^{-1}(-.6)$

21. $\tan^{-1} 5$ **22.** $\tan^{-1}(-5)$ **23.** $\cos^{-1} 1.5$ **24.** $\sin^{-1}(-2.3)$

Draw by hand the graph of each of the following as defined in the text, and give the domain and the range.

25. $y = \cot^{-1} x$ **26.** $y = \csc^{-1} x$ **27.** $y = \sec^{-1} x$

Find the exact real number value of each of the following, using the definitions given in the text.

28. $\cot^{-1}(-1)$ **29.** $\sec^{-1}(-\sqrt{2})$ **30.** $\csc^{-1}(-2)$

31. $\cot^{-1}(-\sqrt{3})$ **32.** $\csc^{-1}(-1)$ **33.** $\sec^{-1}(-2)$

34. The following expressions were used by the mathematicians who computed the value of π to 100,000 decimal places. Use a calculator in radian mode to verify that each is (approximately) correct.

(a) $\pi = 16 \tan^{-1} \dfrac{1}{5} - 4 \tan^{-1} \dfrac{1}{239}$

(b) $\pi = 24 \tan^{-1} \dfrac{1}{8} + 8 \tan^{-1} \dfrac{1}{57} + 4 \tan^{-1} \dfrac{1}{239}$

(c) $\pi = 48 \tan^{-1} \dfrac{1}{18} + 32 \tan^{-1} \dfrac{1}{57} - 20 \tan^{-1} \dfrac{1}{239}$

35. Explain why attempting to find $\sin^{-1} 1.003$ on your calculator will result in an error message.

36. Explain why you are able to find $\tan^{-1} 1.003$ on your calculator. Why is this situation different from the one described in Exercise 35?

Decide whether the statement is true for all real numbers x in the given interval. If it is not true, say why.

37. $\sin(\sin^{-1} x) = x,\quad -1 \le x \le 1$ **38.** $\cos(\cos^{-1} x) = x,\quad -1 \le x \le 1$

39. $\sin^{-1}(\sin x) = x,\quad x \text{ in } (-\infty, \infty)$ **40.** $\cos^{-1}(\cos x) = x,\quad x \text{ in } (-\infty, \infty)$

41. $\tan(\tan^{-1} x) = x,\quad x \text{ in } (-\infty, \infty)$ **42.** $\tan^{-1}(\tan x) = x,\quad x \text{ in } \left(-\dfrac{\pi}{2}, \dfrac{\pi}{2}\right)$

Give the exact value of each of the following, without using a calculator. You may wish to support your answer by using your calculator as described in the group discussions following Examples 3 and 4 in the text.

43. $\tan\left(\arccos \dfrac{3}{4}\right)$ **44.** $\sin\left(\arccos \dfrac{1}{4}\right)$ **45.** $\cos(\tan^{-1}(-2))$

46. $\sec\left(\sin^{-1}\left(-\dfrac{1}{5}\right)\right)$ **47.** $\cot\left(\arcsin\left(-\dfrac{2}{3}\right)\right)$ **48.** $\cos\left(\arctan \dfrac{8}{3}\right)$

49. $\sin\left(2 \tan^{-1} \dfrac{12}{5}\right)$ **50.** $\cos\left(2 \sin^{-1} \dfrac{1}{4}\right)$ **51.** $\cos\left(2 \arctan \dfrac{4}{3}\right)$

52. $\tan\left(2 \cos^{-1} \dfrac{1}{4}\right)$ **53.** $\sin\left(2 \cos^{-1} \dfrac{1}{5}\right)$ **54.** $\cos(2 \tan^{-1}(-2))$

55. $\sec(\sec^{-1} 2)$ **56.** $\csc(\csc^{-1} \sqrt{2})$

Write each of the following as an algebraic expression in u, u > 0.

57. $\sin(\arccos u)$

58. $\tan(\arccos u)$

59. $\cot(\arcsin u)$

60. $\cos(\arcsin u)$

61. $\sin\left(\sec^{-1}\dfrac{u}{2}\right)$

62. $\cos\left(\tan^{-1}\dfrac{3}{u}\right)$

63. $\tan\left(\sin^{-1}\dfrac{u}{\sqrt{u^2+2}}\right)$

64. $\sec\left(\cos^{-1}\dfrac{u}{\sqrt{u^2+5}}\right)$

65. $\sec\left(\text{arccot}\dfrac{\sqrt{4-u^2}}{u}\right)$

66. $\csc\left(\arctan\dfrac{\sqrt{9-u^2}}{u}\right)$

R*elating Concepts*

To evaluate an expression such as

$$\cos\left(\arctan\sqrt{3}+\arcsin\frac{1}{3}\right)$$

we can let $A = \arctan\sqrt{3}$, $B = \arcsin\frac{1}{3}$, and use the identity $\cos(A+B) = \cos A\cos B - \sin A\sin B$. Work Exercises 67–70 in order.

67. If $A = \arctan\sqrt{3}$, what are the values of $\cos A$ and $\sin A$?

68. If $B = \arcsin\frac{1}{3}$, what are the values of $\cos B$ and $\sin B$?

69. Substitute the values you found in Exercises 67 and 68 into the identity $\cos(A+B) = \cos A\cos B - \sin A\sin B$.

70. Simplify the expression obtained in Exercise 69.

Use the technique described in Exercises 67–70 to evaluate the following. Support your answer with a calculator if you wish.

71. $\cos\left(\tan^{-1}\dfrac{5}{12}+\tan^{-1}\dfrac{3}{4}\right)$

72. $\cos\left(\sin^{-1}\dfrac{3}{5}+\cos^{-1}\dfrac{5}{13}\right)$

73. $\sin\left(\sin^{-1}\dfrac{1}{2}+\tan^{-1}(-3)\right)$

74. $\tan\left(\cos^{-1}\dfrac{\sqrt{3}}{2}-\sin^{-1}\left(-\dfrac{3}{5}\right)\right)$

75. The functions $y = \cot^{-1} x$, $y = \sec^{-1} x$, and $y = \csc^{-1} x$ are not found on graphing calculators. However, they can be graphed using the rules shown here. Use these rules to duplicate the screens shown accompanying them.
   (a) For $y = \cot^{-1} x$: $y = \frac{\pi}{2} - \tan^{-1} x$

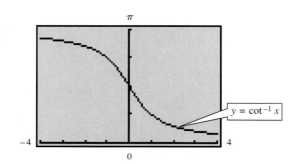

(b) For $y = \sec^{-1} x$: $y = \frac{\pi}{2} - ((x > 0) - (x < 0))(\frac{\pi}{2} - \tan^{-1}(\sqrt{x^2 - 1}))$

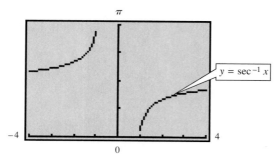

(c) For $y = \csc^{-1} x$: $y = ((x > 0) - (x < 0))(\frac{\pi}{2} - \tan^{-1}(\sqrt{x^2 - 1}))$

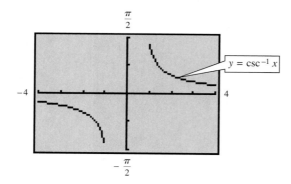

3.6 APPLICATIONS OF THE CIRCULAR FUNCTIONS

Solar Energy ▮ Temperature ▮ Electricity ▮ Music ▮ Sound ▮ Projectile Motion and Trajectory

In this section we examine a variety of applications of circular functions. Many natural phenomena are periodic, and because the circular functions are periodic, they can be used to model and illustrate many occurrences in nature.

Solar Energy

Finding sources of energy has been an important concern since the beginning of civilization. During the past 100 years, people have relied on fossil fuels for a large portion of their energy requirements. Fossils fuels are finite and limited. The heavy use of fossil fuels has caused irreversible damage to our environment and may be accelerating a greenhouse effect. Nuclear energy as an alternative has a potential for providing almost unlimited amounts of energy. Unfortunately, it creates health risks and dangerous nuclear wastes. Currently there is no completely safe disposal method for nuclear wastes. As a result, no new nuclear power plants have been ordered in the United States since 1978.

Over the past twenty-five years the production of solar energy has evolved from a mere kilowatt of electricity to hundreds of megawatts. Solar energy has many advantages over traditional energy sources in that it does not pollute and has the potential of being an unlimited, cheap source of energy. Its use and production is not limited to a small number of countries but is readily available throughout the United States and the world. The North American Southwest has some of the brightest sunlight in the world with a potential to provide up to 2500 kilowatt hours per square meter.

In the design of solar power plants, engineers need to position solar panels perpendicular to the sun's rays so that maximum energy can be collected. Understanding the movement and position of the sun at any time and date are fundamental concepts for solar energy collection.*

EXAMPLE 1

Finding the Angle of Elevation of the Sun

As mentioned in the introduction, knowing the position of the sun in the sky is essential for solar power plants. Solar panels need to be positioned perpendicular to the sun's rays for maximum efficiency. The angle of elevation of the sun (that is, the angle that the sun makes with the horizontal) θ in the sky at any latitude L can be calculated using the formula

$$\sin \theta = \cos D \cos L \cos \omega + \sin D \sin L$$

where $\theta = 0$ corresponds to sunrise and $\theta = \frac{\pi}{2}$ occurs if the sun is directly overhead. ω is the number of radians that the earth has rotated through since noon when $\omega = 0$. ω can be calculated using the formula $\omega = \frac{2\pi t}{24}$ where t is the number of hours past noon. D is the declination of the sun which varies because the earth is tilted on its axis. D can be calculated in radians using the formula

$$D = .409 \sin\{P[N - 82.3 + 1.93 \sin(NP - 2.4P)]\}$$

where $P = \frac{2\pi}{365.25}$ and N is the day number covering a four-year period, where $N = 1$ corresponds to January 1 of a leap year and $N = 1461$ corresponds to December 31 of the fourth year.

Sacramento, California, has a latitude of 38.5°. Find the angle of elevation θ of the sun at 3:00 P.M. on February 29, 2000.

SOLUTION Since 2000 is a leap year, $N = 31 + 29 = 60$ and $P = \frac{2\pi}{365.25} \approx$ .0172.

$$
\begin{aligned}
D &= .409 \sin \{P[N - 82.3 + 1.93 \sin(NP - 2.4P)]\} \\
&= .409 \sin \{.0172[60 - 82.3 + \\
&\quad 1.93 \sin(60 \times .0172 - 2.4 \times .0172)]\} \\
&\approx -.1425
\end{aligned}
$$

At 3:00 P.M., $\omega = \frac{2\pi(3)}{24} \approx .7854$. Also, $L = 38.5° = .6720$ (radian). It now follows that

$$
\begin{aligned}
\sin \theta &= \cos D \cos L \cos \omega + \sin D \sin L \\
&= \cos(-.1425) \cos(.6720) \cos(.7854) + \\
&\quad \sin(-.1425) \sin(.6720) \\
&= .4593.
\end{aligned}
$$

Therefore, $\theta = \sin^{-1} .4593 \approx .4773$ radian or approximately 27.3°. ∎

---

* Winter, C., R. Sizmann, and L. L. Vant-Hunt (Editors), *Solar Power Plants* (New York: Springer-Verlag, 1991).

Temperature

Many cities throughout the world experience seasons. Although each city has its own unique weather patterns, there are also similarities between cities as to when seasons occur and how corresponding temperatures vary. Temperature changes are a primary cause of seasons. The table lists the average monthly temperatures at Vancouver, Canada.

Month	Temperature
Jan	36
Feb	39
Mar	43
Apr	48
May	55
June	59
July	64
Aug	63
Sept	57
Oct	50
Nov	43
Dec	39

Do these temperatures have a pattern? Can we use graphs of the circular functions to model average monthly temperatures in Vancouver? Can temperatures for other cities like Phoenix, Arizona, and Buenos Aires, Argentina, be modeled using a common mathematical technique? As we shall see, the answer to these questions is *yes*.

The average temperatures in Vancouver are coldest in January and warmest in July. These temperatures cycle yearly and may change only slightly after many years. Seasonal temperature changes occur periodically because the earth's axis is tilted and its orbit around the sun is nearly circular. When a phenomenon such as temperature results from circular, periodic motion, the circular functions are often used to mathematically model the data. The graphs of these functions are essential in describing things like world temperatures and seasonal carbon dioxide levels. Their graphs will provide us with both a picture and a better understanding of periodic phenomena.*

EXAMPLE 2

Interpreting a Sine Function Model

The average temperature (in °F) at Mould Bay, Canada, can be approximated by the circular function

$$f(x) = 34 \sin\left[\frac{\pi}{6}(x - 4.3)\right],$$

where x is the month and $x = 1$ corresponds to January.

---

*Miller, A. and J. Thompson, *Elements of Meteorology* (Columbus, Ohio: Charles E. Merrill Publishing Company, 1975).

(a) Graph f over the interval $1 \le x \le 25$. Determine the amplitude and period of the graph.

(b) What is the average temperature during the month of May?

(c) What would be an approximation for the average *yearly* temperature in Mould Bay?

SOLUTION

(a) The graph is shown in Figure 37. Its amplitude is 34 and the period is

$$\frac{2\pi}{\frac{\pi}{6}} = 12.$$

The function f has a period of 12 months or 1 year which agrees with the changing of the seasons.

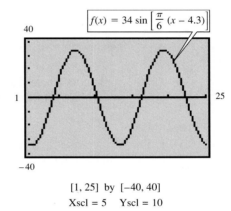

$$f(x) = 34 \sin\left[\frac{\pi}{6}(x - 4.3)\right]$$

40

1   25

−40

[1, 25] by [−40, 40]
Xscl = 5   Yscl = 10

FIGURE 37

(b) May is the 5th month, so the average temperature during the month of May is

$$f(5) = 34 \sin\left[\frac{\pi}{6}(5 - 4.3)\right] \approx 12°F.$$

(c) From the graph it appears that the average yearly temperature is about 0°F, since the graph is centered vertically about the line $y = 0$.   ∎

The next example illustrates how a model such as the one in Example 2 can be found using a set of data points.

EXAMPLE 3

Modeling Temperature with a Sine Function

The maximum average monthly temperature in New Orleans is 82°F and the minimum is 54°. The table on page 228 shows the average monthly temperatures.

(a) Using only these two temperatures, determine a circular function of the form $f(x) = a \sin b(x - d) + c$, where $a, b, c,$ and d are constants, that models the average monthly temperature in New Orleans.

Month	Temperature
Jan	54
Feb	55
Mar	61
Apr	69
May	73
June	79
July	82
Aug	81
Sept	77
Oct	71
Nov	59
Dec	55

(b) On the same coordinate axes, graph f for a two-year period together with the actual data values found in the table.

SOLUTION

(a) Let x represent the month, with January corresponding to $x = 1$. We can use the maximum and minimum average monthly temperatures to find the amplitude a.

$$a = \frac{82 - 54}{2} = 14$$

The average of the maximum and minimum temperatures is a good choice for c. The average is

$$\frac{82° + 54°}{2} = 68°.$$

Since the coldest month is January, when $x = 1$, we should choose d to be about 4. The table shows that temperatures are actually a little warmer after July than before, so we try $d = 4.2$. Since temperatures repeat every 12 months, b is $\frac{2\pi}{12} = \frac{\pi}{6}$. Thus,

$$f(x) = a \sin b(x - d) + c = 14 \sin\left[\frac{\pi}{6}(x - 4.2)\right] + 68.$$

(b) We show a graphing calculator graph of f together with the data points in Figure 38.

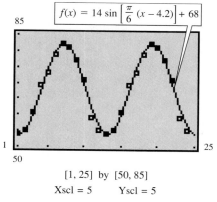

$$f(x) = 14 \sin\left[\frac{\pi}{6}(x - 4.2)\right] + 68$$

[1, 25] by [50, 85]

Xscl = 5   Yscl = 5

FIGURE 38

The function models the data quite accurately.   ❑

Electricity

In 1831 Michael Faraday discovered that when a wire is passed by a magnet, a small electric current is produced in that wire. This phenomenon became known as Faraday's Law. Since then people have used this property to generate electric current for homes and businesses by simultaneously rotating thousands of wires near large electromagnets in order to produce massive amounts of electricity. In 1992 U.S. utilities generated 2,796 billion killowatt-hours of electricity. This amount of electricity could power a 100-watt light bulb for about 3.2 billion years! Electricity has become an important modern convenience in our society.

The electricity that is supplied to most homes is produced by electric generators that rotate at 60 cycles per second. Because of this rotation, electric current alternates its direction in electrical wires and can be modeled accurately using either the sine or cosine functions. Household current is often rated at 115 volts. If the current is alternating direction in the wires, is the voltage always 115 volts or does it actually vary with time? Electric companies charge customers according to the wattage that an electrical device uses and how long it is turned on. Given both the voltage and current supplied to a light bulb, how can its wattage be determined?

In order to understand phenomena such as electric current, sound waves, or stress on your back muscles when you bend at the waist, we will need to use not only circular functions but also the many identities that relate the circular functions to each other. The study of electricity, noise control, and biophysics are all fascinating subjects that require knowledge of the identities introduced earlier in this chapter.*

---

* Weidner, R. and R. Sells, *Elementary Classical Physics,* Vol. 2 (Boston: Allyn and Bacon, 1973). Wright, J. (editor), *The Universal Almanac 1995* (Kansas City: Universal Press Syndicate Company, 1994).

<table>
<tr><td>

EXAMPLE 4

Analyzing Voltage Using a Circular Function

</td></tr>
</table>

Common household electrical current is called alternating current because the current alternates direction within the wires. The voltage V in a typical 115-volt outlet can be expressed using the equation $V = 163 \sin \omega t$ where ω is the angular velocity (in radians per second) of the rotating generator at the electrical plant and t is time measured in seconds.*

(a) It is essential for electrical generators to rotate at precisely 60 cycles per second so that household appliances and computers will function properly. Determine ω for these electrical generators. (Alternating current that cycles 60 times per second is often listed as 60 hertz. 1 hertz is equal to 1 cycle per second and is abbreviated 1 Hz.)

(b) Graph V on the interval $0 \le t \le .05$.

(c) For what value of ϕ will the graph of $V = 163 \cos(\omega t + \phi)$ be the same as the graph of $V = 163 \sin \omega t$?

SOLUTION

(a) Since each cycle is 2π radians, at 60 cycles per second, $\omega = 60(2\pi) = 120\pi$ radians per second.

(b) $V = 163 \sin \omega t = 163 \sin 120\pi t$. Because the amplitude is 163 here, we choose $-200 \le V \le 200$ for the range. See Figure 39.

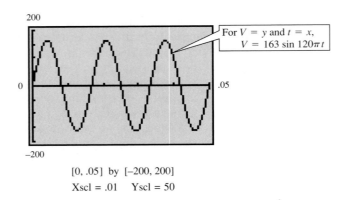

$[0, .05]$ by $[-200, 200]$
Xscl = .01   Yscl = 50

FIGURE 39

(c) The graph of $V = 163 \cos(\omega t + \phi)$ must be translated right so that the maximum voltage occurs after $\frac{1}{4}$ of a cycle. A complete cycle takes 2π units, so we must translate $(\frac{1}{4})(2\pi) = \frac{\pi}{2}$ units. To translate to the right, ϕ must be negative so $\phi = -\frac{\pi}{2}$. ∎

<table>
<tr><td>

EXAMPLE 5

Determining Wattage Consumption

</td></tr>
</table>

If a toaster is plugged into a common household outlet, the wattage consumed is not constant. Instead, it varies at a high frequency according to the equation $W = \frac{V^2}{R}$ where V is the voltage and R is a constant that measures the resistance of the toaster in ohms.* Graph the wattage W consumed by a typical toaster with $R = 15$ and $V = 163 \sin 120\pi t$ over the interval $0 \le t \le .05$. How many oscillations are there? (Think of an *oscillation* as "going from maximum to minimum.")

---

*Bell, D., *Fundamentals of Electric Circuits* 2nd ed. (Reston, Virginia: Reston Publishing Company, Inc., 1981).

SOLUTION   By substituting the given values into the wattage equation, we get

$$W = \frac{V^2}{R} = \frac{(163 \sin 120\pi t)^2}{15}.$$

The graph is shown in Figure 40. To determine the range for W, we note that $\sin 120\pi t$ has a maximum value of 1, so the expression for W has a maximum value of $\frac{163^2}{15} \approx 1771$. The minimum value is 0.

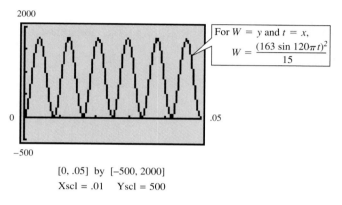

For $W = y$ and $t = x$,

$$W = \frac{(163 \sin 120\pi t)^2}{15}$$

[0, .05] by [−500, 2000]
Xscl = .01   Yscl = 500

FIGURE 40

The graph shows that there are 6 oscillations.   ∎

Music

Music is both art and science. During the Greek and Roman eras, music played an important role in philosophy and science. Although Pythagoras is usually associated with the Pythagorean theorem, in 500 B.C. he also discovered the mathematical relationships between lengths of strings and musical intervals. This discovery of the mathematical ratios that govern pitch and motion was the beginning of the science of musical sound. In the Middle Ages, music was studied together with arithmetic, geometry, and astronomy as part of the liberal arts curriculum. Later in 1862 the psychologist and scientist Hermann von Helmholtz published a classic work that opened a new direction for music using mathematics and technology. This direction has played its biggest role in the recording and reproduction of music. Max Mathews first created complex musical sounds using a computer in 1957. Since then computers have created new sounds that have been possible only through a technical knowledge of music.

When musicians tune instruments, they are able to compare like tones and accurately determine whether their pitches are the same frequency simply by listening, even though these tones vibrate hundreds or thousands of times per second. Some radios and telephones have small speakers that cannot vibrate slower than 200 times per second—yet 35 keys on a piano have frequencies below 200 and all of them can be clearly heard on these speakers. How can we explain these phenomena? What is the advantage of having larger speakers for a stereo if small speakers are capable of reproducing the lower tones?

Explanations of musical phenomena like these require a mathematical understanding of sound. Music is made up of sound waves that cause rapid increases and decreases in air pressure on a person's eardrum. Sound often involves periodic motion through the air. This periodic motion can be modeled using circular functions. Using equations and graphs, important aspects of music can be analyzed. Knowledge about waves and vibrations is necessary in order to create different musical sounds. This knowledge should continue to provide new sounds that will appeal to large and diverse audiences.*

EXAMPLE 6

Describing a
Musical Tone From
a Graph

A basic component of music is a pure tone. The graph in Figure 41 shows the sinusoidal pressure P in pounds per square foot from a pure tone at time t in seconds.

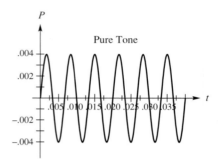

FIGURE 41

(a) The frequency of a pure tone is often measured in a unit called hertz. As mentioned in Example 4, one hertz is defined to be equal to one cycle per second and is abbreviated as Hz. What is the frequency f in hertz of the pure tone shown in the graph?

(b) The time for the tone to produce one complete cycle is called the *period*. Approximate the period T in seconds of the pure tone.

SOLUTION

(a) From the graph we can see that there are 6 cycles in .04 sec. This is equivalent to $\frac{6}{.04} = 150$ cycles per second. The pure tone has a frequency of $f = 150$ Hz.

(b) Six periods cover a time of .04 second. 1 period would be equal to $T = \frac{.04}{6} = \frac{1}{150}$ second. ∎

Sound

EXAMPLE 7

Analyzing the
Concept of Mach
Number of an
Airplane

An airplane flying faster than sound sends out sound waves that form a cone, as shown in Figure 42. The cone intersects the ground to form a hyperbola. As this hyperbola passes over a particular point on the ground, a sonic boom is heard at that point. If α is the angle at the vertex of the cone, then

$$\sin\frac{\alpha}{2} = \frac{1}{m},$$

*Benade, Arthur, *Fundamentals of Musical Acoustics* (New York: Oxford University Press, 1976).

Pierce, John, *The Science of Musical Sound* (New York: Scientific American Books, 1992).

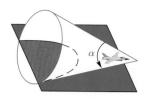

Figure 42

where m is the Mach number for the speed of the plane. (We assume $m > 1$.) The Mach number is the ratio of the speed of the plane to the speed of sound. Thus, a speed of Mach 1.4 means that the plane is flying at 1.4 times the speed of sound.

(a) Find the measure of α, if $m = 1.5$.

(b) Find the Mach number if $\alpha = 60°$.

SOLUTION

(a) Substituting 1.5 for m in the equation and solving yields the following.

$$\sin \frac{\alpha}{2} = \frac{1}{1.5}$$

$$\sin \frac{\alpha}{2} = \frac{2}{3}$$

$$\frac{\alpha}{2} = \sin^{-1}\left(\frac{2}{3}\right)$$

$$\frac{\alpha}{2} \approx 41.8° \qquad \text{Use degree mode.}$$

$$\alpha \approx 83.6°$$

When the Mach number is 1.5, α is about 83.6°.

(b) Let $\alpha = 60°$ and solve for m.

$$\sin \frac{60°}{2} = \frac{1}{m}$$

$$\sin 30° = \frac{1}{m}$$

$$\frac{1}{2} = \frac{1}{m}$$

$$m = 2$$

When $\alpha = 60°$, the Mach number is 2. ∎

Projectile Motion and Trajectory

The final example shows how identities and equation-solving techniques can be used to solve a problem involving trajectory.

EXAMPLE 8

Solving a Problem Involving Altitude of a Projectile

The altitude of a projectile in feet (neglecting air resistance) is given by

$$y = (\tan \theta)x - \frac{16}{v_0{}^2 \cos^2 \theta}x^2,$$

where x is the range (horizontal distance covered) in feet and v_0 is the initial velocity of the projectile at an angle θ from the horizontal. See Figure 43. A projectile is fired with an initial velocity of 100 feet per second. Find the firing angle of the projectile so that it strikes the ground 312.5 feet from the firing point.

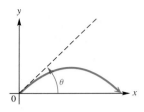

FIGURE 43

SOLUTION We want to find the value of θ so that $y = 0$ when $x = 312.5$ and $v_0 = 100$. Substitute these values into the given equation.

$$y = (\tan \theta)x - \frac{16}{v_0^2 \cos^2 \theta}x^2$$

$$0 = (\tan \theta)(312.5) - \frac{16}{100^2 \cos^2 \theta}(312.5)^2$$

$$0 = \tan \theta - \frac{16}{10{,}000 \cos^2 \theta}(312.5) \qquad \text{Divide both sides by 312.5.}$$

$$0 = \tan \theta - \frac{1}{2 \cos^2 \theta} \qquad \text{Simplify.}$$

$$0 = 2 \cos^2 \theta \tan \theta - 1 \qquad \begin{array}{l}\text{Multiply both}\\\text{sides by}\\ 2 \cos^2 \theta.\end{array}$$

$$0 = 2 \cos^2 \theta\left(\frac{\sin \theta}{\cos \theta}\right) - 1 \qquad \tan \theta = \frac{\sin \theta}{\cos \theta}$$

$$0 = 2 \cos \theta \sin \theta - 1 \qquad \frac{\cos^2 \theta}{\cos \theta} = \cos \theta$$

$$0 = \sin 2\theta - 1 \qquad \begin{array}{l} 2 \cos \theta \sin \theta \\ = \sin 2\theta \end{array}$$

$$\sin 2\theta = 1$$
$$2\theta = 90°$$
$$\theta = 45°$$

The projectile should be fired at an angle of 45° to meet the requirements of the problem. Note that θ must be in the interval $(0°, 90°)$ in this situation. ∎

3.6 EXERCISES

Solve the problem involving solar energy.

1. The solar constant S is the amount of energy per unit area that reaches Earth's atmosphere from the sun. It is equal to 1367 watts per square meter but varies slightly throughout the seasons. This fluctuation ΔS in S is calculated using the formula $\Delta S = .034S \cdot \sin[\frac{2\pi(82.5 - N)}{365.25}]$. As in Example 1, N is the day number covering a four-year period where $N = 1$ corresponds to January 1 of a leap year and $N = 1461$ corresponds to December 31 of the fourth year.*
   (a) Calculate ΔS for $N = 80$ which is the spring equinox in the first year.
   (b) Calculate ΔS for $N = 1268$ which is the summer solstice in the fourth year.
   (c) What is the maximum value of ΔS?
   (d) Find a value for N where ΔS is equal to zero.

2. Refer to Example 1.
   (a) Repeat the example for New Orleans which has a latitude of $L = 30°$.

   (b) Compare your answer to that in the example. Does it agree with your intuition?

3. The ability to calculate the number of daylight hours H at any location is important in estimating the potential solar energy production. H can be calculated using the formula

$$\cos(.1309H) = -\tan D \tan L$$

where D and L are defined in Example 1. Use this equation to calculate the shortest and longest days in Minneapolis, Minnesota, if its latitude is $L = 44.88°$. (*Hint:* The shortest day occurs when $D = -23.44°$ and the longest day occurs when $D = 23.44°$.)*

4. Refer to Exercise 3. Calculate the number of daylight hours at Hartford, Connecticut, on August 12, 2001. (*Hint:* The latitude is $L = 41.93°$.)*

---

* Winter, C., R. Sizmann, and L. L. Vant-Hunt (Editors), *Solar Power Plants* (New York: Springer-Verlag, 1991).

Solve the problem involving temperature.

5. The average temperature (in °F) at Austin, Texas, can be modeled using the function $f(x) = 17.5 \sin[\frac{\pi}{6}(x - 4)] + 67.5$ where x is the month and $x = 1$ corresponds to January.*
   (a) Graph f in the window $[1, 25]$ by $[45, 90]$. Determine the amplitude, period, phase shift, and vertical translation of f.
   (b) What is the average monthly temperature for the month of December?
   (c) Determine the maximum and minimum average monthly temperatures and the months when they occur.
   (d) What would be an approximation for the average *yearly* temperature in Austin?

6. The temperature in Fairbanks is given by

   $$T(x) = 37 \sin\left[\left(\frac{2\pi}{365}\right)(x - 101)\right] + 25,$$

   where $T(x)$ is the temperature (in degrees Celsius) on day x, with $x = 1$ corresponding to January 1 and $x = 365$ corresponding to December 31.[†] On what days was the temperature 0°C? Below 0°C? (Round answers to the nearest whole day.)

7. Refer to the table of average monthly temperatures at Vancouver, Canada, given just prior to Example 2 in this section.
   (a) Plot the average monthly temperature over a two-year period by letting $x = 1$ correspond to the month of January during the first year. Does the data seem to outline a translated sine graph?
   (b) The highest average monthly temperature is 64°F in July and the lowest average monthly temperature is 36°F in January. Their average is 50°F. Graph the data together with the line $y = 50$. What does this line represent with regard to temperature in Vancouver?
   (c) Approximate the amplitude, period, and phase shift of the translated sine wave outlined by the data.
   (d) Determine a model circular function of the form

$f(x) = a \sin b(x - d) + c$ where a, b, c, and d are constants.
   (e) Graph f together with the data on the same coordinate axes.

8. The average monthly temperature (in °F) in Phoenix, Arizona, is shown in the table.*

Month	Temperature
Jan	51
Feb	55
Mar	63
Apr	67
May	77
June	86
July	90
Aug	90
Sept	84
Oct	71
Nov	59
Dec	52

(a) Predict the average yearly temperature and compare it to the actual value of 70°F.
(b) Plot the average monthly temperature over a two-year period by letting $x = 1$ correspond to January of the first year.
(c) Determine a model circular function of the form $f(x) = a \cos b(x - d) + c$ where a, b, c, and d are constants.
(d) Graph f together with the data on the same coordinate axes.

Solve the problems involving electricity. (In Exercises 9 and 10, refer to Example 4.)

9. How many times does the current oscillate in .05 second?

10. What are the maximum and minimum voltages in this outlet? Is the voltage always equal to 115 volts?

11. The voltage E in an electrical circuit is given by $E = 5 \cos 80\pi t$, where t is time measured in seconds.
    (a) Find the amplitude and period.

---

* Miller, A. and J. Thompson, *Elements of Meteorology* (Columbus, Ohio: Charles E. Merrill Publishing Company, 1975).

† From "Is the Graph of Temperature a Sine Curve?" by Barbara Lando and Clifton Lando, *The Mathematics Teacher*, Vol. 70, September 1977, pp. 534–537. Copyright © 1977 by the National Council of Teachers of Mathematics. Reprinted with the permission of The Mathematics Teacher.

(b) The reciprocal of the period, called the *frequency*, is the number of periods completed in one second. Find the frequency.
(c) Find E when $t = 0, .03, .06, .09, .12$.

12. For another electrical circuit, the current E is given by $E = 3.8 \sin 40\pi t$, where t is time measured in seconds.
    (a) Find the amplitude and the period.
    (b) Find the frequency. See Exercise 11(b).
    (c) Find E when $t = .02, .04, .08, .12, .14$.

13. In an electric circuit, let V represent the electromotive force in volts at t seconds. Assume $V = \cos 2\pi t$. Find the smallest positive value of t where $0 \le t \le \frac{1}{2}$ for each of the following values of V.
    (a) $V = 0$ **(b)** $V = .5$ **(c)** $V = .25$

14. A coil of wire rotating in a magnetic field induces a voltage given by

$$e = 20 \sin\left(\frac{\pi t}{4} - \frac{\pi}{2}\right),$$

where t is time in seconds. Find the smallest positive time to produce the following voltages.
    (a) 0 **(b)** $10\sqrt{3}$

Solve the problem involving music.

17. No musical instrument can generate a true pure tone. A pure tone has a unique, constant frequency and amplitude that sounds rather dull and uninteresting. The pressures caused by pure tones on the eardrum are sinusoidal. The change in pressure P in pounds per square foot on a person's eardrum from a pure tone at time t in seconds can be modeled using the equation $P = A \sin(2\pi f t + \phi)$. f is the frequency in cycles per second and ϕ is the phase angle. When P is positive there is an increase in pressure and when P is negative there is a decrease in pressure.[‡]
    (a) Middle C has a frequency of 261.63 cycles per second. Graph this tone with $A = .004$ and $\phi = \frac{\pi}{7}$ in the window $[0, .005]$ by $[-.005, .005]$.
    (b) Determine graphically when $P \le 0$ on $[0, .005]$.

18. A piano string can vibrate at more than one frequency when it is struck. It produces a complex wave that can mathematically be modeled by a sum of several pure tones. If a piano key with a frequency of f_1 is played, then the corresponding string will not only vibrate at f_1 but it will also vibrate at the higher

15. When the two voltages $V_1 = 30 \sin 120\pi t$ and $V_2 = 40 \cos 120\pi t$ are applied to the same circuit, the resulting voltage V will be equal to their sum.[*]
    (a) Graph $V = V_1 + V_2$ in the window $[0, .05]$ by $[-60, 60]$.
    (b) Use graphing to estimate values for a and ϕ so that $V = a \sin(120\pi t + \phi)$.
    (c) Use identities to verify that your expression for V is valid.

16. **Amperage** is a measure of the amount of electricity that is moving through a circuit whereas **voltage** is a measure of the force pushing the electricity. The **wattage** W consumed by an electrical device can be determined by calculating the product of the amperage I and voltage V.[†]
    (a) A household circuit has a voltage of $V = 163 \sin 120\pi t$ when an incandescent light bulb is turned on with an amperage of $I = 1.23 \cdot \sin 120\pi t$. Graph the wattage $W = VI$ consumed by the light bulb in the window $[0, .05]$ by $[-50, 300]$.
    (b) Determine the maximum and minimum wattages used by the light bulb.

frequencies of $2f_1, 3f_1, 4f_1, \ldots, nf_1$. f_1 is called the **fundamental frequency** of the string and higher frequencies are called the **upper harmonics**. The human ear will hear the sum of these frequencies as one complex tone.[‡]
    (a) Suppose that the A key above middle C is played. Its fundamental frequency is $f_1 = 440$ Hz and let its associated pressure be expressed as $P_1 = .002 \sin 880\pi t$. The string will also vibrate at $880, 1320, 1760, \ldots$ hertz. Let the corresponding pressures of these upper harmonics be

$P_2 = \frac{.002}{2} \sin 1760\pi t$, $P_3 = \frac{.002}{3} \sin 2640\pi t$,
$P_4 = \frac{.002}{4} \sin 3520\pi t$, and $P_5 = \frac{.002}{5} \sin 4400\pi t$.
Graph each of the following expressions for P in the window $[0, .01]$ by $[-.005, .005]$.
    (i) $P = P_1$
    (ii) $P = P_1 + P_2$
    (iii) $P = P_1 + P_2 + P_3$
    (iv) $P = P_1 + P_2 + P_3 + P_4$
    (v) $P = P_1 + P_2 + P_3 + P_4 + P_5$

---

[*] Bell, D. *Fundamentals of Electric Circuits 2nd ed.* (Reston: Reston Publishing Company, Inc., 1981).

[†] Wilcox, G. and C. Hesselberth, *Electricity For Engineering Technology* (Boston: Allyn and Bacon, Inc., 1970).

[‡] Roederer, Juan, *Introduction to the Physics and Psychophysics of Music* (London: The English Universities Press Ltd., 1973).

(b) Describe the final graph of P.
(c) What is the maximum pressure of $P = P_1 + P_2 + P_3 + P_4 + P_5$? When does this maximum occur on $[0, .01]$?

19. If a string with a fundamental frequency of 110 hertz is plucked in the middle, it will vibrate at the odd harmonics of 110, 330, 550, . . . hertz but not at the even harmonics of 220, 440, 660, . . . hertz. The resulting pressure P caused by the string can be approximated using the equation

$$P = .003 \sin 220\pi t + \frac{.003}{3} \sin 660\pi t +$$

$$\frac{.003}{5} \sin 1100\pi t + \frac{.003}{7} \sin 1540\pi t.*$$

(a) Graph P in the window $[0, .03]$ by $[-.005, .005]$.
(b) Use the graph to describe the shape of the sound wave that is produced.
(c) At lower frequencies, the inner ear will hear a tone only when the eardrum is moving outward. Determine the times on the interval $[0, .03]$ when this will occur (that is, when $P < 0$).

Solve the problem involving sound.

21. Sound is a result of waves applying pressure to a person's eardrum. For a pure sound wave radiating outward in a spherical shape, the function $P = \frac{a}{r} \cos \left[\frac{2\pi r}{\lambda} - ct \right]$ can be used to express the sound pressure at a radius of r feet from the source. t is the time in seconds, λ is the length of the sound wave in feet, c is the speed of sound in feet per second, and a is the maximum sound pressure at the source measured in pounds per square foot.[‡]
(a) Let $a = .4$ lb/ft^2, $\lambda = 4.9$ ft, and $c = 1026$ ft/sec. Graph the sound pressure at a distance of 10 feet from its source in the window $[0, .05]$ by $[-.05, .05]$. Describe P at this distance.
(b) Now let $a = 3$ and $t = 10$. Graph the sound pressure in the window $[0, 20]$ by $[-2, 2]$. What happens to the pressure P as the radius r increases?
(c) Suppose a person stands at a radius r so that $r = n\lambda$ where n is a positive integer. Use the difference identity for cosine to simplify P in this situation.

22. Small speakers like those found in older radios and telephones often cannot vibrate slower than 200

20. Musicians sometimes tune instruments by playing the same tone on two different instruments and listening for a phenomenon known as **beats.** Beats occur when two tones vary in frequency by only a few hertz. When the two instruments are in tune the beats will disappear. The ear hears beats because the pressure slowly rises and falls as a result of this slight variation in the frequency. This phenomenon can be seen using a graphing calculator.[†]
(a) Consider two tones with frequencies of 220 and 223 hertz and pressures $P_1 = .005 \sin 440\pi t$ and $P_2 = .005 \sin 446\pi t$, respectively. Graph the pressure $P = P_1 + P_2$ felt by an eardrum over the one-second interval $[.15, 1.15]$, using the window $[.15, 1.15]$ by $[-.01, .01]$. How many beats are there in one second?
(b) Repeat part (a) with frequencies of 220 and 216.
(c) Determine a simple way to find the number of beats per second if the frequency of each tone is given.

hertz—yet 35 keys on a piano have frequencies below 200 hertz. When a musical instrument creates a tone of 110 hertz it also creates tones at 220, 330, 440, 550, 660, . . . hertz. A small speaker cannot reproduce the 110-hertz vibration but it can reproduce the higher frequencies which are called the upper harmonics. The low tones can still be heard because the speaker produces **difference tones** of the upper harmonics. The difference between consecutive frequencies is 110 hertz and this difference tone will be heard by a listener. We can see this phenomenon using a graphing calculator.*
(a) Graph the upper harmonics represented by the pressure

$$P = \frac{1}{2} \sin [2\pi(220)t] + \frac{1}{3} \sin [2\pi(330)t] +$$

$$\frac{1}{4} \sin [2\pi(440)t]$$

in the window $[0, .03]$ by $[-1, 1]$.
(b) Estimate all t-coordinates where P is maximum.
(c) Approximate the frequency of these maximum values. What does a person hear in addition to the frequencies of 220, 330, and 440 hertz?

* Benade, Arthur, *Fundamentals of Musical Acoustics* (New York: Oxford University Press, 1976). Roederer, Juan, *Introduction to the Physics and Psychophysics of Music* (London: The English Universities Press Ltd., 1973).

† Pierce, John, *The Science of Musical Sound* (New York: Scientific American Books, 1992).

‡ Beranek, L., *Noise and Vibration Control* (Washington, DC: Institute of Noise Control Engineering, 1988).

Suppose that an airplane flying faster than sound goes directly over you. Assume that the plane is flying level. At the instant that you feel the sonic boom from the plane, the angle of elevation to the plane is given by

$$\alpha = 2 \arcsin \frac{1}{m},$$

where m is the Mach number of the plane's speed. (See Example 7.) Find α to the nearest degree for each of the following values of m.

23. $m = 1.2$ **24.** $m = 1.5$ **25.** $m = 2$ **26.** $m = 2.5$

Solve the problem involving projectile motion and trajectory.

An equation for the curve describing the altitude of a projectile is

$$y = (\tan \theta)x - \frac{16}{v_0^2 \cos^2 \theta}x^2,$$

where v_0 is the initial velocity of the projectile and θ is the angle from the horizontal at which it is fired.

27. A projectile is fired with an initial velocity of 400 ft per sec at an angle of 45° with the horizontal. Find each of the following: **(a)** the range (horizontal distance covered), **(b)** the maximum altitude.

28. Repeat Exercise 27 if the projectile is fired at 800 ft per sec at an angle of 30° with the horizontal.

Solve the following problems on miscellaneous applications of circular functions.

29. The British nautical mile is defined as the length of a minute of arc of a meridian. Since the earth is flat at its poles, the nautical mile, in feet, is given by

$$L = 6077 - 31 \cos 2\theta,$$

where θ is the latitude in degrees.* See the figure.

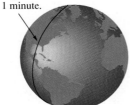

A nautical mile is the length on any of these meridians cut by a central angle of measure 1 minute.

(a) Find the latitude between 0° and 90° at which the nautical mile is 6074 feet.

(b) At what latitude between 0° and 180° is the nautical mile 6108 feet?

(c) In the United States the nautical mile is defined everywhere as 6080.2 feet. At what latitude between 0° and 90° does this agree with the British nautical mile?

30. The equation $.342D \cos \theta + h \cos^2 \theta = \frac{16D^2}{V^2}$ is used in reconstructing accidents in which a vehicle vaults into the air after hitting an obstruction. V is the velocity in feet per second of the vehicle when it hits the obstruction, D is the distance (in feet) from the obstruction to the vehicle's landing point, and h is the difference in height (in feet) between the landing point and the takeoff point. Angle θ is the takeoff angle, the angle between the horizontal and the path of the vehicle. Find θ to the nearest degree if $V = 60$, $D = 80$, and $h = 2$.

If an object is dropped in a vacuum, then the distance, d, the object falls in t seconds is given by

$$d = \frac{1}{2}gt^2,$$

where g is the acceleration due to gravity. At any particular point on the earth's surface, the value of g is a constant, roughly 978 cm per sec per sec. A more exact value of

*Adapted from *A Sourcebook of Applications of School Mathematics* by Donald Bushaw et al. Copyright © 1980 by The Mathematical Association of America. Reprinted by permission.

g at any point on the earth's surface is given by

$$g = 978.0524(1 + .005297 \sin^2 \phi - .0000059 \sin^2 2\phi) - .000094h$$

in cm per second per second, where ϕ is the latitude of the point and h is the altitude of the point in feet. Find g, rounding to the nearest thousandth, given the following.

31. $\phi = 47° 12'$, $h = 387.0$ ft

32. $\phi = 68° 47'$, $h = 1145$ ft

33. A runner's arm swings rhythmically according to the equation

$$y = \left(\frac{\pi}{8}\right) \cos 3\pi\left(t - \frac{1}{3}\right),$$

where y represents the angle between the actual position of the upper arm and the downward vertical position (as shown in the figure*) and where t represents time in seconds. At what times in $[0, 2\pi)$ is the angle y equal to 0?

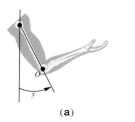

(a)   **(b)**

34. The height in feet above ground of a weight hanging on a spring is given by

$$h = 100 + 10 \cos\left[\left(\frac{2\pi}{3}\right)(t - .03)\right],$$

where t is the number of seconds after the weight is released.

(a) Find h for $t = 0$, 1, and 2 sec.
(b) What are the maximum and minimum distances above ground? What is the period?

35. The graphs of equations of the form $y = \left(\frac{1}{n}\right) \cos(nt - \theta)$ are called *harmonic waves* and are important in the study of music. Let $y_1 = \cos(t - .5\pi)$, $y_2 = \left(\frac{1}{3}\right) \cos(3t - .5\pi)$, $y_3 = \left(\frac{1}{5}\right) \cos(5t - .5\pi)$, $y_4 = \left(\frac{1}{7}\right) \cos(7t - .5\pi)$, $y_5 = \left(\frac{1}{9}\right) \cos(9t - .5\pi)$. Use your calculator to duplicate the following screens for the function defined. Use the window $[0, 4\pi]$ by $[-2, 2]$, with Xscl $= \frac{\pi}{2}$ and Yscl $= 1$.

(a) y_1

(b) $y_1 + y_2$

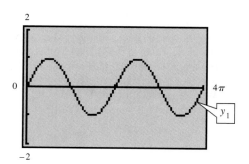
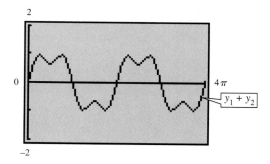

* From *Calculus for Life Sciences* by Rodolfo De Sapio. Copyright © 1978 by Rodolfo De Sapio. Reprinted by permission of the author.

(c) $y_1 + y_2 + y_3$

(d) $y_1 + y_2 + y_3 + y_4$

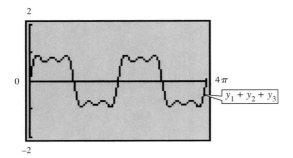

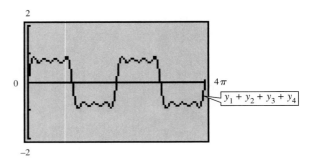

(e) $y_1 + y_2 + y_3 + y_4 + y_5$

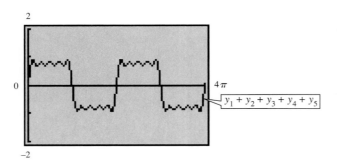

36. If a person bends at the waist with a straight back making an angle of θ degrees with the horizontal, then the force F in pounds exerted on the back muscles can be approximated by the equation $F = \frac{.6W \sin(\theta + 90°)}{\sin 12°}$ where W is the weight of the person.*
   (a) Calculate F when $W = 170$ lb and $\theta = 30°$.
   (b) Use an identity to show that F is approximately equal to $2.9W \cos \theta$.
   (c) For what value of θ is F maximum?

37. At Mauna Loa, Hawaii, atmospheric carbon dioxide levels in parts per million (ppm) have been measured regularly since 1958. The function

$$L(x) = .022x^2 + .55x + 316 + 3.5 \sin(2\pi x)$$

can be used to model these levels where x is in years and $x = 0$ corresponds to 1960.†
   (a) Graph L in the window $[15, 35]$ by $[325, 365]$.

(b) For what x do the seasonal maximum and minimum carbon dioxide levels occur?

38. (Refer to the previous exercise.) The carbon dioxide content in the atmosphere at Barrow, Alaska, in parts per million (ppm) can be modeled using the function

$$C(x) = .04x^2 + .6x + 330 + 7.5 \sin(2\pi x)$$

where $x = 0$ corresponds to 1970.‡
   (a) Graph C in the window $[5, 20]$ by $[320, 380]$, and compare it with the graph for L in the previous exercise.
   (b) Discuss possible reasons why the amplitude of the oscillations in the graph of C are larger than the amplitude of the oscillations in the graph of L which models Hawaii.
   (c) Write a new C function that is valid if x represents the actual year where $1970 \leq x \leq 1995$.

---

*Metcalf, H., *Topics in Classical Biophysics* (Englewood Cliffs: Prentice-Hall, Inc., 1980).

†Nilsson, A., *Greenhouse Earth* (New York: John Wiley & Sons, 1992).

‡Zeilik, M., S. Gregory, and E. Smith, *Introductory Astronomy and Astrophysics* (Philadelphia: Saunders College Publishing, 1992).

Chapter 3

SUMMARY

The identities developed for the trigonometric functions also apply to the circular functions. The negative number identities give the function of the negative of a number in terms of the function of the number. One use of these identities, along with the reciprocal, quotient, and Pythagorean identities, is to express one circular (or trigonometric) function in terms of another. Verifying identities helps to develop this skill which is useful in more advanced work in mathematics.

Equations involving circular functions that are linear or quadratic in, for example, $\sin x$, can be solved with the usual linear or quadratic methods. Their corresponding inequalities can then be solved by observing where the graph is above or below the x-axis.

Identities for the sum or difference of two numbers, such as $\cos(A \pm B)$, twice a number, such as $\sin 2A$, or half a number, such as $\tan \frac{A}{2}$, are derived algebraically using properties from geometry and identities given earlier in this book. These identities are used to find exact function values for certain real numbers. For example, the identities are used to find exact values for $\sin(\frac{\pi}{8})$, using the fact that $\frac{\pi}{8}$ is half of $\frac{\pi}{4}$, or $\cos(\frac{7\pi}{12})$, from the fact that $\frac{7\pi}{12} = \frac{\pi}{3} + \frac{\pi}{4}$. These new identities are also used in verifying more complicated identities.

The double number and half number identities often make it possible to rewrite equations involving circular functions in a different form so that they can be solved by earlier methods. Equations or inequalities that cannot be solved analytically may be solved graphically using methods introduced in Chapter 1.

The inverse circular functions are found, like other inverse functions, by reversing the roles of x and y. For example, $y = \tan^{-1} x$ means that $x = \tan y$ for an appropriate range of y-values. The range of $\sin^{-1} x$ is $[-\frac{\pi}{2}, \frac{\pi}{2}]$, the range of $\cos^{-1} x$ is $[0, \pi]$, and the range of $\tan^{-1} x$ is $(-\frac{\pi}{2}, \frac{\pi}{2})$. The domain and range of the inverse circular functions are part of their definitions and must be memorized. The graph of an inverse circular function is found by reflecting the graph of the corresponding circular function, over a suitably restricted domain, across the line $y = x$.

Applications of the circular functions occur in many fields, such as the study of solar energy, fluctuation of temperature in a particular location, electricity, music, sound waves, and projectile motion and trajectory. The periodic nature of the circular functions allows them to provide models for phenomena that exhibit periodic cycles.

Key Terms

SECTION 3.1

fundamental identities

SECTION 3.2

x-intercept method
intersection-of-graphs method

SECTION 3.3

sum and difference identities
double number identities
half number identities

SECTION 3.5

inverse sine ($\sin^{-1}$)
arcsin x

inverse cosine ($\cos^{-1}$)
arccos x
inverse tangent ($\tan^{-1}$)
arctan x
inverse cotangent
inverse secant
inverse cosecant

SECTION 3.6

Mach number

Chapter 3   REVIEW EXERCISES

For each item in Column I, give the letter of the item in Column II that completes an identity.

Column I   Column II

1. $\sec x$ **(a)** $\dfrac{1}{\sin x}$

2. $\csc x$ **(b)** $\dfrac{1}{\cos x}$

3. $\tan x$ **(c)** $\dfrac{\sin x}{\cos x}$

4. $\cot x$ **(d)** $\dfrac{1}{\cot^2 x}$

5. $\sin^2 x$ **(e)** $\dfrac{1}{\cos^2 x}$

6. $\tan^2 x + 1$ **(f)** $\dfrac{\cos x}{\sin x}$

7. $\tan^2 x$ **(g)** $\dfrac{1}{\sin^2 x}$

8. $1 + \cot^2 x$ **(h)** $1 - \cos^2 x$

Verify analytically each of the following identities.

9. $\sin^2 x - \sin^2 y = \cos^2 y - \cos^2 x$

10. $2 \cos^3 x - \cos x = \dfrac{\cos^2 x - \sin^2 x}{\sec x}$

11. $-\cot \dfrac{x}{2} = \dfrac{\sin 2x + \sin x}{\cos 2x - \cos x}$

12. $\dfrac{\sin^2 x}{2 - 2 \cos x} = \cos^2 \dfrac{x}{2}$

13. $\dfrac{\sin 2x}{\sin x} = \dfrac{2}{\sec x}$

14. $2 \cos A - \sec A = \cos A - \dfrac{\tan A}{\csc A}$

15. $\dfrac{2 \tan B}{\sin 2B} = \sec^2 B$

16. $\tan 4\theta = \dfrac{2 \tan 2\theta}{2 - \sec^2 2\theta}$

17. $1 + \tan^2 \alpha = 2 \tan \alpha \csc 2\alpha$

18. $\dfrac{\sin t}{1 - \cos t} = \cot \dfrac{t}{2}$

19. $\sin 2\alpha = \dfrac{2(\sin \alpha - \sin^3 \alpha)}{\cos \alpha}$

20. $\dfrac{2 \cot x}{\tan 2x} = \csc^2 x - 2$

21. $\tan \theta \cos^2 \theta = \dfrac{2 \tan \theta \cos^2 \theta - \tan \theta}{1 - \tan^2 \theta}$

22. $\tan \theta \sin 2\theta = 2 - 2 \cos^2 \theta$

23. $2 \tan x \csc 2x - \tan^2 x = 1$

24. $2 \sin^3 x - \sin x = \dfrac{\sin^2 x - \cos^2 x}{\csc x}$

Consider the function $f(x) = \cos^2 x \sin x - \sin x$.

25. Solve $f(x) = 0$ analytically over the interval $[0, 2\pi)$, and give solutions in exact form.

26. Graph $y = f(x)$ in the window $[0, 2\pi]$ by $[-4, 4]$.

27. (a) Use the results of Exercise 25 and the graph in Exercise 26 to give the exact solution set of $f(x) > 0$ over $[0, 2\pi)$.
   (b) Repeat part (a) for $f(x) < 0$.

28. Use the quadratic formula to solve the equation $2 \sin^2 x - 3 \sin x - 3 = 0$ over the interval $[0, 2\pi)$. Express solutions to the nearest hundredth. Support your answer with a graph.

Solve each equation over the interval $[0, 2\pi)$. When a solution cannot be easily expressed as a rational multiple of π, give an approximation to the nearest hundredth.

29. $\sin 2x = \cos 2x + 1$

30. $2 \sin 2x = 1$

31. $\sin 2x + \sin 4x = 0$

32. $\cos x - \cos 2x = 2 \cos x$

33. $\tan 2x = \sqrt{3}$

34. $\cos^2 \dfrac{x}{2} - 2 \cos \dfrac{x}{2} + 1 = 0$

35. Use the results of Exercise 29 and a calculator graph of the function $f(x) = \sin 2x - \cos 2x - 1$ to find the solution set of each. Give the solution set over the interval $[0, 2\pi)$. **(a)** $\sin 2x > \cos 2x + 1$ and **(b)** $\sin 2x < \cos 2x + 1$.

36. Graph the function $f(x) = \dfrac{1 - \cos 2x}{\sin 2x}$ in the trig viewing window of your calculator. It is equivalent to one of the six basic circular functions. Identify it, and complete this identity:

$$\frac{1 - \cos 2x}{\sin 2x} = \underline{\hspace{3cm}}.$$

Then verify the identity using analytic methods.

Suppose that $\sin A = -\dfrac{3}{5}$ and $\sin B = \dfrac{12}{13}$, with $\pi < A < \dfrac{3\pi}{2}$ and $0 < B < \dfrac{\pi}{2}$. Use an appropriate identity to find the exact value of each of the following.

37. $\sin(A + B)$

38. $\cos(A + B)$

39. $\tan(A + B)$

40. $\sin(A - B)$

41. $\cos(A - B)$

42. $\tan(A - B)$

43. $\sin 2A$

44. $\cos 2B$

45. $\tan 2A$

46. $\sin \dfrac{A}{2}$

47. $\cos \dfrac{A}{2}$

48. $\tan \dfrac{A}{2}$

Give the exact value of each of the following.

49. $\sin^{-1} \dfrac{\sqrt{2}}{2}$

50. $\arccos\left(-\dfrac{1}{2}\right)$

51. $\tan^{-1}(-\sqrt{3})$

52. $\arcsin(-1)$

53. $\cos^{-1}\left(-\dfrac{\sqrt{2}}{2}\right)$

54. $\arctan \dfrac{\sqrt{3}}{3}$

55. $\sec^{-1}(-2)$

56. $\text{arccsc} \dfrac{2\sqrt{3}}{3}$

57. $\cot^{-1}(-1)$

58. Explain why $\sin^{-1}(-3)$ is not defined.

Find each of the following without the use of a calculator.

59. $\sin\left(\sin^{-1} \dfrac{1}{2}\right)$

60. $\sin\left(\cos^{-1} \dfrac{3}{4}\right)$

61. $\cos(\arctan 3)$

62. $\sec\left(2 \sin^{-1}\left(-\dfrac{1}{3}\right)\right)$

63. $\cos^{-1}\left(\cos \dfrac{3\pi}{2}\right)$

64. $\tan\left(\sin^{-1} \dfrac{3}{5} + \cos^{-1} \dfrac{5}{7}\right)$

Write each of the following as an algebraic expression in u, $u > 0$.

65. $\sin(\tan^{-1} u)$

66. $\tan\left(\arccos \dfrac{u}{\sqrt{u^2 + 1}}\right)$

67. The average monthly temperature (in °F) in Chicago, Illinois, is shown in the table.*

Month	Temperature
Jan	25
Feb	28
Mar	36
Apr	48
May	61
June	72
July	74
Aug	75
Sept	66
Oct	55
Nov	39
Dec	28

(a) Plot the average monthly temperature over a two-year period by letting $x = 1$ correspond to January of the first year.

(b) Determine a model function of the form $f(x) = a \sin b(x - d) + c$ where $a, b, c,$ and d are constants.

(c) Graph f together with the data on the same coordinate axes.

68. The seasonal variation in the length of daylight can be modeled by a sine function. For example, the number of hours of daylight in New Orleans x days after March 21 (disregarding leap year) is approximated by the model

$$y = \frac{35}{3} + \frac{7}{3} \sin \frac{2\pi x}{365}.^{\dagger}$$

Two views of the graph of this function are shown here.

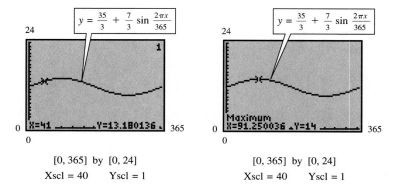

[0, 365] by [0, 24]   [0, 365] by [0, 24]
Xscl = 40   Yscl = 1   Xscl = 40   Yscl = 1

---

*Miller, A. and J. Thompson, *Elements of Meteorology* (Columbus: Charles E. Merrill Publishing Company, 1975).

†Adapted from *A Sourcebook of Applications of School Mathematics* by Donald Bushaw et al. Copyright © 1980 by The Mathematical Association of America. Reprinted by permission.

(a) Interpret the display shown at the bottom of the screen on the left.

(b) Interpret the display shown at the bottom of the screen on the right.

69. In the study of alternating electric current, instantaneous voltage is given by

$$e = E_{max} \sin 2\pi ft,$$

where f is the number of cycles per second, E_{max} is the maximum voltage, and t is time in seconds.

(a) Solve the equation for t.

(b) Find the smallest positive value of t if $E_{max} = 12$, $e = 5$, and $f = 100$.

70. One model for seasonal growth is $f(t) = 1000e^{2 \sin t}$, where t is time in months.* For what months in $[0, 12]$ is the growth equal to 4000 units? 0 units?

71. The amount of pollution in the air fluctuates with the seasons. It is lower after heavy spring rains and higher after periods of little rain. In addition to this seasonal fluctuation, the long-term trend is upward. An idealized graph of this situation is shown in the figure. Circular functions can be used to describe the fluctuating part of the pollution levels. Powers of e can be used to show the long-term growth. In fact, the pollution level in a certain area might be given by

$$P(t) = 7(1 - \cos 2\pi t)(t + 10) + 100e^{.2t},$$

where t is time in years, with $t = 0$ representing January 1 of the base year. Thus, July 1 of the same year would be represented by $t = .5$, and October 1 of the following year would be represented by $t = 1.75$. Find the pollution levels on the following dates:

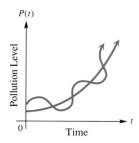

(a) January 1, base year   (b) July 1, base year   (c) January 1, following year

(d) July 1, following year.

72. The figure shows the population of lynx and hares in Canada for the years 1847–1903. The hares are food for the lynx. An increase in hare population causes an increase in lynx population some time later. The increasing lynx population then causes a decline in hare population.

(a) Estimate the length of one period.

(b) Estimate maximum and minimum hare populations.

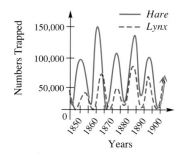

---

* The irrational number $e \approx 2.718281828$ is the base of the natural logarithm function.

73. Suppose that the equation below describes the motion formed by a rhythmically moving arm:

$$y = \frac{1}{3} \sin \frac{4\pi t}{3}.$$

Here t is time (in seconds) and y is the angle formed.
(a) Solve the equation for t.
(b) At what time(s) does the arm form an angle of .3 radian?

74. Sales of snowblowers are seasonal. Suppose the sales of snowblowers in one region of Canada are approximated by

$$S(t) = 500 + 500 \cos\left[\left(\frac{\pi}{6}\right)t\right],$$

where t is time in months, with $t = 0$ corresponding to November. For what months are sales equal to 0?

75. (Refer to Exercise 15 in Section 3.6.) Suppose that for an electric heater the voltage is given by $V = a \sin(2\pi\omega t)$ and the amperage by $I = b \sin(2\pi\omega t)$ where t is time in seconds. Find the period for both V and I.

CHAPTER 4

riangles and
Applications

4.1 RIGHT TRIANGLE TRIGONOMETRY

Introduction to Right Triangle Trigonometry ▮ Significant Digits and
Accuracy ▮ Solving Right Triangles ▮ Applications

Introduction to Right Triangle Trigonometry

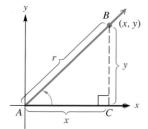

FIGURE 1

Figure 1 shows an acute angle A in standard position. The definitions of the
trigonometric function values of angle A require x, y, and r. As drawn in Figure 1,
x and y are the lengths of the two legs of right triangle ABC, and r is the length of
the **hypotenuse,** the side opposite the right angle.

The side of length y is called the **side opposite** angle A, and the side of length x
is called the **side adjacent** to angle A. The lengths of these sides can be used to
replace x and y in the definition of the trigonometric functions, with r replaced by
the length of the hypotenuse, to get the following right triangle-based definitions.

**RIGHT TRIANGLE-BASED DEFINITIONS OF
TRIGONOMETRIC FUNCTIONS**

For any acute angle A in standard position,

$$\sin A = \frac{y}{r} = \frac{\text{side opposite}}{\text{hypotenuse}} \qquad \csc A = \frac{r}{y} = \frac{\text{hypotenuse}}{\text{side opposite}}$$

$$\cos A = \frac{x}{r} = \frac{\text{side adjacent}}{\text{hypotenuse}} \qquad \sec A = \frac{r}{x} = \frac{\text{hypotenuse}}{\text{side adjacent}}$$

$$\tan A = \frac{y}{x} = \frac{\text{side opposite}}{\text{side adjacent}} \qquad \cot A = \frac{x}{y} = \frac{\text{side adjacent}}{\text{side opposite}}.$$

247

As seen in the first example, a coordinate system is not essential in applying these definitions.

EXAMPLE 1

Finding
Trigonometric
Function Values of
an Acute Angle in a
Right Triangle

Find the values of the trigonometric functions for angles A and B in the right triangle in Figure 2.

SOLUTION   The length of the side opposite angle A is 7. The length of the side adjacent to angle A is 24, and the length of the hypotenuse is 25. Using the relationships given above,

$$\sin A = \frac{\text{side opposite}}{\text{hypotenuse}} = \frac{7}{25} \qquad \csc A = \frac{\text{hypotenuse}}{\text{side opposite}} = \frac{25}{7}$$

$$\cos A = \frac{\text{side adjacent}}{\text{hypotenuse}} = \frac{24}{25} \qquad \sec A = \frac{\text{hypotenuse}}{\text{side adjacent}} = \frac{25}{24}$$

$$\tan A = \frac{\text{side opposite}}{\text{side adjacent}} = \frac{7}{24} \qquad \cot A = \frac{\text{side adjacent}}{\text{side opposite}} = \frac{24}{7}.$$

FIGURE 2

The length of the side opposite angle B is 24, while the length of the side adjacent to B is 7, making

$$\sin B = \frac{24}{25} \qquad \tan B = \frac{24}{7} \qquad \sec B = \frac{25}{7}$$

$$\cos B = \frac{7}{25} \qquad \cot B = \frac{7}{24} \qquad \csc B = \frac{25}{24}. \qquad ∎$$

Because the two acute angles in a right triangle are complementary, the cofunction identities introduced earlier will often be useful when studying right triangle trigonometry. We repeat them in the following box.

COFUNCTION IDENTITIES

If A is an acute angle measured in degrees,

$$\sin A = \cos(90° - A) \qquad \csc A = \sec(90° - A)$$

$$\cos A = \sin(90° - A) \qquad \sec A = \csc(90° - A)$$

$$\tan A = \cot(90° - A) \qquad \cot A = \tan(90° - A).$$

Notice in Example 1 that we have

$$\sin A = \cos B$$
$$\cos A = \sin B$$
$$\tan A = \cot B$$
$$\cot A = \tan B$$
$$\sec A = \csc B$$

and
$$\csc A = \sec B,$$

illustrating these cofunction relationships, since A and B are complements.

Significant Digits and Accuracy

In our discussion of applications of trigonometry in the rest of this chapter, we will be performing many calculations with trigonometric function values. Before doing so we will present a brief explanation of accuracy with such numbers.

Suppose that a wall is measured to the nearest foot and is found to be 18 feet long. Actually this means that the wall has a length between 17.5 feet and 18.5 feet. If the wall is measured more accurately and found to be 18.3 feet long, then its length is really between 18.25 feet and 18.35 feet. A measurement of 18.00 feet would indicate that the length of the wall is between 17.995 feet and 18.005 feet. The measurement 18 feet is said to have two **significant digits** of accuracy; 18.0 has three significant digits, and 18.00 has four.

A significant digit is a digit obtained by actual measurement. A number that represents the result of counting, or a number that results from theoretical work and is not the result of a measurement, is an **exact number.** For example, $\frac{\sqrt{2}}{2}$ is the *exact* value of cos 45°, while a calculator might give an *approximation* of cos 45° as .7071067812 to ten significant digits.

To perform calculations on such approximate numbers, follow the rules given below.

CALCULATION WITH SIGNIFICANT DIGITS

For *adding* and *subtracting*, round the answer so that the last digit you keep is in the right-most column in which all the numbers have significant digits.

For *multiplying* or *dividing*, round the answer to the least number of significant digits found in any of the given numbers.

In our work to follow, we will use the following table for deciding on significant digits in angle measures.

SIGNIFICANT DIGITS FOR ANGLES

Number of Significant Digits	Angle Measure to Nearest:
2	Degree
3	Ten minutes, or nearest tenth of a degree
4	Minute, or nearest hundredth of a degree
5	Tenth of a minute, or nearest thousandth of a degree

For example, an angle measuring 52° 30′ has three significant digits (assuming that 30′ is measured to the nearest ten minutes).

Solving Right Triangles

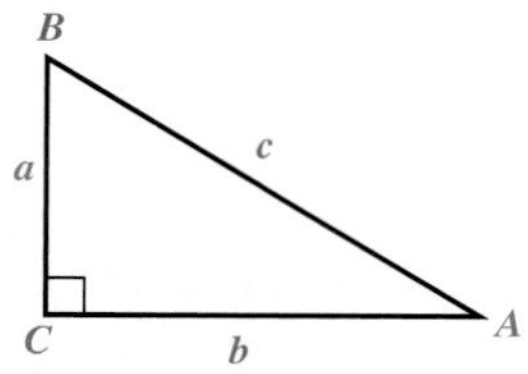

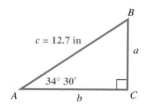

FIGURE 3

With the background that we now have, it is possible to find the measures of all angles and all sides of a right triangle if we know the measures of one acute angle and one side, or if we know the measures of two sides. Finding all measures is called **solving the right triangle.**

In using trigonometry to solve triangles, it is convenient to use a to represent the length of the side opposite angle A, b for the length of the side opposite angle B, and so on. As mentioned earlier, in a right triangle the letter c is reserved for the hypotenuse. Figure 3 shows the labeling of a typical right triangle.

EXAMPLE 2

Solving a Right Triangle Given an Angle and a Side

FIGURE 4

Solve right triangle ABC, with $A = 34°\ 30'$ and $c = 12.7$ in. (See Figure 4.)

SOLUTION   To solve the triangle, find the measures of the remaining sides and angles. The value of a can be found with a trigonometric function involving the known values of angle A and side c. Since the sine of angle A is given by the quotient of the side opposite A and the hypotenuse, use sin A.

$$\sin A = \frac{a}{c}$$

Substituting known values gives

$$\sin 34°\ 30' = \frac{a}{12.7},$$

or, upon multiplying both sides by 12.7,

$$a = 12.7 \sin 34°\ 30'$$
$$a \approx 12.7(.56640624) \qquad \text{Use a calculator.}$$
$$a \approx 7.19 \text{ in.}$$

The value of b could be found with the Pythagorean theorem. It is better, however, to use the information given in the problem rather than a result just calculated. If a mistake were to be made in finding a, then b also would be incorrect. Also, rounding more than once may cause the result to be less accurate. Using cos A gives

$$\cos A = \frac{\text{side adjacent}}{\text{hypotenuse}} = \frac{b}{c}$$
$$\cos 34°\ 30' = \frac{b}{12.7}$$
$$b = 12.7 \cos 34°\ 30'$$
$$b \approx 10.5 \text{ in.}$$

Once b has been found, the Pythagorean theorem could be used as a check. All that remains to solve triangle ABC is to find the measure of angle B. Since $A + B = 90°$ and $A = 34°\ 30'$,

$$A + B = 90°$$
$$B = 90° - A$$
$$B = 89°\ 60' - 34°\ 30'$$
$$B = 55°\ 30'. \qquad ∎$$

NOTE   In Example 2 we could have started by finding the measure of angle B and then used the trigonometric function values of B to find the unknown sides. The process of solving a right triangle (like many problems in mathematics) can usually be done in several ways, each resulting in the correct answer. However, in order to retain as much accuracy as can be expected, always use given information as much as possible, and avoid rounding off in intermediate steps.

EXAMPLE **3**

Solving a Right
Triangle Given Two
Sides

TECHNOLOGICAL NOTE
Once you have mastered
the material on solving
right triangles, you may
wish to write a program
that will accomplish this
goal. You will need to
consider the various
cases of what is given
and what must be found.

Solve right triangle ABC if $a = 29.43$ cm and $c = 53.58$ cm.

SOLUTION   Draw a sketch showing the given information, as in Figure 5. One way to begin is to find angle A by using the sine.

$$\sin A = \frac{\text{side opposite}}{\text{hypotenuse}}$$

$$\sin A = \frac{29.43}{53.58}$$

FIGURE 5

Using the inverse sine function on a calculator, we find that $A \approx 33.32°$. The measure of B is $90° - 33.32° \approx 56.68°$.

We now find b from the Pythagorean theorem, using $a^2 + b^2 = c^2$, or $b^2 = c^2 - a^2$. Since $c = 53.58$ and $a = 29.43$,

$$b^2 = 53.58^2 - 29.43^2$$

giving $\qquad b \approx 44.77$ cm.   ▮

Applications

The process of solving right triangles is easily adapted to solving applied problems. A crucial step in such applications involves sketching the triangle and labeling the given parts correctly. Then we can use the methods described in the earlier examples to find the unknown value or values.

Many applications of right triangles involve the angle of elevation or the angle of depression. The **angle of elevation** from point X to point Y (above X) is the angle made by ray XY and a horizontal ray with endpoint at X. The angle of elevation is always measured from the horizontal. See Figure 6(a). The **angle of depression** from point X to point Y (below X) is the angle made by ray XY and a horizontal ray with endpoint X. See Figure 6(b).

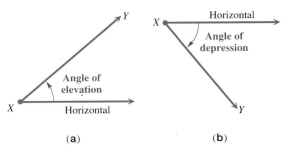

(a)   (b)

FIGURE 6

CAUTION Errors are often made in interpreting the angle of depression. Remember that both the angle of elevation *and* the angle of depression are measured *from* the horizontal *to* the line of sight.

EXAMPLE 4

Finding a Length When the Angle of Elevation Is Known

Donna Garbarino knows that when she stands 123 feet from the base of a flagpole, the angle of elevation to the top is 26° 40′. If her eyes are 5.30 feet above the ground, find the height of the flagpole.

SOLUTION The length of the side adjacent to Donna is known and the length of the side opposite her is to be found. See Figure 7. The ratio that involves these two values is the tangent.

$$\tan A = \frac{\text{side opposite}}{\text{side adjacent}}$$

$$\tan 26° 40′ = \frac{a}{123}$$

$$a = 123 \tan 26° 40′$$

$$a \approx 61.8 \text{ feet}$$

Since Donna's eyes are 5.30 feet above the ground, the height of the flagpole is approximately

$$61.8 + 5.30 \approx 67.1 \text{ feet.} \quad \blacksquare$$

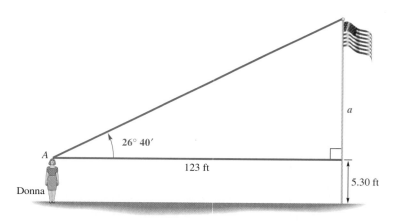

26° 40′

A

123 ft

5.30 ft

a

Donna

FIGURE 7

EXAMPLE 5

Finding the Angle of Elevation When Lengths Are Known

A building that is 34.09 meters tall casts a shadow 37.62 meters long. Find the angle of elevation of the sun.

SOLUTION As shown in Figure 8, the angle of elevation of the sun is angle *B*. Since the side opposite *B* and the side adjacent to *B* are known, use the tangent ratio to find *B*.

$$\tan B = \frac{34.09}{37.62}$$

$$B \approx 42.18° \qquad \text{Use the inverse tangent function.}$$

The angle of elevation of the sun is approximately 42.18°. $\quad \blacksquare$

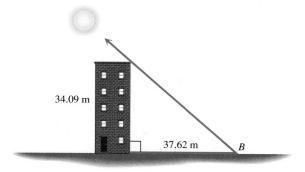

FIGURE 8

Some applications of right triangles involve **bearing,** an important idea in navigation. There are two common ways to express bearing. *When a single angle is given, such as 164°, it is understood that the bearing is measured in a clockwise direction from due north.* Several sample bearings using this first type of system are shown in Figure 9.

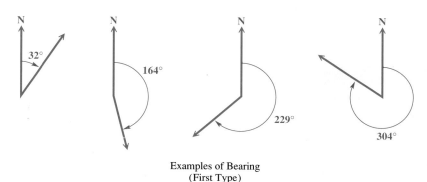

Examples of Bearing
(First Type)

FIGURE 9

EXAMPLE 6

Solving a Problem
Involving Bearing
(First Type)

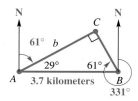

FIGURE 10

Radar stations A and B are on an east-west line, 3.7 kilometers apart. Station A detects a plane at C, on a bearing of 61°. Station B simultaneously detects the same plane, on a bearing of 331°. Find the distance from A to C.

SOLUTION   Draw a sketch showing the given information, as in Figure 10. Since a line drawn due north is perpendicular to an east-west line, right angles are formed at A and B, so that angles CAB and CBA can be found. Angle C is a right angle because angles CAB and CBA are complementary. (If C were not a right angle, the methods of later sections would be needed.) Find distance b by using the cosine function.

$$\cos 29° = \frac{b}{3.7}$$

$$3.7 \cos 29° = b$$

$$b \approx 3.2 \text{ kilometers}$$

Use a calculator and round to the nearest tenth.   ∎

CAUTION   It would be foolish to attempt to solve the problem in Example 6 without drawing a sketch. The importance of a correctly labeled sketch in applications such as this cannot be overemphasized, as some of the necessary information is often not given directly in the statement of the problem, and can only be determined from the sketch.

The second common system for expressing bearing starts with a north-south line and uses an acute angle to show the direction, either east or west, from this line. Figure 11 shows several sample bearings using this system. Either N or S always comes first, followed by an acute angle, and then E or W.

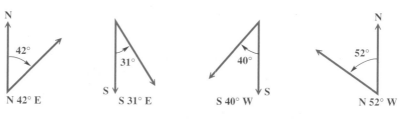

Examples of Bearing
(Second Type)

FIGURE 11

EXAMPLE 7

Solving a Problem
Involving Bearing
(Second Type)

The bearing from A to C is S 52° E. The bearing from A to B is N 84° E. The bearing from B to C is S 38° W. A plane flying at 250 miles per hour takes 2.4 hours to go from A to B. Find the distance from A to C.

SOLUTION   Make a sketch of the situation. First draw the two bearings from point A. Choose a point B on the bearing N 84° E from A and draw the bearing to C. Point C will be located where the bearing lines from A and B intersect as shown in Figure 12.

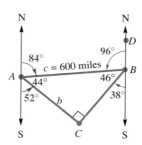

FIGURE 12

Since the bearing from A to B is N 84° E, angle ABD is 180° − 84° = 96°. Thus, angle ABC is 46°. Also, angle BAC is 180° − (84° + 52°) = 44°. Angle C is 180° − (44° + 46°) = 90°. From the statement of the problem, a plane flying at 250 miles per hour takes 2.4 hours to go from A to B. The distance from A to B is the product of rate and time, or

$$c = \text{rate} \times \text{time} = 250(2.4) = 600 \text{ miles.}$$

To find b, the distance from A to C, use the sine. (The cosine could also have been used.)

$$\sin 46° = \frac{b}{c}$$

$$\sin 46° = \frac{b}{600}$$

$$600 \sin 46° = b$$

$$b \approx 430 \text{ miles} \qquad \text{Use a calculator.} \quad \blacksquare$$

EXAMPLE 8

Solving a Problem Involving Angle of Elevation

Francisco needs to know the height of a tree. From a given point on the ground he finds that the angle of elevation to the top of the tree is $36° \, 40'$. He then moves back 50 feet. From the second point, the angle of elevation to the top of the tree is $22° \, 10'$. See Figure 13. Find the height of the tree.

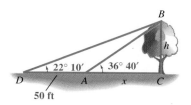

FIGURE 13

SOLUTION The figure shows two unknowns; x, the distance from the center of the trunk of the tree to the point where the first observation was made, and h, the height of the tree. Since nothing is given about the length of the hypotenuse of either triangle ABC or triangle BCD, use a ratio that does not involve the hypotenuse—the tangent.

In triangle ABC, $\tan 36° \, 40' = \dfrac{h}{x}$ or $h = x \tan 36° \, 40'$.

In triangle BCD, $\tan 22° \, 10' = \dfrac{h}{50 + x}$ or $h = (50 + x) \tan 22° \, 10'$.

Since each of these expressions equals h, these expressions must be equal. Thus,

$$x \tan 36° \, 40' = (50 + x) \tan 22° \, 10'.$$

Now use algebra to solve for x.

$$x \tan 36° \, 40' = 50 \tan 22° \, 10' + x \tan 22° \, 10' \qquad \text{Distributive property}$$

$$x \tan 36° \, 40' - x \tan 22° \, 10' = 50 \tan 22° \, 10' \qquad \text{Get } x \text{ terms on one side.}$$

$$x(\tan 36° \, 40' - \tan 22° \, 10') = 50 \tan 22° \, 10' \qquad \text{Factor out } x \text{ on the left.}$$

$$x = \frac{50 \tan 22° \, 10'}{\tan 36° \, 40' - \tan 22° \, 10'} \qquad \text{Divide by the coefficient of } x.$$

We saw that $h = x \tan 36° 40'$. Substituting for x,

$$h = \left(\frac{50 \tan 22° 10'}{\tan 36° 40' - \tan 22° 10'} \right)(\tan 36° 40').$$

From a calculator,

$$\tan 36° 40' \approx .74447242$$
$$\tan 22° 10' \approx .40741394,$$

so

$$\tan 36° 40' - \tan 22° 10' \approx .74447242 - .40741394 \approx .33705848,$$

and

$$h \approx \left(\frac{50(.40741394)}{.33705848} \right)(.74447242) \approx 45 \text{ (rounded)}.$$

The height of the tree is approximately 45 feet.   ∎

NOTE   In practice we usually do not write down the intermediate calculator approximation steps. However, we have done this in Example 8 so that the reader may follow the steps more easily.

FOR GROUP DISCUSSION

An alternate approach to solving the problem in Example 8 uses the intersection-of-graphs capability of the graphing calculator. This approach is based on a similar solution proposed by a student, John Cree, as explained in a letter to the editor in the January 1995 issue of *Mathematics Teacher* from Cree's teacher, Robert Ruzich.*

1. Since the tangent of the angle formed by the graph of $y = mx + b$ with the x-axis is the slope of the line (m), the segment BD lies along the graph of $y_1 = (\tan 22° 10')x$. Enter this equation into your calculator.
2. By similar reasoning and using the fact that A lies 50 feet to the *right* of D, enter the equation $y_2 = (\tan 36° 40')(x - 50)$ into your calculator to represent AB.
3. Using a window of $[0, 200]$ by $[0, 100]$, find the point of intersection of the graphs of y_1 and y_2, using the intersection-of-graphs capability of your calculator.
4. Compare the y-coordinate of the point of intersection to the answer found analytically in Example 8. It should be the same.

Figure 14 illustrates the screen you should obtain when following the procedure described in the previous "For Group Discussion" activity.

---

*Adapted with permission from "Letter to the Editor," by Robert Ruzich (*Mathematics Teacher*, Volume 88, Number 1). Copyright © 1995 by the National Council of Teachers of Mathematics.

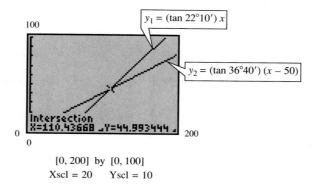

$$y_1 = (\tan 22°10') x$$

$$y_2 = (\tan 36°40') (x - 50)$$

Intersection
X=110.43668 Y=44.993444

[0, 200] by [0, 100]

Xscl = 20   Yscl = 10

FIGURE 14

4.1 EXERCISES

To the Student: Calculator Considerations

Be sure that you know how to put your calculator in degree or radian mode. Also, be aware of the difference between inverse trigonometric functions (such as $\sin^{-1} x$) and reciprocals of trigonometric functions (such as $\csc \theta$, which is equal to $\frac{1}{\sin \theta}$).

You should already be familiar with all of the necessary keystrokes needed to work the exercises in this section. If your answers seem to be "a bit off from the ones given in the back of the book," remember that we give the answers with accuracy based on the guidelines presented at the beginning of this section. In some cases, it may be necessary to convert between decimal degrees and degrees, minutes, and seconds as well.

Find the exact values of the six trigonometric functions of angle A.

1.

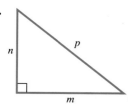

2.

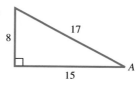

3.

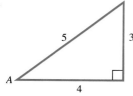

4.

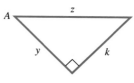

Solve each right triangle.

5.

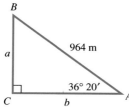

6.

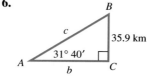

7.

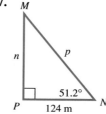

8.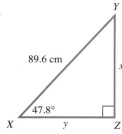

9.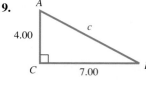

10.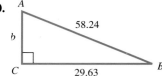

Solve the right triangle ABC, with C = 90°.

11. $A = 28° 00'$, $c = 17.4$ ft

12. $B = 46° 00'$, $c = 29.7$ m

13. $B = 73° 00'$, $b = 128$ in

14. $A = 61° 00'$, $b = 39.2$ cm

15. $a = 76.4$ yd, $b = 39.3$ yd

16. $a = 18.9$ cm, $c = 46.3$ cm

17. Can a right triangle be solved if we are given the measures of its two acute angles and no side lengths? Explain.

18. If we are given an acute angle and a side in a right triangle, what unknown part of the triangle requires the least work to find?

Consider the figure here in Exercises 19 and 20.

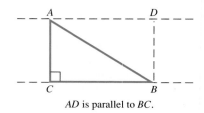

AD is parallel to BC.

19. Explain why the angle of depression *DAB* has the same measure as the angle of elevation *ABC*.

20. Why is angle *CAB not* an angle of depression?

Solve each problem.

21. A 13.5-m fire truck ladder is leaning against a wall. Find the distance the ladder goes up the wall if it makes an angle of $43° 50'$ with the ground.

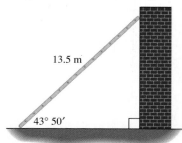

22. To measure the height of a flagpole, Kitty Pellissier finds that the angle of elevation from a point 24.73 ft from the base to the top is $38° 12'$. Find the height of the flagpole.

23. A guy wire 80.1 m long is attached to the top of an antenna mast that is 71.3 m high. Find the angle that the wire makes with the ground.

24. Find the length of a guy wire that makes an angle of $45° 30'$ with the ground if the wire is attached to the top of a tower 63.0 m high.

25. To find the distance *RS* across a lake, a surveyor lays off $RT = 53.1$ m, with angle $T = 32° 10'$, and angle $S = 57° 50'$. Find length *RS*.

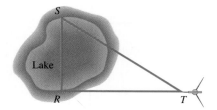

26. A surveyor must find the distance *QM* across a depressed freeway. She lays off $QN = 769$ ft along one side of the freeway, with angle $N = 21° 50'$, and with angle $M = 68° 10'$. Find *QM*.

27. The length of the base of an isosceles triangle is 42.36 in. Each base angle is 38.12°. Find the length of each of the two equal sides of the triangle. (*Hint:* Divide the triangle into two right triangles.)

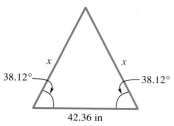

28. Find the altitude of an isosceles triangle having a base of 184.2 cm if the angle opposite the base is 68° 44′.

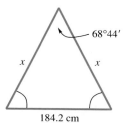

29. Suppose that the angle of elevation of the sun is 23.4°. Find the length of the shadow cast by Cindy Newman, who is 5.75 ft tall.

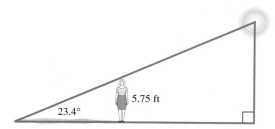

30. The shadow of a vertical tower is 40.6 m long when the angle of elevation of the sun is 34.6°. Find the height of the tower.

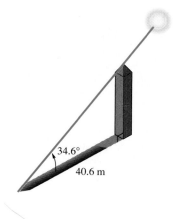

31. Find the angle of elevation of the sun if a 48.6-ft flagpole casts a shadow 63.1 ft long.

32. The angle of depression from the top of a building to a point on the ground is 32° 30′. How far is the point on the ground from the top of the building if the building is 252 m high?

33. An airplane is flying 10,500 feet above the level ground. The angle of depression from the plane to the base of a tree is 13° 50′. How far horizontally must the plane fly to be directly over the tree?

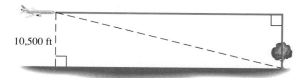

34. The angle of elevation from the top of a small building to the top of a nearby taller building is 46° 40′, while the angle of depression to the bottom is 14° 10′. If the smaller building is 28.0 m high, find the height of the taller building.

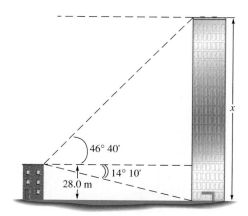

35. A television camera is to be mounted on a bank wall so as to have a good view of the head teller. (See the figure.) Find the angle of depression that the lens should make with the horizontal.

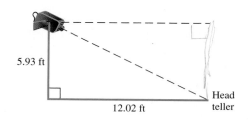

36. A company safety committee has recommended that a floodlight be mounted in a parking lot so as to illuminate the employee exit. (See the figure.) Find the angle of depression of the light.

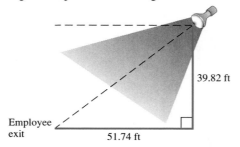

39.82 ft

Employee
exit

51.74 ft

37. A tunnel is to be dug from *A* to *B*. (See the figure.) Both *A* and *B* are visible from *C*. If *AC* is 1.4923 mi and *BC* is 1.0837 mi, and if *C* is 90°, find the measures of angles *A* and *B*.

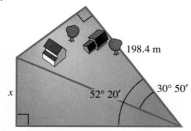

38. A piece of land has the shape shown in the figure. Find *x*.

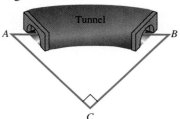

198.4 m

x

52° 20′

30° 50′

39. Find the value of *x* in the figure.

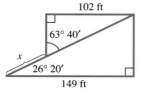

102 ft

63° 40′

x

26° 20′

149 ft

40. The leaning tower of Pisa is approximately 179 ft in height and is approximately 16.5 ft out of plumb (that is, tilted away from the vertical). Find the angle at which it deviates from the vertical.*

41. A regular pentagon (five-sided polygon with equal lengths of sides and equal angles) is inscribed in a circle of radius 7 cm. Find the length of a side of the pentagon. Give your answer to the nearest thousandth.*

42. Find *h* as indicated in the figure.

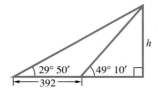

29° 50′   49° 10′   *h*

392

43. Find *h* as indicated in the figure.

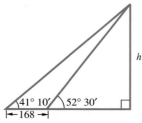

h

41° 10′   52° 30′

168

44. The angle of elevation from a point on the ground to the top of a pyramid is 35° 30′. The angle of elevation from a point 135 ft farther back to the top of the pyramid is 21° 10′. Find the height of the pyramid.

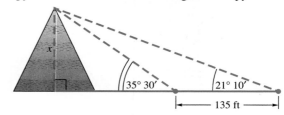

x

35° 30′   21° 10′

135 ft

45. Debbie Glockner, a whale researcher, is watching a whale approach directly toward her as she observes from the top of a lighthouse. When she first begins watching the whale, the angle of depression of the whale is 15° 50′. Just as the whale turns away from the lighthouse, the angle of depression is 35° 40′. If the height of the lighthouse is 68.7 m, find the distance traveled by the whale as it approaches the lighthouse.

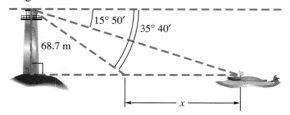

15° 50′   35° 40′

68.7 m

x

*Exercises 40 and 41 are excerpts from *Plane Trigonometry*. Revised Edition by Frank A. Rickey and J. P. Cole, copyright © 1964 by Holt, Rinehart and Winston, Inc., and renewed 1992 by Coleen C. Salley, T. E. Cole, Robert E. Cole, James P. Cole, Jr., Frank A. Rickey, Jr., Mary Ellen Rickey, Mrs. Mary E. Rickey, and W. P. Rickey, reprinted by permission of the publisher.

46. A scanner antenna is on top of the center of a house. The angle of elevation from a point 28.0 m from the center of the house to the top of the antenna is 27° 10′, and the angle of elevation to the bottom of the antenna is 18° 10′. Find the antenna height.

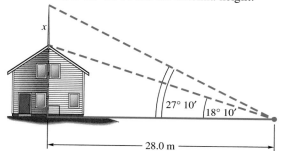

47. The angle of elevation from Lone Pine to the top of Mt. Whitney is 10° 50′. Van Dong Le, traveling 7.00 km from Lone Pine along a straight, level road toward Mt. Whitney, finds the angle of elevation to be 22° 40′. Find the height of the top of Mt. Whitney above the level of the road.

48. A plane flies 1.3 hr at 110 mph on a bearing of 40°. It then turns and flies 1.5 hr at the same speed on a bearing of 130°. How far is the plane from its starting point?

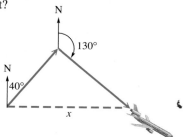

49. A ship travels 50 km on a bearing of 27°, and then travels on a bearing of 117° for 140 km. Find the distance between the starting and ending points.

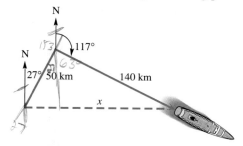

50. Two ships leave a port at the same time. The first ship sails on a bearing of 40° at 18 knots (nautical

miles per hour) and the second at a bearing of 130° at 26 knots. How far apart are they after 1.5 hr?

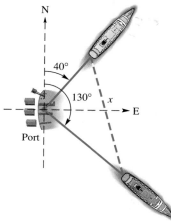

51. Two lighthouses are located on a north-south line. From lighthouse *A* the bearing of a ship 3742 m away is 129° 43′. From lighthouse *B* the bearing of the ship is 39° 43′. Find the distance between the lighthouses.

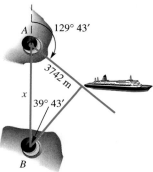

52. A ship leaves its home port and sails on a bearing of N 28° 10′ E. Another ship leaves the same port at the same time and sails on a bearing of S 61° 50′ E. If the first ship sails at 24.0 mph and the second sails at 28.0 mph, find the distance between the two ships after 4 hr.

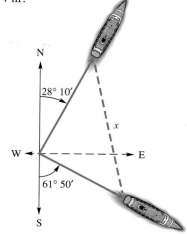

53. Radio direction finders are set up at points A and B, which are 2.50 mi apart on an east-west line. From A it is found that the bearing of the signal from a radio transmitter is N 36° 20′ E, while from B the bearing of the same signal is N 53° 40′ W. Find the distance of the transmitter from B.

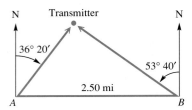

54. The figure shows a magnified view of the threads of a bolt. Find x if d is 2.894 mm.

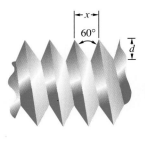

4.2 THE LAW OF SINES

Introduction ∎ Congruency and Oblique Triangles ∎ Derivation of the Law of Sines ∎ Applications ∎ The Ambiguous Case

Introduction

Until now, our applied work with trigonometry has been limited to right triangles. However, the concepts developed earlier in this chapter can be extended so that our work can apply to *all* triangles. Every triangle has three sides and three angles, and we will show that if any three of the six measures of a triangle are known (provided that at least one is that of a side), then the other three measures can be found. Again, this is called *solving the triangle*.

Congruency and Oblique Triangles

Recall from geometry the following axioms that allow us to prove that two triangles are congruent (that is, their corresponding sides and angles are equal).

CONGRUENCE AXIOMS

Side-Angle-Side (SAS)
If two sides and the included angle of one triangle are equal, respectively, to two sides and the included angle of a second triangle, then the triangles are congruent.

Angle-Side-Angle (ASA)
If two angles and the included side of one triangle are equal, respectively, to two angles and the included side of a second triangle, then the triangles are congruent.

Side-Side-Side (SSS)
If three sides of one triangle are equal, respectively, to three sides of a second triangle, then the triangles are congruent.

Throughout this section and the next, keep in mind that whenever any of the groups of data described above are given, the triangle is uniquely determined; that is, all other data in the triangle are given by one and only one set of measures.

A triangle that is not a right triangle is called an **oblique triangle.** The measures of the three sides and the three angles of a triangle can be found if at least one side and any other two measures are known. There are four possible cases.

DATA REQUIRED FOR SOLVING OBLIQUE TRIANGLES

1. One side and two angles are known.
2. Two sides and one angle not included between the two sides are known. This case may lead to more than one triangle.
3. Two sides and the angle included between the two sides are known.
4. Three sides are known.

NOTE   If we know three angles of a triangle, we cannot find unique side lengths, since AAA assures us only of similarity, not congruence. For example, there are infinitely many triangles ABC with $A = 35°$, $B = 65°$, and $C = 80°$.

The first two cases require the use of the *law of sines*, which is introduced in this section. The last two cases require the use of the *law of cosines*, introduced in the next section.

Derivation of the Law of Sines

To derive the law of sines, start with an oblique triangle as in Figure 15 (on p. 264). (The discussion to follow applies to either of the triangles in Figures 15(a) or 15(b).) First, construct the perpendicular line from B to side AC or its extension. Let h be the length of this perpendicular line. Then c is the hypotenuse of right triangle ADB, and a is the hypotenuse of right triangle BDC. By results from the previous section,

$$\text{in triangle } ADB, \quad \sin A = \frac{h}{c} \quad \text{or} \quad h = c \sin A,$$

$$\text{in triangle } BDC, \quad \sin C = \frac{h}{a} \quad \text{or} \quad h = a \sin C.$$

Since $h = c \sin A$ and $h = a \sin C$,

$$a \sin C = c \sin A,$$

or, upon dividing both sides by $\sin A \sin C$,

$$\frac{a}{\sin A} = \frac{c}{\sin C}.$$

In a similar way, by constructing the perpendicular lines from other vertices, it can be shown that

$$\frac{a}{\sin A} = \frac{b}{\sin B} \quad \text{and} \quad \frac{b}{\sin B} = \frac{c}{\sin C}.$$

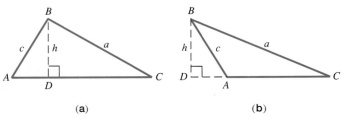

(a)   **(b)**

FIGURE 15

This discussion proves the following theorem.

LAW OF SINES

In any triangle ABC, with sides a, b, and c,

$$\frac{a}{\sin A} = \frac{b}{\sin B}, \quad \frac{a}{\sin A} = \frac{c}{\sin C}, \quad \text{and} \quad \frac{b}{\sin B} = \frac{c}{\sin C}.$$

This can be written in compact form as

$$\frac{a}{\sin A} = \frac{b}{\sin B} = \frac{c}{\sin C}.$$

Sometimes an alternate form of the law of sines,

$$\frac{\sin A}{a} = \frac{\sin B}{b} = \frac{\sin C}{c},$$

is convenient to use.

Applications

If two angles and the side opposite one of the angles are known, the law of sines can be used directly to solve for the side opposite the other known angle. The triangle can then be solved completely, as shown in the first example.

EXAMPLE 1

Using the Law of Sines to Solve a Triangle

Solve triangle ABC if $A = 32.0°$, $B = 81.8°$, and $a = 42.9$ centimeters. (See Figure 16.)

SOLUTION   Start by drawing a triangle, roughly to scale, and labeling the given parts as in Figure 16. Since the values of A, B, and a are known, use the part of the law of sines that involves these variables.

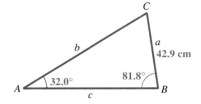

FIGURE 16

$$\frac{a}{\sin A} = \frac{b}{\sin B}$$

Substituting the known values gives

$$\frac{42.9}{\sin 32.0°} = \frac{b}{\sin 81.8°}.$$

Multiply both sides of the equation by sin 81.8°.

$$b = \frac{42.9 \sin 81.8°}{\sin 32.0°}$$

$b \approx$ **80.1** centimeters.   Use a calculator.

Find C from the fact that the sum of the angles of any triangle is 180°.

$$A + B + C = 180°$$
$$C = 180° - A - B$$
$$C = 180° - 32.0° - 81.8° = 66.2°$$

Now use the law of sines again to find c. (Why does the Pythagorean theorem not apply?)

$$\frac{a}{\sin A} = \frac{c}{\sin C}$$

$$\frac{42.9}{\sin 32.0°} = \frac{c}{\sin 66.2°}$$

$$c = \frac{42.9 \sin 66.2°}{\sin 32.0°}$$

$$c \approx 74.1 \text{ centimeters} \quad \blacksquare$$

CAUTION In applications of oblique triangles, such as the one that follows in Example 2, a properly labeled sketch is essential in order to set up the correct equation.

EXAMPLE 2
Using the Law of Sines in an Application

Tri Nguyen wishes to measure the distance across the Big Muddy River. (See Figure 17.) He finds that $C = 112° \, 53'$, $A = 31° \, 06'$, and $b = 347.6$ feet. Find the required distance.

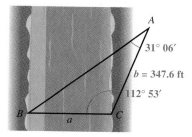

$$A$$
$$31° \, 06'$$
$$b = 347.6 \text{ ft}$$
$$112° \, 53'$$
$$B \quad a \quad C$$

FIGURE 17

SOLUTION To use the law of sines, one side and the angle opposite it must be known. Since the only side whose length is given is b, angle B must be found before the law of sines can be used.

$$B = 180° - A - C = 180° - 31° \, 06' - 112° \, 53' = 36° \, 01'$$

Now the required distance a can be found. Use the form of the law of sines involving A, B, and b.

$$\frac{a}{\sin A} = \frac{b}{\sin B}$$

Substitute the known values.

$$\frac{a}{\sin 31° \, 06'} = \frac{347.6}{\sin 36° \, 01'}$$

$$a = \frac{347.6 \sin 31° \, 06'}{\sin 36° \, 01'}$$

$$a \approx \textbf{305.3} \text{ feet} \quad\quad\quad \text{Use a calculator.} \quad \blacksquare$$

The next example involves the use of bearing, first dicussed in Section 4.1.

EXAMPLE **3**

Using the Law of
Sines in an
Application

Two tracking stations are on an east-west line 110 miles apart. A forest fire is located on a bearing of N 42° E from the western station at *A* and also on a bearing of N 15° E from the eastern station at *B*. How far is the fire from the western station? (See Figure 18.)

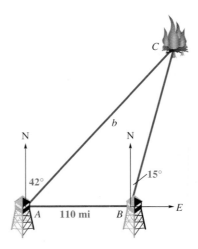

FIGURE 18

SOLUTION   Figure 18 shows the two stations at points *A* and *B* and the fire at point *C*. Angle *BAC* = 90° − 42° = 48°, the obtuse angle at *B* equals 90° + 15° = 105°, and the third angle, *C*, equals 180° − 105° − 48° = 27°. Using the law of sines to find side *b* gives

$$\frac{b}{\sin 105°} = \frac{110}{\sin 27°}$$

$$b \approx 234,$$

or approximately 230 miles (to two significant digits).   ◼

The Ambiguous Case

The law of sines can be used when given two angles and the side opposite one of these angles. Also, if two angles and the included side are known, then the third angle can be found by using the fact that the sum of the angles of a triangle is 180°, and then applying the law of sines. However, if we are given the lengths of two sides and the angle opposite one of them, it is possible that 0, 1, or 2 such triangles exist. (Recall that there is no "SSA" congruence theorem.)

To illustrate these facts, suppose that the measure of acute angle *A* of triangle *ABC*, the length of side *a*, and the length of side *b* are given. Draw angle *A* having a terminal side of length *b*. Now draw a side of length *a* opposite angle *A*. The following chart shows that there might be more than one possible outcome. This situation is called the **ambiguous case of the law of sines.**

Number of Possible Triangles	Sketch	Condition Necessary for Case to Hold
0	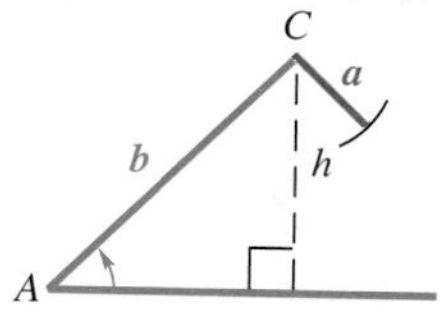	$a < h$ ($h = b \sin A$)
1	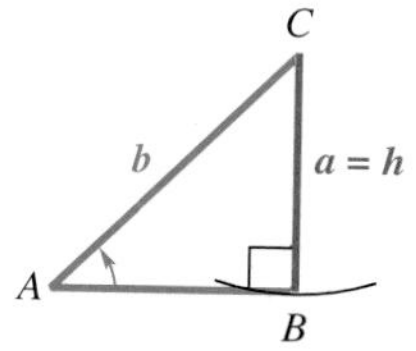	$a = h$
1	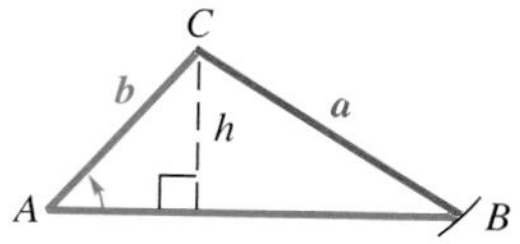	$a > b$
2	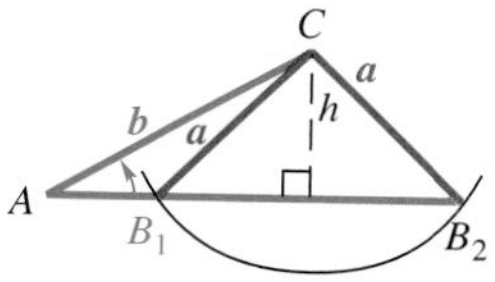	$b > a > h$

If angle A is obtuse, there are two possible outcomes, as shown in the next chart.

Number of Possible Triangles	Sketch	Condition Necessary for Case to Hold
0	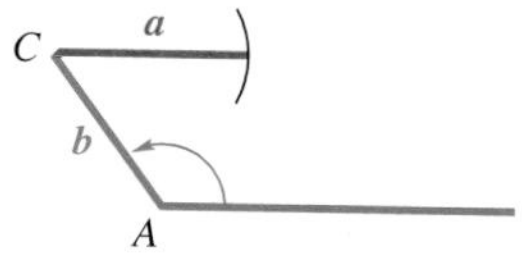	$a \leq b$
1	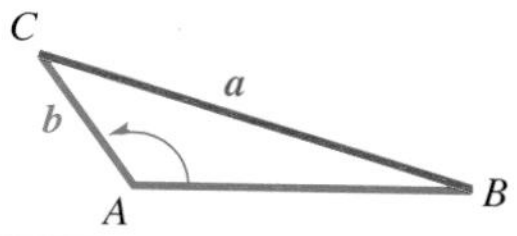	$a > b$

As the remaining examples of this section will illustrate, applying the law of sines to the values of a, b, and A and some basic properties of geometry and trigonometry will allow us to determine which of these cases applies. The following basic facts should be kept in mind.

FACTS TO REMEMBER WHEN USING THE LAW OF SINES

1. For any angle θ of a triangle, $0 < \sin \theta \leq 1$. If $\sin \theta = 1$, then $\theta = 90°$ and the triangle is a right triangle.
2. $\sin \theta = \sin(180° - \theta)$ (That is, supplementary angles have the same sine value.)
3. The smallest angle is opposite the shortest side, the largest angle is opposite the longest side, and the middle-valued angle is opposite the medium side (assuming the triangle has sides that are all of different lengths).

EXAMPLE 4

Solving a Triangle Using the Law of Sines (No Such Triangle)

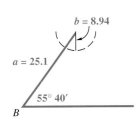

FIGURE 19

Solve triangle ABC if $B = 55° \, 40'$, $b = 8.94$ meters, and $a = 25.1$ meters.

SOLUTION   Since we are given B, b, and a, use the law of sines to find A.

$$\frac{\sin A}{a} = \frac{\sin B}{b}$$

Substitute the given values.

$$\frac{\sin A}{25.1} = \frac{\sin 55° \, 40'}{8.94}$$

$$\sin A = \frac{25.1 \sin 55° \, 40'}{8.94}$$

$$\sin A \approx 2.3184379$$

Since $\sin A$ cannot be greater than 1, there can be no such angle A and thus no triangle with the given information. An attempt to sketch such a triangle leads to the situation seen in Figure 19.   ◗

EXAMPLE 5

Solving a Triangle Using the Law of Sines (Two Triangles)

Solve triangle ABC if $A = 55° \, 20'$, $a = 22.8$ feet, and $b = 24.9$ feet.

SOLUTION   To begin, use the law of sines to find angle B.

$$\frac{a}{\sin A} = \frac{b}{\sin B}$$

$$\frac{22.8}{\sin 55° \, 20'} = \frac{24.9}{\sin B}$$

$$\sin B = \frac{24.9 \sin 55° \, 20'}{22.8}$$

$$\sin B \approx .89822938$$

Since $\sin B \approx .89822938$, to the nearest ten minutes we have one value of B as

$$B \approx 64° \, 00'$$

using the inverse sine function of a calculator. However, since supplementary angles have the same sine value, another *possible* value of B is

$$B \approx 180° - 64° \, 00' \approx 116° \, 00'.$$

To see if $B \approx 116° \, 00'$ is a valid possibility, simply add $116° \, 00'$ to the measure of the given value of A, $55° \, 20'$. Since $116° \, 00' + 55° \, 20' = 171° \, 20'$, and this sum is less than $180°$ (the sum of the angles of a triangle), we know that it is a valid angle measure for this triangle.

To keep track of these two different values of B, let

$$B_1 \approx 116° \, 00' \quad \text{and} \quad B_2 \approx 64° \, 00'.$$

Now separately solve triangles $AB_1 C_1$ and $AB_2 C_2$ shown in Figure 20.

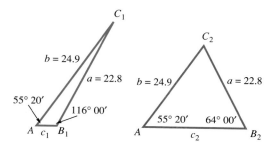

FIGURE 20

Let us begin with $AB_1 C_1$. Find C_1 first.

$$C_1 = 180° - A - B_1 \approx 8° \, 40'.$$

Now, use the law of sines to find c_1.

$$\frac{a}{\sin A} = \frac{c_1}{\sin C_1}$$

$$\frac{22.8}{\sin 55° \, 20'} = \frac{c_1}{\sin 8° \, 40'}$$

$$c_1 = \frac{22.8 \sin 8° \, 40'}{\sin 55° \, 20'}$$

$$c_1 \approx 4.18 \text{ feet}$$

To solve triangle $AB_2 C_2$, first find C_2.

$$C_2 = 180° - A - B_2 \approx 60° \, 40'$$

By the law of sines,

$$\frac{22.8}{\sin 55° \, 20'} = \frac{c_2}{\sin 60° \, 40'}$$

$$c_2 = \frac{22.8 \sin 60° \, 40'}{\sin 55° \, 20'}$$

$$c_2 \approx 24.2 \text{ feet.} \quad \blacksquare$$

CAUTION When solving a triangle using the type of data given in Example 5, do not forget to find the possible obtuse angle. The inverse sine function of the calculator will not give it directly. As we shall see in the next example, it is possible that the obtuse angle will not be a valid measure.

EXAMPLE 6

Solving a Triangle Using the Law of Sines (One Triangle)

Solve triangle ABC given $A = 43.5°$, $a = 10.7$ inches, and $b = 7.2$ inches.

SOLUTION To find angle B use the law of sines.

$$\frac{\sin B}{7.2} = \frac{\sin 43.5°}{10.7}$$

$$\sin B = \frac{7.2 \sin 43.5°}{10.7} \approx .46319186$$

The inverse sine function of the calculator gives us

$$B \approx 27.6°$$

as the acute angle. The other possible value of B is $180° - 27.6° \approx 152.4°$. However, when we add this possible obtuse angle to the given angle $A = 43.5°$, we get $152.4° + 43.5° = 195.9°$, which is greater than $180°$. So there can be only one triangle. Then angle $C \approx 180° - 27.6° - 43.5° = 108.9°$, and side c can be found with the law of sines.

$$\frac{c}{\sin 108.9°} = \frac{10.7}{\sin 43.5°}$$

$$c = \frac{10.7 \sin 108.9°}{\sin 43.5°}$$

$$c \approx 14.7 \text{ inches} \quad \blacksquare$$

TECHNOLOGICAL NOTE
Graphing calculators can be programmed to use the law of sines to solve a triangle, given the appropriate data. You may wish to explore this further.

EXAMPLE 7

Analyzing Data Involving an Obtuse Angle

Without using the law of sines, explain why the data

$$A = 104°, \ a = 26.8 \text{ meters}, \ b = 31.3 \text{ meters}$$

cannot be valid for a triangle ABC.

SOLUTION Since A is an obtuse angle, the largest side of the triangle must be a, the side opposite A. However, we are given $b > a$, which is impossible if A is obtuse. Therefore, no such triangle ABC exists. $\quad \blacksquare$

4.2 EXERCISES

To the Student: Calculator Considerations
You should not need to use any functions on your calculator that you have not already learned. For the sake of accuracy, when making approximations, do not round off during intermediate steps—wait until the final step of your calculation to give the appropriate approximation.

Solve each of the following triangles.

1. $A = 37°$, $B = 48°$, $c = 18$ m

2. $B = 52°$, $C = 29°$, $a = 43$ cm

3. $A = 46° \ 30'$, $B = 52° \ 50'$, $b = 87.3$ mm

4. $A = 59° \ 30'$, $B = 48° \ 20'$, $b = 32.9$ m

5. $C = 74.08°$, $B = 69.38°$, $c = 45.38$ m

6. $A = 87.2°$, $b = 75.9$ yd, $C = 74.3°$

7. $B = 38° 40'$, $a = 19.7$ cm, $C = 91° 40'$

8. $B = 20° 50'$, $C = 103° 10'$, $AC = 132$ ft

9. $A = 35.3°$, $B = 52.8°$, $AC = 675$ ft

10. Explain why the law of sines cannot be used to solve a triangle if we are given the lengths of the three sides of a triangle.

11. In Example 1, we ask the question "Why does the Pythagorean theorem not apply?" Answer this question.

12. Kala Wanersdorfer, a perceptive trigonometry student, makes the statement "If we know *any* two angles and one side of a triangle, then the triangle is uniquely determined." Is this a valid statement? Explain, referring to the congruence axioms given in this section.

Solve each problem.

13. To find the distance AB across a river, a distance $BC = 354$ m is laid off on one side of the river. It is found that $B = 112° 10'$ and $C = 15° 20'$. Find AB.

14. To determine the distance RS across a deep canyon, Joanna lays off a distance $TR = 582$ yd. She then finds that $T = 32° 50'$ and $R = 102° 20'$. Find RS.

15. Radio direction finders are placed at points A and B, which are 3.46 mi apart on an east-west line, with A west of B. From A the bearing of a certain radio transmitter is 47.7°, and from B the bearing is 302.5°. Find the distance of the transmitter from A.

16. A ship is sailing due north. At a certain point the bearing of a lighthouse 12.5 km distant is N 38.8° E. Later on, the captain notices that the bearing of the lighthouse has become S 44.2° E. How far did the ship travel between the two observations of the lighthouse?

17. A folding chair is to have a seat 12.0 in deep with angles as shown in the figure. How far down from the seat should the crossing legs be joined? (Find x in the figure.)

18. Mark notices that the bearing of a tree on the opposite bank of a river flowing north is 115.45°. Lisa is on the same bank as Mark, but 428.3 m away. She notices that the bearing of the tree is 45.47°. The two banks are parallel. What is the distance across the river?

19. Three gears are arranged as shown in the figure. Find angle θ.

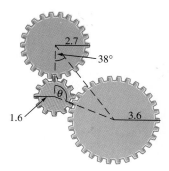

20. Three atoms with atomic radii of 2.0, 3.0, and 4.5 are arranged as in the figure. Find the distance between the centers of atoms A and C.

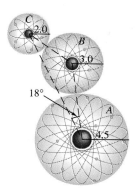

21. The bearing of a lighthouse from a ship was found to be N 37° E. After the ship sailed 2.5 miles due south, the new bearing was N 25° E. Find the distance between the ship and the lighthouse at each location.

22. A balloonist is directly above a straight road 1.5 miles long that joins two villages. She finds that the town closer to her is at an angle of depression of 35° and the farther town is at an angle of depression of 31°. How high above the ground is the balloon?

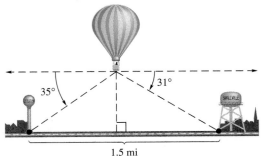

1.5 mi

23. From shore station A, a ship C is observed in the direction N 22.4° E. The same ship is observed to be in the direction N 10.6° W from shore station B, located at a distance of 25.5 km exactly southeast of A. Find the distance of the ship from station A.

24. A helicopter is sighted at the same time by two ground observers who are 3 mi apart on the same side of the helicopter. (See the figure.) They report the angles of elevation as 20.5° and 27.8°. How high is the helicopter?

20.5° 27.8°
3 mi

25. A rocket tracking station has two telescopes T_1 and T_2, placed 1.73 km apart, that lock onto the rocket and continuously transmit the angles of elevation to a computer. Find the distance to the rocket from T_1 at the moment when the angles of elevation are 28.1° and 79.5°, as shown in the figure.

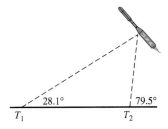

28.1°   79.5°
T_1 T_2

26. A surveyor standing 48.0 m from the base of a building measures the angle to the top of the building and finds it to be 37.4°. (See the figure.) The surveyor then measures the angle to the top of a clock tower on the building, finding that it is 45.6°. Find the height of the clock tower.

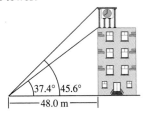

37.4° 45.6°
48.0 m

27. A slider crank mechanism is shown in the figure. Find the distance between the wrist pin W and the connecting rod center C.

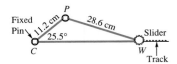

Fixed Pin — 11.2 cm — P — 28.6 cm — Slider
25.5°
C W
Track

28. A satellite S is traveling in a circular orbit 1600 km above the earth. It will pass directly over a tracking station T at noon. The satellite takes 2 hours to make a complete orbit. Assume that the radius of the earth is 6400 km. The tracking antenna is aimed 30° above the horizon. At what time will the satellite pass through the beam of the antenna?*

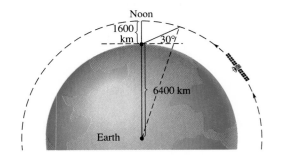

Noon
1600 km   30°
6400 km
Earth

29. Since the moon is a relatively close celestial object, its distance can be measured directly using trigonometry. To find this distance, two different photographs of the moon are taken at precisely the same time in two different locations with a known distance between them. The moon will have a

*From *Space Mathematics* by Bernice Kastner, Ph.D. Copyright © 1972 by the National Aeronautics and Space Administration. Courtesy of NASA.

different angle of elevation at each location. On April 29, 1976, at 11:35 A.M., the lunar angles of elevation during a partial solar eclipse at Bochum in upper Germany and at Donaueschingen in lower Germany were measured as 52.6997° and 52.7430°, respectively. The two cities are 398 kilometers apart.* Calculate the distance to the moon from Bochum on this day. Disregard the curvature of the earth in this calculation.

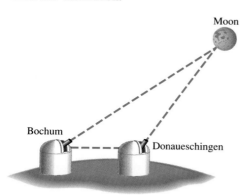

30. The distance covered by an aerial photograph is determined by both the focal length of the camera and the tilt of the camera from the perpendicular to the ground. Although the tilt is usually small, both archaeological and Canadian photographs often use larger tilts. A camera lens with a 12-inch focal length will have an angular coverage of 60°. If an aerial photograph is taken with this camera tilted $\theta = 35°$ at an altitude of 5000 feet, calculate the ground distance d in feet that will be shown in this photograph.† See the figure. (The 60° angle is bisected.)

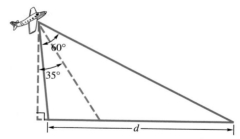

31. (See Exercise 30.) A camera lens with a 6-inch focal length has an angular coverage of 86°. Suppose an aerial photograph is taken vertically with no tilt at an altitude of 3500 feet over the horizon line with an increasing slope of 5° as shown in the figure. Calculate the ground distance CB that would appear in the resulting photograph.

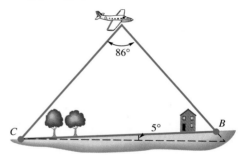

32. Repeat Exercise 31 if the camera lens has an 8.25-inch focal length with an angular coverage of 72°.

In each set of data for triangle ABC, two sides and an angle opposite one of them are given. As explained in Examples 4–6, such a set of data may lead to 0, 1, or 2 triangles. Solve for all possible triangles.

33. $A = 42.5°$, $a = 15.6$ ft, $b = 8.14$ ft

34. $C = 52.3°$, $a = 32.5$ yd, $c = 59.8$ yd

35. $B = 72.2°$, $b = 78.3$ m, $c = 145$ m

36. $C = 68.5°$, $c = 258$ cm, $b = 386$ cm

37. $A = 38° 40'$, $a = 9.72$ km, $b = 11.8$ km

38. $C = 29° 50'$, $a = 8.61$ m, $c = 5.21$ m

39. $B = 32° 50'$, $a = 7540$ cm, $b = 5180$ cm

40. $C = 22° 50'$, $b = 159$ mm, $c = 132$ mm

41. $A = 96.80°$, $b = 3.589$ ft, $a = 5.818$ ft

42. $C = 88.70°$, $b = 56.87$ yd, $c = 112.4$ yd

43. Apply the law of sines to the following: $a = \sqrt{5}$, $c = 2\sqrt{5}$, $A = 30°$. What is the value of sin C? What is the measure of C? Based on its angle measures, what kind of triangle is triangle ABC?

44. In your own words, explain the condition that must exist to determine that there is no triangle satisfying the given values of a, b, and B, once the value of sin B is found.

---

*Schlosser, W., T. Schmidt-Kaler, and E. Milone, *Challenges of Astronomy* (New York: Springer-Verlag, 1991).

†Brooks, R. and Dieter Johannes, *Phytoarchaeology* (Portland, OR: Dioscorides Press, 1990). Moffitt, F., *Photogrammetry* (Scranton: International Textbook Company, 1967).

45. Without using the law of sines, explain why no triangle ABC exists satisfying $A = 103° 20'$, $a = 14.6$ ft, $b = 20.4$ ft.

46. Apply the law of sines to the data given in Example 7. Describe in your own words what happens when you try to find the measure of angle B using a calculator.

47. A surveyor reported the following data about a piece of property: "The property is triangular in shape, with dimensions as shown in the figure." Use the law of sines to see whether such a piece of property could exist.

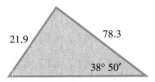

Can such a triangle exist?

48. The surveyor tries again: "A second triangular piece of property has dimensions as shown." This time it turns out that the surveyor did not consider every possible case. Use the law of sines to show why.

When a light ray travels from one medium, such as air, to another medium, such as water or glass, the speed of the light changes, and the direction that the ray is traveling changes. (This is why a fish under water is in a different position than it appears to be.) These changes are given by Snell's law

$$\frac{c_1}{c_2} = \frac{\sin \theta_1}{\sin \theta_2},$$

where c_1 is the speed of light in the first medium, c_2 is the speed of light in the second medium, and θ_1 and θ_2 are the angles shown in the figure. In the following exercises, assume that $c_1 = 3 \times 10^8$ m per sec. Find the speed of light in the second medium.

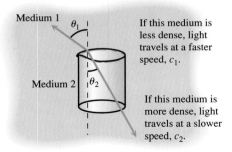

Medium 1 — If this medium is less dense, light travels at a faster speed, c_1.

Medium 2 — If this medium is more dense, light travels at a slower speed, c_2.

49. $\theta_1 = 46°$, $\theta_2 = 31°$

50. $\theta_1 = 39°$, $\theta_2 = 28°$

Find θ_2 for the following values of θ_1 and c_2. Round to the nearest degree.

51. $\theta_1 = 40°$, $c_2 = 1.5 \times 10^8$ m per sec

52. $\theta_1 = 62°$, $c_2 = 2.6 \times 10^8$ m per sec

53. The figure shows a fish's view of the world above the surface of the water.*

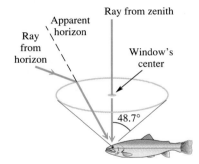

Suppose that a light ray comes from the horizon, enters the water, and strikes the fish's eye. Let us assume that this ray gives a value of 90° for angle θ_1 in the formula for Snell's law. (In a practical situation this angle would probably be a little less than 90°.) The speed of light in water is about 2.254×10^8 m per sec. Find angle θ_2.

54. What is wrong with this statement, given by a student as the law of sines?

$$\frac{a}{A} = \frac{b}{B} = \frac{c}{C}$$

4.3 THE LAW OF COSINES

Introduction ▍ Derivation of the Law of Cosines ▍ Applications

Introduction

Recall from the previous section that if we are given two sides and the included angle or three sides of a triangle, a unique triangle is formed. These are the SAS and SSS cases, respectively. In both cases, however, we cannot begin the solution of the triangle by using the law of sines. Both of these cases require the use of the law of cosines, introduced in this section.

It will be helpful to remember the following property of triangles when applying the law of cosines.

RESTRICTION ON TRIANGLE SIDE LENGTHS

In any triangle, the sum of the lengths of any two sides must be greater than the length of the remaining side.

For example, it would be impossible to construct a triangle with sides of lengths 3, 4, and 10. See Figure 21.

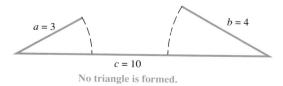

No triangle is formed.

FIGURE 21

Derivation of the Law of Cosines

To derive the law of cosines, let *ABC* be any oblique triangle. Choose a coordinate system so that vertex *B* is at the origin and side *BC* is along the positive *x*-axis. See Figure 22.

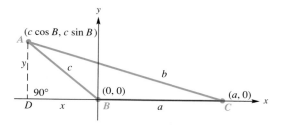

FIGURE 22

Let (x, y) be the coordinates of vertex A of the triangle. Verify that for angle B, whether obtuse or acute,

$$\sin B = \frac{y}{c} \quad \text{and} \quad \cos B = \frac{x}{c}.$$

(Here we assume that x is negative if B is obtuse.) From these results

$$y = c \sin B \quad \text{and} \quad x = c \cos B,$$

so that the coordinates of point A become

$$(c \cos B, c \sin B).$$

Point C has coordinates $(a, 0)$, and AC has length b. By the distance formula,

$$b = \sqrt{(c \cos B - a)^2 + (c \sin B)^2}.$$

Squaring both sides and simplifying gives

$$
\begin{aligned}
b^2 &= (c \cos B - a)^2 + (c \sin B)^2 \\
&= c^2 \cos^2 B - 2ac \cos B + a^2 + c^2 \sin^2 B \\
&= a^2 + c^2(\cos^2 B + \sin^2 B) - 2ac \cos B \\
&= a^2 + c^2(1) - 2ac \cos B \\
&= a^2 + c^2 - 2ac \cos B.
\end{aligned}
$$

This result is one form of the law of cosines. In the work above, we could just as easily have placed A or C at the origin. This would have given the same result, but with the variables rearranged. These various forms of the law of cosines are summarized in the following theorem.

LAW OF COSINES

In any triangle ABC, with sides a, b, and c,

$$
\begin{aligned}
a^2 &= b^2 + c^2 - 2bc \cos A \\
b^2 &= a^2 + c^2 - 2ac \cos B \\
c^2 &= a^2 + b^2 - 2ab \cos C.
\end{aligned}
$$

The law of cosines says that the square of a side of a triangle is equal to the sum of the squares of the other two sides, minus twice the product of the two sides and the cosine of the angle included between them.

NOTE If we let $C = 90°$ in the third form of the law of cosines given above, we have $\cos C = \cos 90° = 0$, and the formula becomes

$$c^2 = a^2 + b^2,$$

the familiar equation of the Pythagorean theorem. Thus, the Pythagorean theorem is a special case of the law of cosines.

Applications

We will now investigate how the law of cosines is used in applied situations.

EXAMPLE **1**

Using the Law of
Cosines in an
Application

A surveyor wishes to find the distance between two inaccessible points A and B on opposite sides of a lake. While standing at point C, she finds that $AC = 259$ meters, $BC = 423$ meters, and angle ACB measures $132°\ 40'$. Find the distance AB. (See Figure 23.)

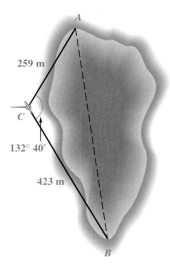

259 m

C

$132°\ 40'$

423 m

A

B

FIGURE 23

SOLUTION   The law of cosines can be used here, since we know the lengths of two sides of the triangle and the measure of the included angle.

$$AB^2 = 259^2 + 423^2 - 2(259)(423) \cos 132°\ 40'$$
$$AB^2 \approx 394{,}510.6 \qquad \text{Use a calculator.}$$
$$AB \approx 628 \qquad \text{Take the square root and round to 3 significant digits.}$$

The distance between the points is approximately 628 meters.   ◻

EXAMPLE **2**

Using the Law of
Cosines to Solve a
Triangle

Solve triangle ABC if $A = 42.3°$, $b = 12.9$ meters, and $c = 15.4$ meters. (See Figure 24.)

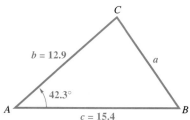

C

$b = 12.9$

a

$42.3°$

A

$c = 15.4$

B

FIGURE 24

SOLUTION   Start by finding a with the law of cosines.

$$a^2 = b^2 + c^2 - 2bc \cos A$$
$$a^2 = 12.9^2 + 15.4^2 - 2(12.9)(15.4) \cos 42.3°$$
$$a^2 \approx 109.7$$
$$a \approx \sqrt{109.7} \text{ meters} \qquad \text{Leave in this form for now.}$$

We now must find the measures of angles B and C. There are several approaches that can be used at this point. Let us use the law of sines to find one of these angles. Of the two remaining angles, B must be the smaller since it is opposite the shorter of the two sides b and c. Therefore, it cannot be obtuse, and we will avoid any ambiguity when we find its sine.

$$\frac{\sin 42.3°}{\sqrt{109.7}} = \frac{\sin B}{12.9}$$

$$\sin B = \frac{12.9 \sin 42.3°}{\sqrt{109.7}}$$

$$B \approx 56.0° \qquad \text{Use the inverse sine function of a calculator.}$$

The easiest way to find C is to subtract the sum of A and B from $180°$.

$$C = 180° - (A + B) \approx 81.7°.$$

$\sqrt{109.7} \approx 10.5$, so $a \approx 10.5$, and the triangle is solved.   ∎

CAUTION   Had we chosen to use the law of sines to find C rather than B in Example 2, we would not have known whether C equals $81.7°$ or its supplement, $98.3°$.

EXAMPLE 3

Using the Law of Cosines to Solve a Triangle

Solve triangle ABC if $a = 9.47$ feet, $b = 15.9$ feet, and $c = 21.1$ feet.

SOLUTION   We are given the lengths of three sides of the triangle, so we may use the law of cosines to solve for any angle of the triangle. Let us solve for C, the largest angle, using the law of cosines. We will be able to tell if C is obtuse if $\cos C < 0$. Use the form of the law of cosines that involves C.

$$c^2 = a^2 + b^2 - 2ab \cos C,$$

or $\qquad\qquad \cos C = \dfrac{a^2 + b^2 - c^2}{2ab}.$

Using the inverse cosine function of the calculator, we get the obtuse angle C.

$$C \approx 109.9°$$

We can use either the law of sines or the law of cosines to find $B \approx 45.1°$. (Verify this.) Since $A = 180° - B - C$,

$$A \approx 25.0°. \qquad ∎$$

As shown in this section and the previous one, four possible cases can occur when solving an oblique triangle. These cases are summarized in the chart that follows, along with a suggested procedure for solving in each case. There are other procedures that work, but we give the one that is most efficient. In all four cases, it is assumed that the given information actually produces a triangle.

Case	Suggested Procedure for Solving
One side and two angles are known. (SAA or ASA)	1. Find the remaining angle using the angle sum formula ($A + B + C = 180°$). 2. Find the remaining sides using the law of sines.
Two sides and one angle (not included between the two sides) are known. (SSA)	*Be aware of the ambiguous case; there may be two triangles.* 1. Find an angle using the law of sines. 2. Find the remaining angle using the angle sum formula. 3. Find the remaining side using the law of sines. *If two triangles exist, repeat Steps 1, 2, and 3.*

Case	Suggested Procedure for Solving
Two sides and the included angle are known. (SAS)	1. Find the third side using the law of cosines. 2. Find the smaller of the two remaining angles using the law of sines. 3. Find the remaining angle using the angle sum formula.
Three sides are known. (SSS)	1. Find the largest angle using the law of cosines. 2. Find either remaining angle using the law of sines. 3. Find the remaining angle using the angle sum formula.

4.3 EXERCISES

To the Student: Calculator Considerations

The law of cosines can be programmed into modern graphing calculators. You may wish to attempt to write such a program. Read your owner's manual for guidelines to programming. There are also programs available from user's groups, newsletters on technology in mathematics, etc., and you may wish to investigate these sources.

Solve each of the following triangles using the laws of cosines and sines as necessary. In Exercises 7–12, give answers in degrees.

1. $C = 28.3°$, $b = 5.71$ in, $a = 4.21$ in

2. $A = 41.4°$, $b = 2.78$ yd, $c = 3.92$ yd

3. $C = 45.6°$, $b = 8.94$ m, $a = 7.23$ m

4. $A = 67.3°$, $b = 37.9$ km, $c = 40.8$ km

5. $A = 80° 40'$, $b = 143$ cm, $c = 89.6$ cm

6. $C = 72° 40'$, $a = 327$ ft, $b = 251$ ft

7. $a = 9.3$ cm, $b = 5.7$ cm, $c = 8.2$ cm

8. $a = 28$ ft, $b = 47$ ft, $c = 58$ ft

9. $a = 42.9$ m, $b = 37.6$ m, $c = 62.7$ m

10. $a = 189$ yd, $b = 214$ yd, $c = 325$ yd

11. $AB = 1240$ ft, $AC = 876$ ft, $BC = 965$ ft

12. $AB = 298$ m, $AC = 421$ m, $BC = 324$ m

13. Refer to Figure 21. If you attempt to find any angle of a triangle using the values $a = 3$, $b = 4$, and $c = 10$ with the law of cosines, what happens?

14. A familiar saying is "The shortest distance between two points is a straight line." Explain how this relates to the geometric property that states that the sum of the lengths of any two sides of a triangle must be greater than the remaining side.

Solve each problem.

15. Points A and B are on opposite sides of Lake Yankee. From a third point, C, the angle between the lines of sight to A and B is 46.3°. If AC is 350 m long and BC is 286 m long, find AB.

16. The sides of a parallelogram are 4.0 cm and 6.0 cm. One angle is 58° while another is 122°. Find the lengths of the diagonals of the parallelogram.

17. Airports A and B are 450 km apart, on an east-west line. Tom flies in a northeast direction from A to airport C. From C he flies 359 km on a bearing of 128° 40′ to B. How far is C from A?

18. Two ships leave a harbor together, traveling on courses that have an angle of 135° 40′ between them. If they each travel 402 mi, how far apart are they?

19. A ship is sailing east. At one point, the bearing of a submerged rock is 45° 20′. After sailing 15.2 mi, the bearing of the rock has become 308° 40′. Find the distance of the ship from the rock at the latter point.

20. From an airplane flying over the ocean, the angle of depression to a submarine lying just under the surface is 24° 10′. At the same moment the angle of depression from the airplane to a battleship is 17° 30′. (See the figure.) The distance from the airplane to the battleship is 5120 ft. Find the distance between the battleship and the submarine. (Assume the airplane, submarine, and battleship are in a vertical plane.)

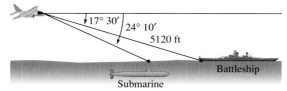

21. Two boats leave a dock together. Each travels in a straight line. The angle between their courses measures 54° 10′. One boat travels 36.2 km per hr, and the other travels 45.6 km per hr. How far apart will they be after 3 hr?

22. Find the lengths of both diagonals of a parallelogram with adjacent sides of 12 cm and 15 cm if the angle between these sides is 33°.

23. A crane with a counterweight is shown in the figure. Find the horizontal distance between points A and B.

24. A weight is supported by cables attached to both ends of a balance beam, as shown in the figure. What angles are formed between the beam and the cables?

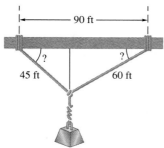

25. To measure the distance through a mountain for a proposed tunnel, a point C is chosen that can be reached from each end of the tunnel. (See the figure.) If AC = 3800 m, BC = 2900 m, and angle C = 110°, find the length of the tunnel.

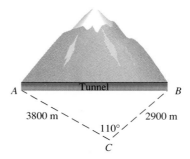

26. A baseball diamond is a square, 90 ft on a side, with home plate and the three bases as vertices. The pitcher's rubber is located 60.5 ft from home plate. Find the distance from the pitcher's rubber to each of the bases.

27. The Vietnam Veterans' Memorial in Washington, D.C., is in the shape of an unenclosed isosceles triangle (that is, V-shaped) with equal sides of length 246.75 feet and the angle between these sides measuring 125° 12′. Find the distance between the ends of the two equal sides.

28. Starting at point A, a ship sails 18.5 km on a bearing of 189°, then turns and sails 47.8 km on a bearing of 317°. Find the distance of the ship from point A.

29. Two towns 21 mi apart are separated by a dense forest. (See the figure.) To travel from town A to town B, a person must go 17 mi on a bearing of 325°, then turn and continue for 9 mi to reach town B. Find the bearing of B from A.

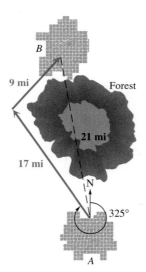

30. Two factories blow their whistles at exactly 5:00. A man hears the two blasts at 3 seconds and 6 seconds after 5:00, respectively. The angle between his lines of sight to the two factories is 42.2°. If sound travels 344 m per sec, how far apart are the factories?

31. A satellite traveling in a circular orbit 1600 km above Earth is due to pass directly over a tracking station at noon.* (See the figure.) Assume that the satellite takes 2 hr to make an orbit and that the radius of Earth is 6400 km. Find the distance between the satellite and the tracking station at 12:03 P.M.

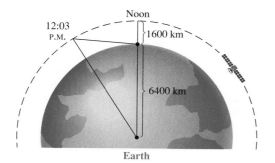

32. A ship sailing due east in the North Atlantic has been warned to change course to avoid a group of icebergs. The captain turns and sails on a bearing of 62° for a while, then changes course again to a bearing of 115° until the ship reaches its original course. (See the figure.) How much farther did the ship have to travel to avoid the icebergs?

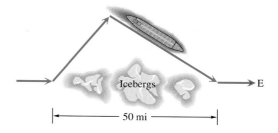

33. A parallelogram has sides of lengths 25.9 cm and 32.5 cm. The longer diagonal has a length of 57.8 cm. Find the angle opposite the diagonal.

34. A person in a plane flying a straight course observes a mountain at a bearing 24.1° to the right of its course. At that time the plane is 7.92 km from the mountain. A short time later, the bearing to the mountain becomes 32.7°. How far is the airplane from the mountain when the second bearing is taken?

To help predict eruptions from the volcano Mauna Loa on the island of Hawaii, scientists keep track of the volcano's movement by using a "super triangle" with vertices on the three volcanoes shown on the map below. (For example, in a recent year, Mauna Loa moved 6 inches, a result of increasing internal pressure.) Refer to the map to work Exercises 35 and 36.

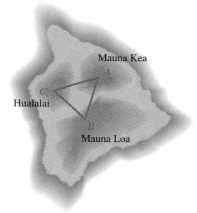

35. $AB = 22.47928$ mi, $AC = 28.14276$ mi, $A = 58.56989°$; find BC

36. $AB = 22.47928$ mi, $BC = 25.24983$ mi, $A = 58.56989°$; find B

*From *Space Mathematics* by Bernice Kastner, Ph.D. Copyright © 1972 by the National Aeronautics and Space Administration. Courtesy of NASA.

37. The layout for a child's playhouse has the dimensions given in the figure. Find x.

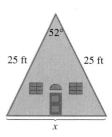

38. To find the distance between two small towns, an Electronic Distance Measuring (EDM) instrument is placed on a hill from which both towns are visible. The distance to each town from the EDM and the angle between the two lines of sight are measured. (See the figure.) Find the distance between the towns.

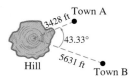

39. *Triangulation* is a technique used by surveyors to find the area of lots that have straight line boundaries. The region being surveyed is divided up into triangles and then the sum of the triangular areas is found.* The figure shows the dimensions in feet of a lake shore lot in Minnesota. The bearings of some of the lot lines are also shown. Angle *BAE* measures 88° 41′ 42″. What is the distance from *B* to *E*?

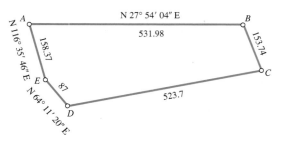

Length measures are in feet.

40. In order to obtain accurate aerial photographs, ground control must determine the coordinates of control points located on the ground that can be identified in the photographs. Using these known control points, the orientation and scale of each photograph can be determined. Then, unknown positions and distances can easily be determined. Before an aerial photograph is taken for highway design, horizontal control points must be determined and the distance between them calculated. The figure shows three consecutive control points *A*, *B*, and *C*. A surveyor measures a baseline distance of 92.13 feet from *B* to an arbitrary point *P*. Angles *BAP* and *BCP* are found to be 2° 22′ 47″, and 5° 13′ 11″, respectively. Then, angles *APB* and *CPB* are determined to be 63° 04′ 25″ and 74° 19′ 49″, respectively. Determine the distance between the control points *A* and *B* and between *B* and *C*.[†]

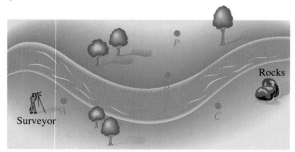

4.4 ADDITIONAL APPLICATIONS

Applying Formulas Involving Trigonometric Functions ▮ Heron's Area Formula ▮ Another Area Formula ▮ Linear and Angular Velocity

Applying Formulas Involving Trigonometric Functions

Many formulas involved in real-life applications include one or more trigonometric functions. The first example illustrates one such formula.

---

* Mueller I. and K. Ramsayer, *Introduction to Surveying* (New York: Frederick Ungar Publishing Co., 1979).

[†] Moffitt, F., *Photogrammetry* (Scranton: International Textbook Company, 1967).

EXAMPLE **1**
Using a Formula from Highway Design

When highway curves are designed, the outside of the curve is often slightly elevated or inclined above the inside of the curve. See Figure 25. This inclination is called **superelevation.** For safety reasons it is important that both the curve's radius and superelevation are correct for a given speed limit. If an automobile is traveling at velocity V (in feet per second), the safe radius R for a curve with superelevation α can be calculated using the formula $R = \frac{V^2}{g(f + \tan \alpha)}$ where f and g are constants, and α is in degrees.* A roadway is being designed for automobiles traveling at 45 miles per hour. If $\alpha = 3°$, $g = 32.2$, and $f = .14$, calculate R.

FIGURE 25

SOLUTION   First we must convert 45 miles per hour to feet per second. Since there are 5280 feet in one mile and 3600 seconds in one hour,

$$V = \frac{45 \cdot 5280}{3600} = 66 \text{ feet per second.}$$

Now substitute into the formula.

$$R = \frac{V^2}{g(f + \tan \alpha)}$$

$$R = \frac{66^2}{32.2(.14 + \tan 3°)} \qquad \text{\small } V = 66, g = 32.2, f = .14, \alpha = 3°$$

$$R \approx 703 \qquad\qquad \text{\small Use a calculator.}$$

The safe radius is approximately 703 feet.   ◼

Heron's Area Formula

The law of cosines can be used to derive a formula for the area of a triangle when only the lengths of the three sides are known. This formula is known as Heron's formula, named after the Greek mathematician Heron of Alexandria, who lived around A.D. 75. It is found in his work *Metrica*. See Exercises 27–32 for an outline of the proof that this formula is valid.

HERON'S AREA FORMULA

If a triangle has sides of lengths a, b, and c, and if the **semiperimeter** is

$$s = \frac{1}{2}(a + b + c),$$

then the area A of the triangle is

$$A = \sqrt{s(s - a)(s - b)(s - c)}.$$

---

*Mannering, F. and W. Kilareski, *Principles of Highway Engineering and Traffic Control* (New York: John Wiley & Sons, Inc., 1990).

EXAMPLE 2
Using Heron's Formula to Find an Area

The distance "as the crow flies" from Los Angeles to New York is 2451 miles, from New York to Montreal is 331 miles, and from Montreal to Los Angeles is 2427 miles. What is the area of the triangular region having these three cities as vertices?

SOLUTION   Figure 26 shows that we can let $a = 2451$, $b = 331$, and $c = 2427$. Then the semiperimeter s is given by

$$s = \frac{1}{2}(2451 + 331 + 2427) = 2604.5.$$

Using Heron's formula, the area A is found as follows.

$$A = \sqrt{s(s - a)(s - b)(s - c)}$$
$$A = \sqrt{2604.5(2604.5 - 2451)(2604.5 - 331)(2604.5 - 2427)}$$
$$A \approx 401{,}700$$

The area of the triangular region is approximately 401,700 square miles.   ∎

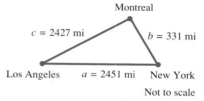

Los Angeles $a = 2451$ mi   New York

Not to scale

FIGURE 26

FOR **GROUP DISCUSSION**

1. If your class is held in a rectangular-shaped room, measure the length and the width, and then multiply them to find the area. Now, measure a diagonal of the room, and use Heron's formula with the length, the width, and the diagonal to find the area of half the room. Double this result. Do your area calculations agree?

2. For a triangle to exist, the sum of the lengths of any two sides must exceed the length of the remaining side. Have half of the class try to calculate the area of a "triangle" with $a = 4$, $b = 8$, and $c = 12$, while the other half tries to calculate the area of the "triangle" with $a = 10$, $b = 20$, and $c = 34$. In both cases, use Heron's formula. Then, discuss the results, drawing diagrams on the chalkboard to support the results obtained.

3. A popular textbook for mathematics survey courses contains the following problem and diagram: *Find the perimeter and area of the shaded region. (See Figure 27.)*

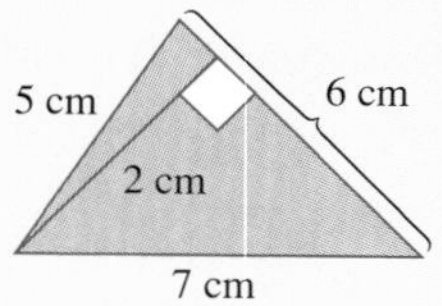

5 cm   6 cm

2 cm

7 cm

FIGURE 27

The perimeter, obviously, is 18 cm. Now, divide the class into two groups. Have one group determine the area using area = $(\frac{1}{2})$(base)(height), and have the other group determine the area using Heron's formula. Then discuss your results. What is the problem with this problem?

If we know the measures of two sides of a triangle and the angle included between them, then we can find the area $\mathcal{A}$ of the triangle using the following formula.

AREA OF A TRIANGLE

In any triangle ABC, the area $\mathcal{A}$ is given by any of the following formulas:

$$\mathcal{A} = \frac{1}{2}\, bc \, \sin A, \qquad \mathcal{A} = \frac{1}{2}\, ab \, \sin C, \qquad \mathcal{A} = \frac{1}{2}\, ac \, \sin B.$$

That is, the area is given by half the product of the lengths of two sides and the sine of the angle included between them.

The derivation of this formula is outlined in Exercises 33–36.

EXAMPLE 3

Finding the Area of a Triangle Using $\mathcal{A} = \frac{1}{2} ab \sin C$

Find the area of triangle ABC if $A = 24° \, 40'$, $b = 27.3$ centimeters, and $C = 52° \, 40'$.

SOLUTION Before we can use the formula given above, we must use the law of sines to find either a or c. Since the sum of the measures of the angles of any triangle is $180°$,

$$B = 180° - 24° \, 40' - 52° \, 40' = 102° \, 40'.$$

Now use the form of the law of sines that relates a, b, A, and B to find a.

$$\frac{a}{\sin A} = \frac{b}{\sin B}$$

$$\frac{a}{\sin 24° \, 40'} = \frac{27.3}{\sin 102° \, 40'}$$

Solve for a to verify that $a \approx 11.7$ centimeters. Now find the area.

$$\mathcal{A} = \frac{1}{2}\, ab \, \sin C \approx 127 \text{ (with } C = 52° \, 40'\text{)}$$

The area of triangle ABC is 127 square centimeters (to three significant digits).

Linear and Angular Velocity

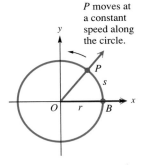

P moves at a constant speed along the circle.

FIGURE 28

Suppose that point P moves at a constant speed along a circle of radius r and center O. See Figure 28. The measure of how fast the position of P is changing is called **linear velocity.**

If v represents linear velocity, then

$$v = \frac{s}{t},$$

where s is the length of the arc traced by point P at time t. (This formula is just a restatement of the familiar result $d = rt$ with s as distance, v as the rate, and t as time.)

Look at Figure 28 again. As point P moves along the circle, ray OP rotates around the origin. Since the ray OP is the terminal side of angle POB, the measure of the angle changes as P moves along the circle. The measure of how fast angle POB is changing is called **angular velocity.** Angular velocity, written ω, can be given as

$$\omega = \frac{\theta}{t}, \quad \theta \text{ in radians,}$$

where θ is the measure of angle POB at time t. As with the earlier formula of this type, θ must be measured in radians, with ω expressed as radians per unit of time. Angular velocity is used in physics and engineering, among other applications.

In Chapter 2 we saw that the length s of the arc intercepted on a circle of radius r by a central angle of measure θ radians was found to be $s = r\theta$. Using this formula, the formula for linear velocity, $v = \frac{s}{t}$, becomes

$$v = \frac{r\theta}{t}$$

or $v = r\omega$

This last formula relates linear and angular velocity.

EXAMPLE **4**

Using the Linear and Angular Velocity Formulas

Suppose that point P is on a circle with a radius of 10 centimeters, and ray OP is rotating with angular velocity of $\frac{\pi}{18}$ radian per second.

(a)   Find the angle generated by P in 6 seconds.

SOLUTION   The velocity of ray OP is $\omega = \frac{\pi}{18}$ radian per second. Since $\omega = \frac{\theta}{t}$, then in 6 seconds

$$\frac{\pi}{18} = \frac{\theta}{6},$$

or $\theta = 6\left(\frac{\pi}{18}\right) = \frac{\pi}{3}$ radians.

(b)   Find the distance traveled by P along the circle in 6 seconds.

SOLUTION   In 6 seconds P generates an angle of $\frac{\pi}{3}$ radians. Since $s = r\theta$,

$$s = 10\left(\frac{\pi}{3}\right) = \frac{10\pi}{3} \text{ centimeters.}$$

(c)   Find the linear velocity of P.

SOLUTION   Since $v = \frac{s}{t}$, in 6 seconds

$$v = \frac{\dfrac{10\pi}{3}}{6} = \frac{5\pi}{9} \text{ centimeters per second.} \quad \blacksquare$$

In practical applications, angular velocity is often given as revolutions per unit of time, which must be converted to radians per unit of time before using the formulas given in this section.

EXAMPLE 5 Using the Linear and Angular Velocity Formulas	A belt runs a pulley of radius 6 centimeters at 80 revolutions per minute. **(a)** Find the angular velocity of the pulley in radians per second.

SOLUTION In one minute, the pulley makes 80 revolutions. Each revolution is 2π radians, for a total of

$$80(2\pi) = 160\pi \text{ radians per minute.}$$

Since there are 60 seconds in a minute, ω, the angular velocity in radians per second, is found by dividing 160π by 60.

$$\omega = \frac{160\pi}{60} = \frac{8\pi}{3} \text{ radians per second}$$

(b) Find the linear velocity of the belt in centimeters per second.

SOLUTION The linear velocity of the belt will be the same as that of a point on the circumference of the pulley. Thus,

$$v = r\omega$$

$$v = 6\left(\frac{8\pi}{3}\right)$$

$$v = 16\pi \text{ centimeters per second}$$

$$v \approx 50.3 \text{ centimeters per second.} \quad \blacksquare$$

4.4 EXERCISES

Solve each problem involving a formula with trigonometric functions.

1. Refer to the formula in Example 1. What should the radius of the curve be if the speed limit is increased to 70 miles per hour?

2. Refer to the formula in Example 1. How would increasing angle α affect the results? Verify your answer by repeating parts of Example 1 with $\alpha = 4°$.

3. A highway curve has a radius of $R = 1150$ feet and a superelevation of $\alpha = 2.1°$. What should the speed limit (in miles per hour) be for this curve?

4. When an automobile travels uphill or downhill on a highway, it experiences a force due to gravity. This force F is called *grade resistance* and is computed using the equation $F = W \sin \theta$, where θ is the grade and W is the weight of the automobile. If the automobile is moving uphill, $\theta > 0$, and if it is moving downhill, $\theta < 0$. See the figure.*
   (a) Calculate F to two significant digits for a 2500-pound car traveling an uphill grade with $\theta = 2.5°$.

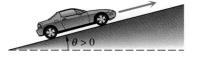

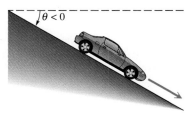

(b) Calculate F to two significant digits for a 5000-pound truck traveling a downhill grade with $\theta = -6.1°$.

(c) Calculate F for $\theta = 0°$ and $\theta = 90°$. Do these answers agree with your intuition?

---

*Mannering, F. and W. Kilareski, *Principles of Highway Engineering and Traffic Control* (New York: John Wiley & Sons, Inc., 1990).

5. A 10-ft-high movie screen is located 2 ft above the eyes of the viewers, all of whom are sitting at the same level. A viewer seated 5 ft from the screen has the maximum viewing angle, given by x in the equation

$$\frac{\tan x + .4}{1 - .4 \tan x} = 2.4.$$

Find the maximum viewing angle (in degrees).

6. Artificial satellites that orbit Earth often use VHF signals to communicate with the ground. VHF signals travel in straight lines. The height h in miles of the satellite above Earth and the time T that the satellite can communicate with a fixed location on the ground are related by the equation

$$h = R\left(\frac{1}{\cos(\frac{180T}{P})} - 1\right),$$

where $R = 3955$ miles is the radius of Earth and P is the period for the satellite to orbit Earth.[*]
   (a) Find h when $T = 25$ min and $P = 140$ min. (Evaluate the cosine function in degree mode.)
   (b) Discuss what must happen to h if one wants T to increase to 30 min.

7. In the Kodak Customer Service Pamphlet AA-26, entitled *Optical Formulas and Their Applications,* the near and far limits of the depth of field (how close or how far away an object can be placed and still be in focus) are given by the following formulas:

$$w_1 = \frac{u^2(\tan \theta)}{L + u(\tan \theta)} \quad \text{and} \quad w_2 = \frac{u^2(\tan \theta)}{L - u(\tan \theta)}.$$

In these equations, θ represents the angle (in degrees) between the lens and the "circle of confusion," which is the circular image on the film of a point that is not exactly in focus. L is the diameter of the lens opening, and u is the distance to the object being photographed. Find the angle so that near and far limits of the depth of field are approximately 2 m and 6 m respectively, when L is .00625 m and the object being photographed is 6 m from the camera.[†]

8. If aerodynamic resistance is ignored, the braking distance D (in feet) for an automobile to change its velocity from V_1 to V_2 (feet per second) can be calculated using the equation

$$D = \frac{1.05(V_1^2 - V_2^2)}{64.4(K_1 + K_2 + \sin \theta)}.$$

K_1 is a constant determined by the efficiency of the brakes and tires, K_2 is a constant determined by the rolling resistance of the automobile, and θ is the grade of the highway.[‡]
   (a) Compute the number of feet required to slow a car from 55 to 30 miles per hour while traveling uphill with a grade of $\theta = 3.5°$. Let $K_1 = .4$ and $K_2 = .02$.
   (b) Repeat part (a) with $\theta = -2°$.
   (c) How is braking distance affected by the grade θ? Does this agree with your driving experience?

9. (Refer to Exercise 8.) An automobile is traveling at 90 miles per hour on a highway with a downhill grade of $\theta = -3.5°$. The driver sees a stalled truck in the road 200 feet away and immediately applies the brakes. Assuming that a collision cannot be avoided, how fast (in miles per hour) is the car traveling when it hits the truck? (Use the same values for K_1 and K_2 as in Exercise 8.)

10. A basic highway curve connecting two straight sections of road is often circular. See the figure. The points P and S mark the beginning and end of the curve. Let Q be the point of intersection where the two straight sections of highway leading into the curve would meet if extended. The radius of the curve is R and the angle θ denotes how many degrees the curve turns.[‡]
   (a) If $R = 965$ ft and $\theta = 37°$, find the distance d between P and Q.
   (b) Find an expression in terms of R and θ for the distance between points M and N.

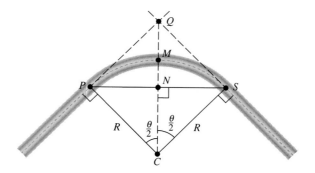

[*] Schlosser, W., T. Schmidt-Kaler, and E. Milone, *Challenges of Astronomy* (New York: Springer-Verlag, 1991).

[†] "Optical Formulas and Their Applications," *Kodak Customer Service Pamphlet AA-26.* Reprinted by permission of Eastman Kodak Company.

[‡] Mannering, F. and W. Kilareski, *Principles of Highway Engineering and Traffic Control* (New York: John Wiley & Sons, Inc., 1990).

11. When a highway goes downhill and then uphill it is called a *sag curve*. Sag curves are designed so that at night, headlights shine sufficiently far down the road to allow for a safe stopping distance. See the figure. The minimum length L of a sag curve is determined by the height h of the car's headlights above the pavement, the downhill grade $\theta_1 < 0°$, the uphill grade $\theta_2 > 0°$, and the safe stopping distance S for a given speed limit. In addition, L is dependent on the vertical alignment of the headlights. Headlights are usually pointed upward at a slight angle α above the horizontal of the car. Using these quantities, L can then be computed using the formula

$$L = \frac{(\theta_2 - \theta_1)S^2}{200(h + S \tan \alpha)}, \quad \text{where } S < L.*$$

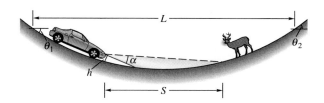

(a) Compute L for a 55-mile-per-hour speed limit where $h = 1.9$ ft, $\alpha = .9°$, $\theta_1 = -3°$, $\theta_2 = 4°$, and $S = 336$ feet.
(b) Repeat part (a) with $\alpha = 1.5°$.

12. When a large-view camera is used to take a picture of an object that is not parallel to the film, the lens board should be tilted so that the planes containing the subject, the lens board, and the film intersect in a line. (See the figure.) This gives the best "depth of field."[†]

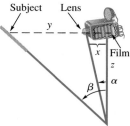

(a) Write two equations, one relating α, x, and z, and the other relating β, x, y, and z.
(b) Eliminate z from the equations in part (a) to get one equation relating α, β, x, and y.
(c) Solve the equation from part (b) for α.
(d) Solve the equation from part (b) for β.

Find the area of each triangle by using one of the area formulas introduced in this section.

13. $A = 42.5°$, $b = 13.6$ m, $c = 10.1$ m

14. $B = 124.5°$, $a = 30.4$ cm, $c = 28.4$ cm

15. $A = 56.80°$, $b = 32.67$ in, $c = 52.89$ in

16. $A = 24° 25'$, $B = 56° 20'$, $c = 78.40$ cm

17. $a = 12$ m, $b = 16$ m, $c = 25$ m

18. $a = 154$ cm, $b = 179$ cm, $c = 183$ cm

19. $a = 76.3$ ft, $b = 109$ ft, $c = 98.8$ ft

20. $a = 22$ in, $b = 45$ in, $c = 31$ in

21. $a = 25.4$ yd, $b = 38.2$ yd, $c = 19.8$ yd

22. $a = 15.89$ in, $b = 21.74$ in, $c = 10.92$ in

Solve each problem.

23. A painter is going to apply a special coating to a triangular metal plate on a new building. Two sides measure 16.1 m and 15.2 m. She knows that the angle between these sides is 125°. What is the area of the surface she plans to cover with the coating?

24. A real estate agent wants to find the area of a triangular lot. A surveyor takes measurements and finds that two sides are 52.1 m and 21.3 m, and the angle between them is 42.2°. What is the area of the lot?

25. A painter needs to cover a triangular region 75 m by 68 m by 85 m. A can of paint covers 75 sq m of area. How many cans (to the next higher number of cans) will be needed?

26. Find the area of the Bermuda Triangle if the sides of the triangle have the approximate lengths 850 miles, 925 miles, and 1300 miles.

* Mannering, F. and W. Kilareski, *Principles of Highway Engineering and Traffic Control* (New York: John Wiley & Sons, Inc., 1990).

† From *A Sourcebook of Applications of School Mathematics* by Donald Bushaw et al. Copyright © 1980 by The Mathematical Association of America. Reprinted by permission.

Relating Concepts

From the law of cosines, in triangle ABC, $\cos A = \frac{b^2 + c^2 - a^2}{2bc}$. Use this equation to show that each of the following is true, and from these exercises, prove Heron's formula. Work Exercises 27–32 in order.

27. $1 + \cos A = \dfrac{(b + c + a)(b + c - a)}{2bc}$

28. $1 - \cos A = \dfrac{(a - b + c)(a + b - c)}{2bc}$

29. $\cos \dfrac{A}{2} = \sqrt{\dfrac{s(s - a)}{bc}}$ $\left(\text{Hint: } \cos \dfrac{A}{2} = \sqrt{\dfrac{1 + \cos A}{2}} \right)$

30. $\sin \dfrac{A}{2} = \sqrt{\dfrac{(s - b)(s - c)}{bc}}$ $\left(\text{Hint: } \sin \dfrac{A}{2} = \sqrt{\dfrac{1 - \cos A}{2}} \right)$

31. The area of a triangle having sides b and c and angle A is given by $(\frac{1}{2})bc \sin A$. Show that this result can be written as

$$\sqrt{\frac{1}{2}bc(1 + \cos A) \cdot \frac{1}{2}bc(1 - \cos A)}.$$

32. Use the results of Exercises 27–31 to prove Heron's area formula.

Use the figures below to derive the area formula $\mathcal{A} = \frac{1}{2}bc \sin A$. Work Exercises 33–36 in order.

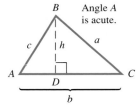
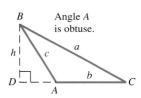

33. Using triangle ABD, find an expression for $\sin A$ in terms of h and c.

34. Solve for h.

35. The familiar area formula for a triangle, area $= \frac{1}{2}$(base)(height), can now be applied. Since the height to base b is h, use the expression for h found in Exercise 34 and write the formula in terms of b and that expression.

36. Explain why, if $A = 90°$, the formula $\mathcal{A} = \frac{1}{2}bc \sin A$ becomes the familiar area formula.

Solve the following problems involving linear and angular velocity.

37. A tire is rotating 600 times per minute. Through how many degrees does a point on the edge of the tire move in $\frac{1}{2}$ second?

38. An airplane propeller rotates 1000 times per minute. Find the number of degrees that a point on the edge of the propeller will rotate in 1 second.

39. A pulley rotates through 75° in one minute. How many rotations does the pulley make in an hour?

40. (a) How many inches will the weight in the figure rise if the pulley is rotated through an angle of 71° 50′?

(b) Through what angle, to the nearest minute, must the pulley be rotated to raise the weight 6 in.?

9.27 in

41. Find the radius of the pulley in the figure if a rotation of 51.6° raises the weight 11.4 cm.

42. The figure shows the chain drive of a bicycle. How far will the bicycle move if the pedals are rotated through 180°? Assume that the radius of the bicycle wheel is 13.6 in.

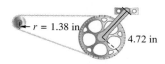

r = 1.38 in.
4.72 in

43. The speedometer of a small pickup truck is designed to be accurate with tires of radius 14 in.
 (a) Find the number of rotations of a tire in 1 hr if the truck is driven at 55 mph.
 (b) Suppose that oversize tires of radius 16 in. are placed on the truck. If the truck is now driven for 1 hr with the speedometer reading 55 mph, how far has the truck gone? If the speed limit is 55 mph, does the driver deserve a speeding ticket?

44. A railroad track is laid along the arc of a circle of radius 1800 ft. The circular part of the track subtends a central angle of 40°. How long (in seconds) will it take a point on the front of a train traveling 30 mph to go around this portion of the track?

45. Two pulleys of diameter 4 m and 2 m, respectively, are connected by a belt. The larger pulley rotates 80 times per min. Find the speed of the belt in meters per second and the angular velocity of the smaller pulley.

46. The earth revolves on its axis once every 24 hr. Assuming that the earth's radius is 6400 km, find the following.
 (a) Angular velocity of the earth in radians per day and radians per hr
 (b) Linear velocity at the North Pole or South Pole
 (c) Linear velocity at Quito, Ecuador, a city on the equator

47. The earth travels about the sun in an orbit that is almost circular. Assume that the orbit is a circle, with a radius of 93,000,000 mi. (See the figure.)
 (a) Assume that a year is 365 days, and find θ, the angle formed by the earth's movement in one day.
 (b) Give the angular velocity in radians per hour.
 (c) Find the linear velocity of the earth in miles per hour.

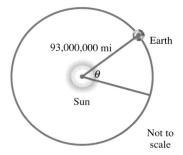

93,000,000 mi
Earth
θ
Sun
Not to scale

48. The pulley shown has a radius of 12.96 cm. Suppose that it takes 18 sec for 56 cm of belt to go around the pulley. Find the angular velocity of the pulley in radians per second.

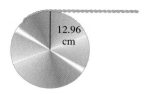

12.96 cm

49. The two pulleys in the figure have radii of 15 cm and 8 cm, respectively. The larger pulley rotates 25 times in 36 sec. Find the angular velocity of each pulley in radians per sec.

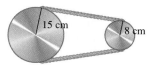

15 cm
8 cm

50. A gear is driven by a chain that travels 1.46 m per sec. Find the radius of the gear if it makes 46 revolutions per min.

51. A thread is being pulled off a spool at the rate of 59.4 cm per sec. Find the radius of the spool if it makes 152 revolutions per min.

A *sector of a circle* is the portion of the interior of a circle intercepted by a central angle. (See the figure.) It can be shown that the area of a sector of a circle with radius r and central angle θ, where θ is in radians, is given by the formula

$$A = \frac{1}{2}r^2\theta.$$

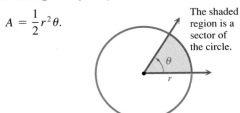

The shaded region is a sector of the circle.

Use this formula to solve the problems in Exercises 52–55.

52. Find the area of a sector of a circle having radius 29.2 m and central angle $\theta = \frac{5\pi}{6}$.

53. Find the area of a sector of a circle having radius 12.7 cm and central angle $\theta = 81°$.

54. The figure shows Medicine Wheel, an Indian structure in northern Wyoming. This circular structure is perhaps 200 years old. There are 32 spokes in the wheel, all equally spaced.

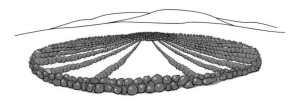

(a) Find the measure of each central angle in degrees and in radians.
(b) If the radius of the wheel is 76 ft, find the circumference.
(c) Find the length of each arc intercepted by consecutive pairs of spokes.
(d) Find the area of each sector formed by consecutive spokes.

55. The unusual corral in the figure (at the top of the next column) is separated into 26 areas, many of which approximate sectors of a circle. Assume that the corral has a diameter of 50 m.
(a) Find the central angle for each region, assuming that the 26 regions are all equal sectors, with the fences meeting at the center.
(b) What is the area of each sector?

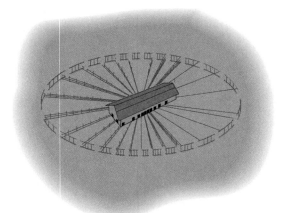

56. Eratosthenes (*ca.* 230 B.C.) made a famous measurement of the earth. He observed at Syene (the modern Aswan) at noon and at the summer solstice that a vertical stick had no shadow, while at Alexandria (on the same meridian as Syene) the sun's rays were inclined $\frac{1}{50}$ of a complete circle to the vertical. See the figure. He then calculated the circumference of the earth from the known distance of 5000 stades between Alexandria and Syene. Obtain Eratosthenes' result of 250,000 stades for the circumference of the earth. There is reason to suppose that a stade is about equal to 516.7 ft. Assuming this, use Eratosthenes' result to calculate the polar diameter of the earth in miles. (The actual polar diameter of the earth is approximately 7900 mi.)*

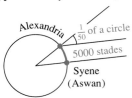

---

*Mathematical exercise and figure from *A Survey of Geometry*, Vol. 1 by Howard Eves. Reprinted by permission of the author.

Chapter 4 SUMMARY

In this chapter we see how the trigonometric functions can be defined in terms of ratios of lengths of the sides of a right triangle. For an acute angle θ in a right triangle, the sides can be identified as *opposite θ*, *adjacent to θ*, and the *hypotenuse*. Ratios of these sides form the six trigonometric functions of θ, as seen in the first section.

Solving a right triangle means finding the measures of all its angles and sides. We do this by using the trigonometric function ratios. The degree of accuracy depends on the accuracy of the given information.

Practical problems often require using or finding the angle of elevation or depression. These angles are always measured from the horizontal. Some applications involve bearing, which is measured in two ways. One way is to measure in a clockwise direction from due north. With another method we measure the angle from due north to the west or east, or from due south to the west or east.

The law of sines and the law of cosines are used to solve oblique triangles. By the law of sines the ratio of any side of a triangle to the sine of the opposite angle equals the corresponding ratio of another side to the sine of its opposite angle. If we are given two sides and the angle opposite one of them, the information may lead to no triangle, one triangle, or two triangles. This situation is called the ambiguous case of the law of sines. It is useful to remember that the largest angle must be opposite the largest side, and the smallest angle must be opposite the smallest side.

The law of cosines says that the square of a side of a triangle is equal to the sum of the squares of the other two sides, minus twice the product of the two sides and the cosine of the included angle. Note the similarity to the Pythagorean theorem. When using the law of cosines, remember that the sum of the lengths of any two sides of a triangle must be greater than the length of the third side.

The trigonometric functions often appear in formulas from mechanics, engineering, and other fields, leading to meaningful applications. Using the laws of sines and cosines, we can derive two formulas for the area of a triangle, one of which is known as Heron's formula. Heron's formula is used when three side lengths of a triangle are known. The other area formula is used when two sides and the included angle are known.

The linear velocity of a point moving along a circle tells how fast the point is moving. As the point moves along the circle, the positive angle in standard position also changes. The rate at which the angle changes is the angular velocity. Linear and angular velocity have many practical applications.

Key Terms

SECTION 4.1

hypotenuse
side opposite an angle
side adjacent to an angle
significant digits
exact number
angle of elevation
angle of depression
bearing

SECTION 4.2

congruent triangles
oblique triangle

SECTION 4.4

linear velocity
angular velocity

Chapter 4   **REVIEW EXERCISES**

Find the exact values of the six trigonometric functions of angle A.

1.

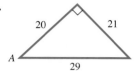

2.

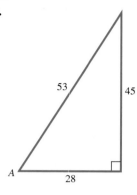

Solve each right triangle.

3.

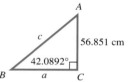

4.

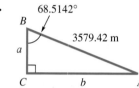

Use the right triangle shown, and give the exact value of the specified function.

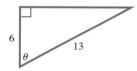

5. $\sin \theta$

6. $\sec(90° - \theta)$

7. Find the degree measure of θ correct to the nearest tenth.

8. What justifies this statement? "If I know the value of $\sin \theta$, then I know the value of $\cos(90° - \theta)$."

Solve the right triangle with C = 90°.

9. $A = 39.72°$, $b = 38.97$ m

10. $a = 270.0$ m, $b = 298.6$ m

Solve the problem.

11. The angle of elevation from a point 93.2 feet from the base of a tower to the top of the tower is 38° 20′. Find the height of the tower.

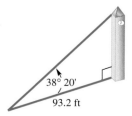

12. The angle of depression of a television tower to a point on the ground 36.0 meters from the bottom of the tower is 29.5°. Find the height of the tower.

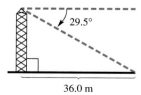

13. A rectangle has adjacent sides measuring 10.93 centimeters and 15.24 centimeters. The angle between the diagonal and the longer side is 35.65°. Find the length of the diagonal.

14. An isosceles triangle has a base of length 49.28 meters. The angle opposite the base is 58.746°. Find the length of each of the two equal sides.

15. The bearing of B from C is 254°. The bearing of A from C is 344°. The bearing of A from B is 32°. The distance from A to C is 780 meters. Find the distance from A to B.

16. A ship leaves a pier on a bearing of S 55° E and travels for 80 kilometers. It then turns and continues on a bearing of N 35° E for 74 kilometers. How far is the ship from the pier?

17. Two cars leave an intersection at the same time. One heads due south at 55 miles per hour. The other travels due west. After two hours, the bearing of the car headed west from the car headed south is 324°. How far apart are they at that time?

18. From the top of a building that overlooks an ocean, an observer watches a boat sailing directly toward the building. If the observer is 150 feet above sea level and if the angle of depression of the boat changes from 27° to 39° during the period of observation, approximate the distance that the boat travels.

19. Find the measure of h to the nearest unit.

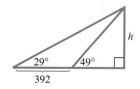

20. The highest mountain peak in the world is Mt. Everest located in the Himalayas. The height of this enormous mountain was determined in 1856 by surveyors using trigonometry long before it was first climbed in 1953. This difficult measurement had to be done from a large distance away. At an altitude of 14,545 feet on a different mountain, the straight line distance to the peak of Mt. Everest is 27.0134 miles and its angle of elevation is $\theta = 5.82°$.* See the figure.
   (a) Approximate the height of Mt. Everest.
   (b) In the actual measurement, Mt. Everest was over a hundred miles away and the curvature of the earth had to be taken into account. Would the curvature of the earth make the peak appear taller or shorter than it actually is?

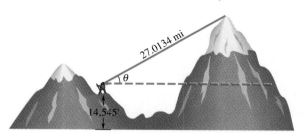

Use the law of sines and/or the law of cosines to find the indicated part of triangle ABC.

21. $C = 74° \ 10'$, $c = 96.3$ m, $B = 39° \ 30'$; find b

22. $a = 86.14$ in, $b = 253.2$ in, $c = 241.9$ in; find A

23. $A = 129° \ 40'$, $a = 127$ ft, $b = 69.8$ ft; find B

24. $B = 120.7°$, $a = 127$ ft, $c = 69.8$ ft; find b

25. $C = 51° \ 20'$, $c = 68.3$ m, $b = 58.2$ m; find B

26. $A = 51° \ 20'$, $c = 68.3$ m, $b = 58.2$ m; find a

27. $a = 165$ m, $A = 100.2°$, $B = 25.0°$; find b

28. $a = 14.8$ m, $b = 19.7$ m, $c = 31.8$ m; find B

29. $B = 39° \ 50'$, $b = 268$ m, $a = 340$ m; find A

30. $A = 46° \ 10'$, $b = 184$ cm, $c = 192$ cm; find a

31. $C = 79° \ 20'$, $c = 97.4$ mm, $a = 75.3$ mm; find A

32. $a = 7.5$ ft, $b = 12.0$ ft, $c = 6.9$ ft; find C

Use the law of sines to find all possible triangles with the given information.

33. $A = 25° \ 10'$, $a = 6.92$ yd, $b = 4.82$ yd

34. $A = 61.7°$, $a = 78.9$ m, $b = 86.4$ m

*Dunham, W., *The Mathematical Universe* (New York: John Wiley & Sons, Inc., 1994).

Solve the problem. Use the law of sines and/or the law of cosines.

35. The angles of elevation of a balloon from two points *A* and *B* on level ground are 24° 50′ and 47° 20′, respectively. As shown in the figure, points *A* and *B* are in the same vertical plane and are 8.4 miles apart. Approximate the height of the balloon above the ground to the nearest tenth of a mile.

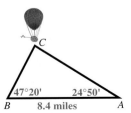

36. The pitcher's mound on a regulation softball field is 46 feet from home plate. The distance between the bases is 60 feet, as shown in the figure. How far is the pitcher's mound at point *M* from third base (point *T*)? Give your answer to the nearest foot.

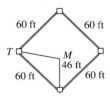

37. The course for a boat race starts at point *X* and goes in the direction S 48° W to point *Y*. It then turns and goes S 36° E to point *Z*, and finally returns back to point *X*. If point *Z* lies 10 kilometers directly south of point *X*, find the distance from *Y* to *Z* to the nearest kilometer.

38. Radio direction finders are placed at points *A* and *B*, which are 3.46 miles apart on an east-west line, with *A* west of *B*. From *A* the bearing of a certain illegal pirate radio transmitter is 48°, and from *B* the bearing is 302°. Find the distance between the transmitter and *A*.

39. To measure the distance *AB* across a canyon for a power line, a surveyor measures angles *B* and *C* and the distance *BC*. (See the figure.) What is the distance from *A* to *B*?

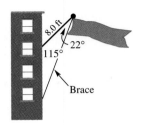

40. A banner on an 8.0-ft pole is to be mounted on a building at an angle of 115°, as shown in the figure. Find the length of the brace.

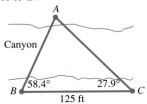

41. A hanging sculpture in an art gallery is to be hung with two wires of lengths 15.0 ft and 12.2 ft so that the angle between them is 70.3°. How far apart should the ends of the wire be placed on the ceiling?

42. A pipeline is to run between points *A* and *B*, which are separated by a protected wetlands area. To avoid the wetlands, the pipe will run from *A* to *C* and then to *B*. The distances involved are *AB* = 150 km, *AC* = 102 km, and *BC* = 135 km. What angle should be used at point *C*?

Solve the following "miscellaneous" problems involving trigonometric functions and formulas.

43. A painting 3 ft high and 6 ft from the floor will cut off an angle *θ* to an observer, where

$$\theta = \tan^{-1}\left(\frac{x}{x^2 + 2}\right).$$

Assume that the observer is *x* ft from the wall where the painting is displayed and that the eyes of the observer are 5 ft above the ground. See the figure. Rounding to the nearest degree, find the values of *θ* for the following values of *x*:

(a) 1   **(b)** 2   **(c)** 3.

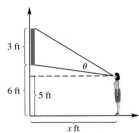

44. The slope of a line is defined as the ratio of the vertical change to the horizontal change. As shown in the figure, the tangent of the *angle of inclination* θ is given by the ratio of the side opposite and the side adjacent. This ratio is the same as that used in finding the slope, m, so that $m = \tan \theta$. In the figure on the right, let the two lines have angles of inclination α and β, and slopes m_1 and m_2, respectively. Let θ be the smallest positive angle between the lines. Show that

$$\tan \theta = \frac{m_2 - m_1}{1 + m_1 m_2}.$$

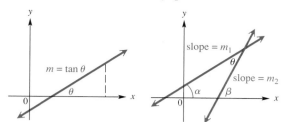

45. Many computer languages such as BASIC and FORTRAN have only the arctangent function available. To use the other inverse trigonometric functions, it is necessary to express them in terms of arctangent. This can be done as follows.
 (a) Let $u = \arcsin x$. Solve the equation for x in terms of u.
 (b) Use the result of part (a) to label the three sides of the triangle of the figure in terms of x.
 (c) Use the triangle from part (b) to write an equation for $\tan u$ in terms of x.
 (d) Solve the equation from part (c) for u.

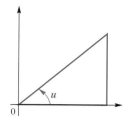

46. Exact values of the trigonometric functions of $15°$ can be found by the following method, an alternative to the use of the half number formulas. Start with a right triangle ABC having a $60°$ angle at A and a $30°$ angle at B. Let the hypotenuse of this triangle have length 2. Extend side BC and draw a semicircle with diameter along BC extended, center at B, and radius AB. Draw segment AE. (See the figure.) Since any angle inscribed in a semicircle is a right angle, triangle AED is a right triangle.

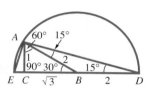

Prove each of the following statements.
 (a) Triangle ABD is isosceles.
 (b) Angle ABD is $150°$.
 (c) Angle DAB and angle ADB are each $15°$.
 (d) DC has length $2 + \sqrt{3}$.
 (e) Since AC has length 1, the length of AD is $AD = \sqrt{1^2 + (2 + \sqrt{3})^2}$. Reduce this to $\sqrt{8 + 4\sqrt{3}}$, and show that this result equals $\sqrt{6} + \sqrt{2}$.
 (f) Use angle ADB of triangle ADE and find $\cos 15°$.
 (g) Show that AE has length $\sqrt{6} - \sqrt{2}$. Then find $\sin 15°$.
 (h) Use triangle ACE and find $\tan 15°$.

47. A method that surveyors use to determine a small distance d between two points P and Q is called the *subtense bar method*. The subtense bar with length b is centered at Q and situated perpendicular to the line of sight between P and Q. See the figure. The angle θ is measured and then the distance d can be determined.*

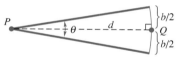

 (a) Find d when $\theta = 1° 23' 12''$ and $b = 2$ meters.
 (b) The angle θ usually cannot be measured more accurately than to the nearest $1''$. How much change would there be in the value of d if θ were measured $1''$ larger?

48. (See Exercise 47.) A variation of the subtense bar method that surveyors use to determine larger distances d between two points P and Q is shown in the accompanying figure. In this case the subtense bar with length b is placed between the points P and Q so that the bar is centered on and perpendicular to the line of sight connecting P and Q. The angles α and β are measured from points P and Q, respectively.*

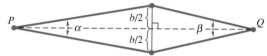

 (a) Find a formula for d involving α, β, and b.
 (b) Use your formula to determine d if $\alpha = 37' 48''$, $\beta = 42' 03''$, and $b = 2$ meters.

---

*Mueller I. and K. Ramsayer, *Introduction to Surveying* (New York: Frederick Ungar Publishing Co., 1979).

49. When an automobile travels along a circular curve, objects like trees and buildings situated on the inside of the curve can obstruct a driver's vision. These obstructions prevent the driver from seeing sufficiently far down the highway to ensure a safe stopping distance. See the accompanying figure. The *minimum* distance d that should be cleared on the inside of the highway is given by the equation $d = R(1 - \cos\frac{\beta}{2})$.*

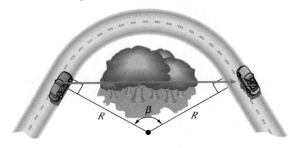

(a) It can be shown that if β is measured in degrees then $\beta \approx \frac{57.3\,S}{R}$, where S is the safe stopping distance for the given speed limit. Compute d for a 55-mile-per-hour speed limit if $S = 336$ feet and $R = 600$ feet.

(b) Compute d for a 65-mile-per-hour speed limit if $S = 485$ feet and $R = 600$ feet.

(c) How does the speed limit affect the amount of land that should be cleared on the inside of the curve?

50. With your calculator in degree mode, follow these steps:
(a) Enter the year of your birth (all four digits).
(b) Subtract the number of years that have elapsed since 1980. For example, if it is 1996, subtract 16.
(c) Find the sine of the display.
(d) Find the inverse sine of the new display.
(e) What do you notice about the result in part (d)? (For an explanation of why this procedure works as it does, see "Sine of the Times: Your Age in a Flash" by E. John Hornsby, Jr., in the October 1985 issue of *Mathematics Teacher*.)

Find the area of the triangle with the given information.

51. $b = 840.6$ m, $c = 715.9$ m, $A = 149° \ 18'$

52. $a = 6.90$ ft, $b = 10.2$ ft, $C = 35° \ 10'$

53. $a = .913$ km, $b = .816$ km, $c = .582$ km

54. $a = 43$ m, $b = 32$ m, $c = 51$ m

Solve each of the following problems involving linear and angular velocity.

55. Find t if $\theta = \frac{5\pi}{12}$ radians and $\omega = \frac{8\pi}{9}$ radians per sec.

56. Find θ if $t = 12$ sec and $\omega = 9$ radians per sec.

57. Find ω if $t = 8$ sec and $\theta = \frac{2\pi}{5}$ radians.

58. Find ω if $s = \frac{12\pi}{25}$ ft, $r = \frac{3}{5}$ ft, $t = 15$ sec.

59. Find s if $r = 11.46$ cm, $\omega = 4.283$ radians per sec, and $t = 5.813$ sec.

60. Find the linear velocity of a point on the edge of a flywheel of radius 7 m if the flywheel is rotating 90 times per sec.

---

* Mannering, F. and W. Kilareski, *Principles of Highway Engineering and Traffic Control* (New York: John Wiley & Sons., 1990).

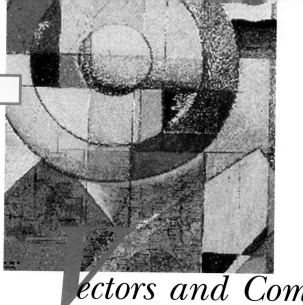

CHAPTER 5

Vectors and Complex Numbers

5.1 VECTORS AND APPLICATIONS

Basic Terminology ▮ Direction Angle and Components ▮ Operations with Vectors ▮ Dot Product and Angle Between Vectors ▮ Applications of Vectors

Basic Terminology

Many quantities in mathematics involve magnitudes, such as 45 lb or 60 mph. These quantities are called **scalars.** Other quantities, called **vector quantities,** involve both magnitude and direction. Typical vector quantities are velocity, acceleration, and force.

A vector quantity is often represented with a directed line segment, called a **vector.** The length of the vector represents the magnitude of the vector quantity. The direction of the vector, indicated with an arrowhead, represents the direction of the quantity. For example, the vector in Figure 1 represents a force of 10 lb applied at an angle of 30° from the horizontal.

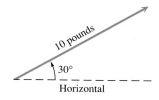

FIGURE 1

The symbol for a vector is often printed in boldface type. To write vectors by hand, it is customary to use an arrow over the letter or letters. Thus **OP** and $\overrightarrow{OP}$

both represent vector **OP**. Vectors may be named with either one lowercase or uppercase letter, or two uppercase letters. When two letters are used, the first indicates the *initial point* and the second indicates the *terminal point* of the vector. Knowing these points gives the direction of the vector. For example, vectors **OP** and **PO** in Figure 2 are not the same vectors. They have the same magnitude, but opposite directions. The magnitude of vector **OP** is written $|\mathbf{OP}|$.

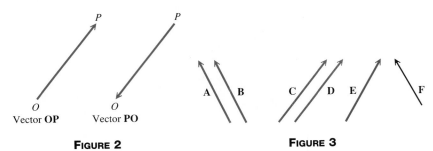

FIGURE 2   **FIGURE 3**

Two vectors are *equal* if and only if they both have the same directions and the same magnitudes. In Figure 3, vectors **A** and **B** are equal, as are vectors **C** and **D**.

To find the **sum** of two vectors **A** and **B**, we place the initial point of vector **B** at the terminal point of vector **A**, as shown in Figure 4. The vector with the same initial point as **A** and the same terminal point as **B** is the sum **A** + **B**. The sum of two vectors is also a vector.

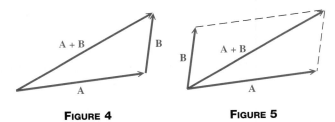

FIGURE 4   **FIGURE 5**

Another way to find the sum of two vectors is to use the **parallelogram rule.** Place vectors **A** and **B** so that their initial points coincide. Then complete a parallelogram that has **A** and **B** as two sides. The diagonal of the parallelogram with the same initial point as **A** and **B** is the same vector sum **A** + **B** found by the definition. Compare Figures 4 and 5.

Parallelograms can be used to show that vector **B** + **A** is the same as vector **A** + **B**, or that

$$\mathbf{A} + \mathbf{B} = \mathbf{B} + \mathbf{A},$$

so that vector addition is *commutative.*

The vector sum **A** + **B** is called the **resultant** of vectors **A** and **B**. Each of the vectors **A** and **B** is a **component** of vector **A** + **B**. In many practical applications, such as surveying, it is necessary to break a vector into its **vertical** and **horizontal components.** These components are two vectors, one vertical and one horizontal, whose resultant is the original vector. As shown in Figure 6(a) vector **OR** is the vertical component and vector **OS** is the horizontal component of **OP**. Figure 6(b) shows an alternative method of indicating a resultant.

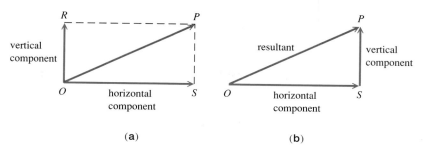

FIGURE 6

For every vector **v** there is a vector −**v** that has the same magnitude as **v** but opposite direction. Vector −**v** is called the **opposite** of **v**. (See Figure 7.) The sum of **v** and −**v** has magnitude 0 and is called the **zero vector.** As with real numbers, to *subtract* vector **B** from vector **A**, find the vector sum **A** + (−**B**). (See Figure 8.)

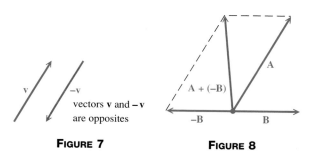

FIGURE 7   **FIGURE 8**

The **scalar product** of a real number (or scalar) k and a vector **u** is the vector $k \cdot \mathbf{u}$ which has magnitude $|k|$ times the magnitude of **u**. As suggested by Figure 9, the vector $k \cdot \mathbf{u}$ has the same direction as **u** if $k > 0$, and opposite direction if $k < 0$.

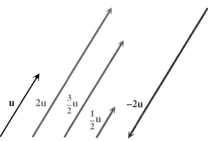

FIGURE 9

EXAMPLE 1

Finding the
Magnitude of
a Resultant

Two forces of 15 and 22 newtons (a *newton* is a unit of force used in physics) act on a point in the plane. If the angle between the forces is 100°, find the magnitude of the resultant force.

SOLUTION   As shown in Figure 10, a parallelogram that has the forces as adjacent sides can be formed. The angles of the parallelogram adjacent to angle P each measure 80°, since adjacent angles of a parallelogram are supplementary. Opposite sides of the parallelogram are equal in length. The resultant force divides the

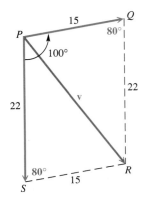

FIGURE 10

parallelogram into two triangles. Use the law of cosines with either triangle to get

$$|\mathbf{v}|^2 = 15^2 + 22^2 - 2(15)(22) \cos 80°$$
$$\approx 225 + 484 - 115$$
$$|\mathbf{v}|^2 \approx 594$$
$$|\mathbf{v}| \approx 24.$$

To the nearest unit, the magnitude of the resultant force is 24 newtons. ∎

Direction Angle and Components

If a vector is placed so that its initial point coincides with the origin of a rectangular coordinate system, then the angle between the x-axis and the vector, measured in a counterclockwise direction, is called a **direction angle** for the vector. In Figure 11, $\mathbf{u}$ has direction angle θ and magnitude r. The following basic rules for vectors are derived from the definition of direction angle and earlier results.

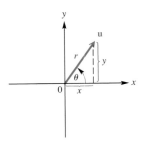

FIGURE 11

TECHNOLOGICAL NOTE
Many graphing calculators have built-in functions that will allow the user to find the horizontal and vertical components when the magnitude and direction angle are known, and also give the magnitude and direction angle when the horizontal and vertical components are known. The procedures for activating these functions vary greatly among models, and you should consult your owner's manual to see whether your model has these functions and how to use them if it does.

BASIC RULES FOR VECTORS

Let a vector have direction angle θ and magnitude r. Then the horizontal component of the vector has magnitude

$$x = r \cos \theta,$$

and the vertical component has magnitude

$$y = r \sin \theta.$$

Also,

$$x^2 + y^2 = r^2 \qquad \text{and} \qquad \tan \theta = \frac{y}{x}, \quad x \neq 0.$$

EXAMPLE 2

Finding Horizontal and Vertical Components

FIGURE 12

Vector $\mathbf{w}$ has magnitude 25.0 and direction angle 41.7°. Find the magnitudes of the horizontal and vertical components of the vector.

SOLUTION In Figure 12 the vertical component is labeled $\mathbf{v}$ and the horizontal component is labeled $\mathbf{u}$. Using the basic rules for vectors,

$$|\mathbf{v}| = 25.0 \sin 41.7° \approx 16.6,$$

and

$$|\mathbf{u}| = 25.0 \cos 41.7° \approx 18.7.$$

To the nearest tenth, the horizontal component is 18.7 and the vertical component is 16.6. ∎

FOR **GROUP DISCUSSION**

Modern graphing calculators are capable of finding resultants, direction angles, and horizontal and vertical components. Refer to your instruction manual to learn how to find (a) the magnitude of the resultant and the direction angle if the horizontal and vertical components are known and (b) the horizontal and vertical components if the magnitude of the resultant and the direction angle are known. To determine whether you have learned how to do these correctly, verify the following statements.

1. If the magnitude of the resultant is 3.5 and the direction angle is 150°, the horizontal component is -3.031088913 and the vertical component is 1.75. (Be sure that your calculator is in degree mode.)
2. If the horizontal component is 4.7 and the vertical component is 7.6, the magnitude of the resultant is 8.935882721 and the direction angle is 58.26648085°. (Again, be sure that your calculator is in degree mode.)

Let vector **u** be placed in a plane so that the initial point of the vector is at the origin, $(0, 0)$, and the endpoint is at the point (a, b). A vector with initial point at the origin is called a **position vector** or (sometimes) a **radius vector.** A position vector having endpoint at the point (a, b) is called the **vector (a, b).** To avoid confusion, the vector (a, b) is written as $\langle a, b \rangle$. The numbers a and b are called the **x-component** and **y-component,** respectively. Figure 13 shows the vector $\mathbf{u} = \langle a, b \rangle$.

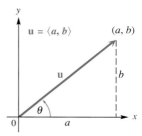

FIGURE 13

LENGTH OR MAGNITUDE OF A VECTOR

The length or magnitude of vector $\mathbf{u} = \langle a, b \rangle$ is given by

$$|\mathbf{u}| = \sqrt{a^2 + b^2}.$$

EXAMPLE **3**

Finding Magnitude and Direction Angle

Figure 14 shows vector $\mathbf{u} = \langle 3, -2 \rangle$. Find the magnitude and direction angle for **u**.

SOLUTION   The magnitude is

$$\sqrt{3^2 + (-2)^2} = \sqrt{13}.$$

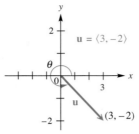

FIGURE 14

To find the direction angle θ, start with

$$\tan \theta = \frac{y}{x} = \frac{-2}{3} = -\frac{2}{3}.$$

Vector **u** has positive x-component and negative y-component, placing the vector in quadrant IV. A graphing calculator gives $\tan^{-1}\left(-\frac{2}{3}\right) \approx -33.7°$. Adding $360°$ yields the direction angle $326.3°$. This angle is shown in Figure 14. ◨

Operations with Vectors

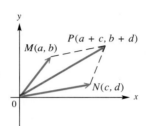

FIGURE 15

Let vector **OM** in Figure 15 be given by $\langle a, b \rangle$, and vector **ON** be given by $\langle c, d \rangle$. Let **OP** be given by $\langle a + c, b + d \rangle$. With facts from geometry points O, N, M, and P can be shown to form the vertices of a parallelogram. Since a diagonal of this parallelogram gives the resultant of **OM** and **ON**, vector **OP** is given by **OP** = **OM** + **ON**, with the resultant of $\langle a, b \rangle$ and $\langle c, d \rangle$ given by $\langle a + c, b + d \rangle$. In the same way, $k \cdot \langle a, b \rangle = \langle ka, kb \rangle$ for any real number k.

VECTOR OPERATIONS

For any real numbers a, b, c, d, and k,

$$\langle a, b \rangle + \langle c, d \rangle = \langle a + c, b + d \rangle$$
$$k \cdot \langle a, b \rangle = \langle ka, kb \rangle.$$

EXAMPLE 4

Performing Vector Operations

Let $\mathbf{u} = \langle -2, 1 \rangle$ and $\mathbf{v} = \langle 4, 3 \rangle$. Find each of the following: **(a)** $\mathbf{u} + \mathbf{v}$, **(b)** $-2\mathbf{u}$, **(c)** $4\mathbf{u} + 3\mathbf{v}$. (See Figure 16.)

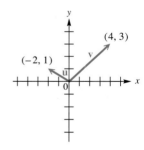

FIGURE 16

SOLUTION

(a) $\mathbf{u} + \mathbf{v} = \langle -2, 1 \rangle + \langle 4, 3 \rangle = \langle -2 + 4, 1 + 3 \rangle = \langle 2, 4 \rangle$

(b) $-2\mathbf{u} = -2 \cdot \langle -2, 1 \rangle = \langle -2(-2), -2(1) \rangle = \langle 4, -2 \rangle$

(c) $4\mathbf{u} + 3\mathbf{v} = 4 \cdot \langle -2, 1 \rangle + 3 \cdot \langle 4, 3 \rangle = \langle -8, 4 \rangle + \langle 12, 9 \rangle$
$$= \langle -8 + 12, 4 + 9 \rangle = \langle 4, 13 \rangle \quad ◨$$

For a vector **u** with magnitude r and direction angle θ, it was shown above that the horizontal and vertical components of **u** have magnitudes

$$|\mathbf{x}| = r \cos \theta \quad \text{and} \quad |\mathbf{y}| = r \sin \theta.$$

This leads to the following result.

> If a vector **u** has direction angle θ and magnitude r, then
> $$\mathbf{u} = \langle r \cos \theta, r \sin \theta \rangle.$$

EXAMPLE 5

Writing Vectors in the Form $\langle a, b \rangle$

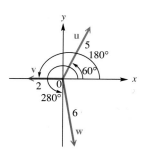

FIGURE 17

Write the vectors in Figure 17 in the form $\langle a, b \rangle$.

SOLUTION Vector **u** in Figure 17 has a magnitude of 5 and a direction angle 60°. By the result above,

$$\mathbf{u} = \langle 5 \cos 60°, 5 \sin 60° \rangle = \left\langle 5 \cdot \frac{1}{2}, 5 \cdot \frac{\sqrt{3}}{2} \right\rangle = \left\langle \frac{5}{2}, \frac{5\sqrt{3}}{2} \right\rangle.$$

Also, $\mathbf{v} = \langle 2 \cos 180°, 2 \sin 180° \rangle = \langle 2(-1), 2(0) \rangle = \langle -2, 0 \rangle.$

Finally, $\mathbf{w} = \langle 6 \cos 280°, 6 \sin 280° \rangle$

or $\mathbf{w} \approx \langle 1.0419, -5.9088 \rangle.$ ◻

FOR GROUP DISCUSSION

Use your calculator's capability to work with vectors to support the answers found in Example 5. Note that for **u**, the calculator will give a decimal approximation for the vertical component. Verify that what the calculator gives is the same result you get if you use the calculator to approximate $\frac{5\sqrt{3}}{2}$. Be sure your calculator is in degree mode.

Dot Product and Angle Between Vectors

The **dot product** of the two vectors $\mathbf{u} = \langle a, b \rangle$ and $\mathbf{v} = \langle c, d \rangle$ is denoted by $\mathbf{u} \cdot \mathbf{v}$, pronounced "**u** dot **v**," and defined by the formula

$$\mathbf{u} \cdot \mathbf{v} = ac + bd.$$

The dot product of two vectors is a real number, not a vector. It is also known as the *inner product* or *scalar product*. Dot products are used to determine the angle between two vectors, to derive geometric theorems, and to solve problems in physics.

EXAMPLE 6

Finding the Dot Product

Find the dot product for **(a)** $\langle 2, 3 \rangle \cdot \langle 4, -1 \rangle$ and **(b)** $\langle 6, 4 \rangle \cdot \langle -2, 3 \rangle$.

SOLUTION

(a) $\langle 2, 3 \rangle \cdot \langle 4, -1 \rangle = (2)(4) + (3)(-1) = 5$

(b) $\langle 6, 4 \rangle \cdot \langle -2, 3 \rangle = (6)(-2) + (4)(3) = 0$ ◻

The following properties of dot products are easily verified.

> **PROPERTIES OF THE DOT PRODUCT**
>
> For all vectors **u**, **v**, and **w** and real numbers k,
>
> **a.** $\mathbf{u} \cdot \mathbf{v} = \mathbf{v} \cdot \mathbf{u}$ **d.** $(k\mathbf{u}) \cdot \mathbf{v} = k(\mathbf{u} \cdot \mathbf{v}) = \mathbf{u} \cdot (k\mathbf{v})$
>
> **b.** $\mathbf{u} \cdot (\mathbf{v} + \mathbf{w}) = \mathbf{u} \cdot \mathbf{v} + \mathbf{u} \cdot \mathbf{w}$ **e.** $\mathbf{0} \cdot \mathbf{u} = 0$
>
> **c.** $(\mathbf{u} + \mathbf{v}) \cdot \mathbf{w} = \mathbf{u} \cdot \mathbf{w} + \mathbf{v} \cdot \mathbf{w}$ **f.** $\mathbf{u} \cdot \mathbf{u} = |\mathbf{u}|^2.$

To prove the first part of (d), let $\mathbf{u} = \langle a, b \rangle$ and $\mathbf{v} = \langle c, d \rangle$. Then

$$(k\mathbf{u}) \cdot \mathbf{v} = (k\langle a, b \rangle) \cdot \langle c, d \rangle = \langle ka, kb \rangle \cdot \langle c, d \rangle = kac + kbd$$
$$= k(ac + bd) = k(\langle a, b \rangle \cdot \langle c, d \rangle) = k(\mathbf{u} \cdot \mathbf{v}).$$

The proofs of the remaining properties are done in a similar manner.

As demonstrated in Example 6, the product of two vectors can be positive or zero. It may also be negative. There is a geometric interpretation to the dot product that explains when each of these cases occurs. This interpretation involves the angle between the two vectors. Consider the vectors $\mathbf{u} = \langle a, b \rangle$ and $\mathbf{v} = \langle c, d \rangle$ drawn from the origin of a rectangular coordinate system as shown in Figure 18. The **angle θ between u and v** is defined to be the angle having the two vectors as its sides for which $0° \leq \theta \leq 180°$. The following theorem relates the dot product to the angle between the vectors.

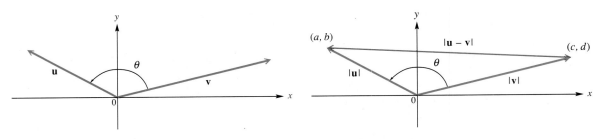

FIGURE 18   **FIGURE 19**

GEOMETRIC INTERPRETATION OF DOT PRODUCT

If θ is the angle between the two nonzero vectors $\mathbf{u}$ and $\mathbf{v}$, where $0° \leq \theta \leq 180°$, then

$$\mathbf{u} \cdot \mathbf{v} = |\mathbf{u}| \, |\mathbf{v}| \cos \theta.$$

First examine the case where $\theta \neq 0°$ and $\theta \neq 180°$. To prove the theorem, consider the triangle shown in Figure 19 with sides of length $|\mathbf{u}|$, $|\mathbf{v}|$, and $|\mathbf{u} - \mathbf{v}|$. Let $\mathbf{u} = \langle a, b \rangle$ and $\mathbf{v} = \langle c, d \rangle$. Then $\mathbf{u} - \mathbf{v} = \langle a - c, b - d \rangle$. By the law of cosines,

$$|\mathbf{u} - \mathbf{v}|^2 = |\mathbf{u}|^2 + |\mathbf{v}|^2 - 2|\mathbf{u}| \, |\mathbf{v}| \cos \theta,$$

which may be rearranged to produce

$$|\mathbf{u}| \, |\mathbf{v}| \cos \theta = \frac{1}{2}(|\mathbf{u}|^2 + |\mathbf{v}|^2 - |\mathbf{u} - \mathbf{v}|^2).$$

Now use the distance formula.

$$|\mathbf{u}| \, |\mathbf{v}| \cos \theta = \frac{1}{2}(a^2 + b^2 + c^2 + d^2 - (a - c)^2 - (b - d)^2)$$

$$= ac + bd$$

$$= \mathbf{u} \cdot \mathbf{v} \qquad \text{Definition of dot product}$$

Let us now derive the result for the two cases $\theta = 0°$ and $\theta = 180°$. When $\theta = 0°$, the vectors have the same direction and therefore $\mathbf{u} = k\mathbf{v}$ for some positive real number k. When $\theta = 180°$, the vectors have opposite directions and therefore $\mathbf{u} = k\mathbf{v}$ for some negative real number k. In either of these two cases,

$$\mathbf{u} \cdot \mathbf{v} = (k\mathbf{v}) \cdot \mathbf{v} = k(\mathbf{v} \cdot \mathbf{v}) = k|\mathbf{v}|^2$$
$$|\mathbf{u}||\mathbf{v}| \cos \theta = |k\mathbf{v}||\mathbf{v}| \cos \theta = |k||\mathbf{v}||\mathbf{v}| \cos \theta$$
$$= |k||\mathbf{v}|^2 \cos \theta.$$

Now, when $\theta = 0°$, $\qquad |k||\mathbf{v}|^2 \cos \theta = k|\mathbf{v}|^2(1) = k|\mathbf{v}|^2$.
When $\theta = 180°$, $\qquad |k||\mathbf{v}|^2 \cos \theta = -k|\mathbf{v}|^2(-1) = k|\mathbf{v}|^2$.

Thus, the theorem holds for both of these cases.

EXAMPLE 7

Finding the Angle
Between Two
Vectors

Find the angle between the two vectors $\mathbf{u} = \langle 3, 4 \rangle$ and $\mathbf{v} = \langle 2, 1 \rangle$.

SOLUTION By the equation discussed above,

$$\cos \theta = \frac{\mathbf{u} \cdot \mathbf{v}}{|\mathbf{u}||\mathbf{v}|} = \frac{\langle 3, 4 \rangle \cdot \langle 2, 1 \rangle}{|\langle 3, 4 \rangle||\langle 2, 1 \rangle|}$$

$$= \frac{3(2) + 4(1)}{\sqrt{9 + 16} \sqrt{4 + 1}} = \frac{10}{5\sqrt{5}} \approx .894427191$$

Therefore, $\qquad \theta = \cos^{-1}(.894427191) \approx 26.57°.$ ∎

For angles θ between $0°$ and $180°$, $\cos \theta$ is positive, zero, or negative when θ is less than, equal to, or greater than $90°$, respectively. Therefore, the dot product is positive, zero, or negative according to this table.

Dot Product	Angle between Vectors
positive	acute
zero	right
negative	obtuse

Applications of Vectors

The law of sines and the law of cosines, introduced in Chapter 4, are often useful in solving applied problems involving vectors. You may wish to review them at this time.

Navigation problems, such as the one in the following example, can often be solved by using vectors.

EXAMPLE 8

Applying Vectors to
a Navigation
Problem

A ship leaves port on a bearing of $28°$ and travels 8.2 miles. The ship then turns due east and travels 4.3 miles. How far is the ship from port? What is its bearing from port?

SOLUTION In Figure 20, vectors **PA** and **AE** represent the ship's path. The magnitude and bearing of the resultant **PE** can be found as follows. Triangle *PNA* is a right triangle, so angle $NAP = 90° - 28° = 62°$. Then angle $PAE = 180° - 62° = 118°$. Use the law of cosines to find $|\mathbf{PE}|$, the magnitude of vector **PE**.

$$|\mathbf{PE}|^2 = 8.2^2 + 4.3^2 - 2(8.2)(4.3) \cos 118°$$
$$|\mathbf{PE}|^2 \approx 118.84$$

Therefore, $\qquad |\mathbf{PE}| \approx 10.9,$

or 11 miles, rounded to two significant digits.

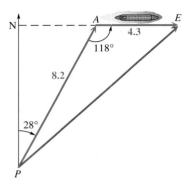

FIGURE 20

To find the bearing of the ship from port, first find angle *APE*. Use the law of sines, along with the value of $|\mathbf{PE}|$ before rounding.

$$\frac{\sin APE}{4.3} = \frac{\sin 118°}{10.9}$$

$$\sin APE = \frac{4.3 \sin 118°}{10.9}$$

$$\text{angle } APE \approx 20.4°$$

After rounding, angle *APE* is 20°, and the ship is 11 miles from port on a bearing of 28° + 20° = 48°. ◨

In air navigation, the **airspeed** of a plane is its speed relative to the air, while the **groundspeed** is its speed relative to the ground. Because of wind, these two speeds are usually different. The groundspeed of the plane is represented by the vector sum of the airspeed and windspeed vectors. See Figure 21.

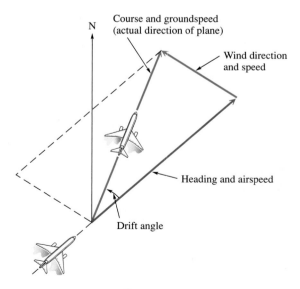

FIGURE 21

EXAMPLE 9

Applying Vectors to a Navigation Problem

A plane with an airspeed of 192 miles per hour is headed on a bearing of 121°. A north wind is blowing (from north to south) at 15.9 miles per hour. Find the groundspeed and the actual bearing of the plane.

SOLUTION   In Figure 22, the groundspeed is represented by $|\mathbf{x}|$. We must find angle α to find the bearing, which will be $121° + \alpha$. From Figure 22, angle BCO equals angle AOC, which equals 121°. Find $|\mathbf{x}|$ by the law of cosines.

$$|\mathbf{x}|^2 = 192^2 + 15.9^2 - 2(192)(15.9) \cos 121°$$
$$|\mathbf{x}|^2 \approx 40{,}261$$

Therefore, $|\mathbf{x}| \approx 200.7$,

or 201 miles per hour. Now find α by using the law of sines. As before, use the value of $|\mathbf{x}|$ before rounding.

$$\frac{\sin \alpha}{15.9} = \frac{\sin 121°}{200.7}$$
$$\sin \alpha \approx .06792320$$
$$\alpha \approx 3.89°$$

After rounding, α is 3.9°. The groundspeed is about 201 miles per hour, on a bearing of approximately 125°.   ◼

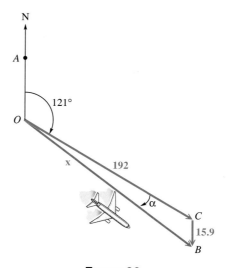

FIGURE 22

5.1   EXERCISES

Exercises 1–4 refer to the following vectors.

1. Name all pairs of vectors that appear to be equal.

2. Name all pairs of vectors that are opposites.

3. Name all pairs of vectors where the first is a scalar multiple of the other, with the scalar positive.

4. Name all pairs of vectors where the first is a scalar multiple of the other, with the scalar negative.

*Exercises 5–22 refer to the vectors pictured. Make a copy or a careful sketch of each, and then draw a sketch to represent the following vectors. For example, find **a** + **e** by placing **a** and **e** so that their initial points coincide. Then use the parallelogram rule to find the resultant, as shown in the figure.*

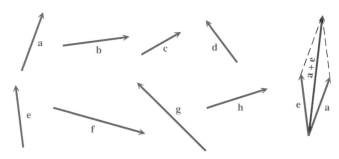

5. −**b**   **6.** −**g**   **7.** 3**a**   **8.** 2**h**   **9.** **a** + **c**

10. **a** + **b**   **11.** **h** + **g**   **12.** **e** + **f**   **13.** **a** + **h**   **14.** **b** + **d**

15. **h** + **d**   **16.** **a** + **f**   **17.** **a** − **c**   **18.** **d** − **e**   **19.** **a** + (**b** + **c**)

20. (**a** + **b**) + **c**   **21.** **c** + **d**   **22.** **d** + **c**

23. From the results of Exercises 19 and 20, do you think vector addition is associative?

24. From the results of Exercises 21 and 22, do you think vector addition is commutative?

*For each pair of vectors **u** and **w** with angle θ between them, sketch the resultant.*

25. $|\mathbf{u}| = 12, |\mathbf{w}| = 20, \theta = 27°$ **26.** $|\mathbf{u}| = 8, |\mathbf{w}| = 12, \theta = 20°$ **27.** $|\mathbf{u}| = 20, |\mathbf{w}| = 30, \theta = 30°$

28. $|\mathbf{u}| = 27, |\mathbf{w}| = 50, \theta = 12°$ **29.** $|\mathbf{u}| = 50, |\mathbf{w}| = 70, \theta = 40°$

*For each of the following, vector **v** has the given magnitude and makes an angle θ with the horizontal. Find the horizontal and vertical components of **v**, first using the relationships x = r cos θ and y = r sin θ. Then use the capabilities of your calculator to support your answers.*

30. $|\mathbf{v}| = 50, \theta = 20°$ **31.** $|\mathbf{v}| = 12, \theta = 38°$ **32.** $|\mathbf{v}| = 150, \theta = 70°$

33. $|\mathbf{v}| = 26, \theta = 50°$ **34.** $|\mathbf{v}| = 47.8, \theta = 35°50'$ **35.** $|\mathbf{v}| = 15.4, \theta = 27.5°$

36. $|\mathbf{v}| = 78.9, \theta = 59°40'$ **37.** $|\mathbf{v}| = 198, \theta = 128.5°$ **38.** $|\mathbf{v}| = 238, \theta = 146.3°$

In each of the following, two forces act at a point in the plane. The angle between the two forces is given. Find the magnitude of the resultant force.

39. Forces of 250 and 450 newtons, forming an angle of 85°

40. Forces of 19 and 32 newtons, forming an angle of 118°

41. Forces of 17.9 and 25.8 pounds, forming an angle of 105°30′

42. Forces of 75.6 and 98.2 pounds, forming an angle of 82°50′

Find the dot product for each of the following pairs of vectors.

43. $\langle 6, -1 \rangle, \langle 2, 5 \rangle$ **44.** $\langle -3, 8 \rangle, \langle 7, -5 \rangle$ **45.** $\langle 2, -3 \rangle, \langle 6, 5 \rangle$

46. $\langle 1, 2 \rangle$, $\langle 3, -1 \rangle$ **47.** $\langle 4, 0 \rangle$, $\langle 5, -9 \rangle$ **48.** $\langle 2, 4 \rangle$, $\langle 0, -1 \rangle$

Find the angles between each of the following pairs of vectors.

49. $\langle 2, 1 \rangle$, $\langle -3, 1 \rangle$ **50.** $\langle 1, 7 \rangle$, $\langle 1, 1 \rangle$ **51.** $\langle 1, 2 \rangle$, $\langle -6, 3 \rangle$
52. $\langle 4, 0 \rangle$, $\langle 2, 2 \rangle$ **53.** $\langle 3, 4 \rangle$, $\langle 0, 1 \rangle$ **54.** $\langle -5, 12 \rangle$, $\langle 3, 2 \rangle$

Let $\mathbf{u} = \langle -2, 1 \rangle$, $\mathbf{v} = \langle 3, 4 \rangle$, *and* $\mathbf{w} = \langle -5, 12 \rangle$. *Evaluate each of the following.*

55. $(3\mathbf{u}) \cdot \mathbf{v}$ **56.** $\mathbf{u} \cdot (\mathbf{v} - \mathbf{w})$ **57.** $\mathbf{u} \cdot \mathbf{v} - \mathbf{u} \cdot \mathbf{w}$ **58.** $\mathbf{u} \cdot (3\mathbf{v})$

Two vectors are said to be *orthogonal* when the angle between them is a right angle. The following result is an important consequence of the table found in this section.

ORTHOGONAL VECTORS

Two nonzero vectors $\mathbf{u}$ and $\mathbf{v}$ are **orthogonal vectors** if and only if $\mathbf{u} \cdot \mathbf{v} = 0$.

Use this property to determine whether the following pairs of vectors are orthogonal.

59. $\langle 1, 2 \rangle$, $\langle -6, 3 \rangle$ **60.** $\langle 3, 4 \rangle$, $\langle 6, 8 \rangle$ **61.** $\langle 1, 0 \rangle$, $\langle \sqrt{2}, 0 \rangle$
62. $\langle 1, 1 \rangle$, $\langle 1, -1 \rangle$ **63.** $\langle \sqrt{5}, -2 \rangle$, $\langle -5, 2\sqrt{5} \rangle$ **64.** $\langle -4, 3 \rangle$, $\langle 8, -6 \rangle$

Solve each of the following problems.

65. A force of 176 lb makes an angle of 78°50′ with a second force. The resultant of the two forces makes an angle of 41°10′ with the first force. Find the magnitude of the second force and of the resultant.

66. A force of 28.7 lb makes an angle of 42°10′ with a second force. The resultant of the two forces makes an angle of 32°40′ with the first force. Find the magnitudes of the second force and of the resultant.

67. A plane flies 650 mph on a bearing of 175.3°. A 25-mph wind, from a direction of 266.6°, blows against the plane. Find the resulting bearing of the plane.

68. A pilot wants to fly on a bearing of 74.9°. By flying due east, he finds that a 42-mph wind, blowing from the south, puts him on course. Find the airspeed and the groundspeed.

69. Starting at point *A*, a ship sails 18.5 km on a bearing of 189°, then turns and sails 47.8 km on a bearing of 317°. Find the distance of the ship from point *A*.

70. A pilot is flying at 168 mph. She wants her flight path to be on a bearing of 57°40′. A wind is blowing from the south at 27.1 mph. Find the bearing the pilot should fly, and find the plane's groundspeed.

71. What bearing and airspeed are required for a plane to fly 400 mi due north in 2.5 hr if the wind is blowing from a direction of 328° at 11 mph?

72. A plane is headed due south with an airspeed of 192 mph. A wind from a direction of 78° is blowing at 23 mph. Find the groundspeed and resulting bearing of the plane.

73. An airplane is headed on a bearing of 174° at an airspeed of 240 km per hr. A 30-km-per-hr wind is blowing from a direction of 245°. Find the groundspeed and resulting bearing of the plane.

74. A ship sailing due east in the North Atlantic has been warned to change course to avoid a group of icebergs. The captain turns and sails on a bearing of 62° for a while, then changes course again to a bearing of 115° until the ship reaches its original course. (See the figure.) How much farther did the ship have to travel to avoid the icebergs?

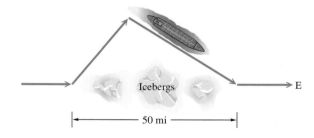

75. In order to find the coordinates of control points for aerial photographs, ground control must first locate basic control monuments established by the U.S. Coast and Geodetic Survey and the U.S. Geological

Survey. These monuments have published x- and y-coordinates called *state plane coordinates*. Using these monuments and common surveying techniques the coordinates of the control points can be determined. Two basic control monuments A and B have coordinates in feet of $x_A = 2,101,345.1$, $y_A = 998,764.3$ and $x_B = 2,131,667.8$, $y_B = 923.541.7$. The location of an unknown control point P is to be determined. If angles PAB and PBA are measured as $37°41'37''$ and $57°52'04''$, respectively, determine the state plane coordinates of control point P.* See the figure.

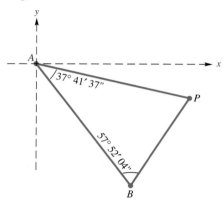

76. Refer to Exercise 75. The state plane coordinates in feet for two basic control monuments A and B are $x_A = 1,651,334.5$, $y_A = 1,102,749.3$ and $x_B = 1,654,639.9$, $y_B = 1,103,431.2$, respectively. The location of an unknown control point C is to be determined. Distance CA is measured as 4834.7 feet and distance CB is 3371.8 feet. Determine the state plane coordinates of the control point C.

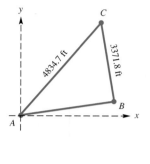

*R*elating Concepts

Consider the two vectors $\mathbf{v}$ *and* $\mathbf{u}$ *shown. Assume all values are exact. Work Exercises 77–82 in order.*

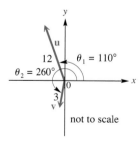

not to scale

77. Use trigonometry alone (without using vector notation) to find the magnitude and direction angle of $\mathbf{u} + \mathbf{v}$. You should use the law of cosines and the law of sines in your work.

78. Find the horizontal and vertical components of $\mathbf{u}$, using your calculator.

79. Find the horizontal and vertical components of $\mathbf{v}$, using your calculator.

---

*Moffitt, F., *Photogrammetry* (Scranton: International Textbook Company, 1967).

80. Find the horizontal and vertical components of **u** + **v** by adding the results you obtained in Exercises 78 and 79.

81. Use your calculator to find the magnitude and direction angle of the vector **u** + **v**.

82. Compare your answers in Exercises 77 and 81. What do you notice? Which method of solution do you prefer?

5.2 TRIGONOMETRIC OR POLAR FORM OF COMPLEX NUMBERS; OPERATIONS AND APPLICATIONS

The Complex Plane and Vector Representation ∎ Trigonometric or Polar Form ∎ Products of Complex Numbers in Polar Form ∎ Quotients of Complex Numbers in Polar Form ∎ Applications

The Complex Plane and Vector Representation

Complex numbers of the form $a + bi$, where a and b are real numbers and $i^2 = -1$, are studied in algebra courses. We will refer to this form as rectangular form, since the real numbers a and b can be represented by a point (a, b) in the rectangular coordinate plane. We can extend the concepts of rectangular coordinates so that a complex number, such as $2 - 3i$, can be represented by a position vector $\langle 2, -3 \rangle$. In this context, the horizontal axis will be called the **real axis,** and the vertical axis the **imaginary axis.** Then complex numbers can be graphed in this **complex plane,** as shown in Figure 23 for the complex number $2 - 3i$.

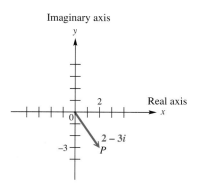

FIGURE 23

TECHNOLOGICAL NOTE
Some graphing calculators are capable for performing arithmetic of complex numbers, often in ordered pair form. Consult your owner's manual to see if your model has this capability.

Each complex number graphed in this way determines a unique position vector.

Recall from algebra that the sum of the two complex numbers $4 + i$ and $1 + 3i$ is found as follows:

$$(4 + i) + (1 + 3i) = 5 + 4i$$

Graphically, the sum of two complex numbers is represented by the vector that is the resultant of the vectors corresponding to the two numbers. The vectors representing the complex numbers $4 + i$ and $1 + 3i$ and the resultant vector that represents their sum, $5 + 4i$, are shown in Figure 24.

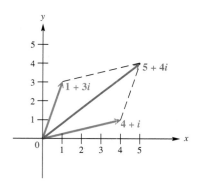

FIGURE 24

EXAMPLE 1

Expressing the Sum
of Complex
Numbers
Graphically

Find the sum of $6 - 2i$ and $-4 - 3i$. Graph both complex numbers and their resultant.

SOLUTION   The sum is found by adding the two numbers.

$$(6 - 2i) + (-4 - 3i) = 2 - 5i$$

The graphs are shown in Figure 25.   ◼

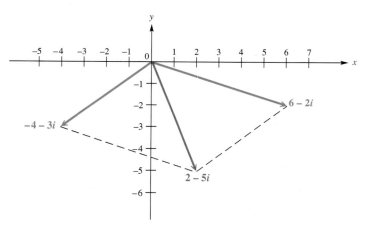

FIGURE 25

Trigonometric or Polar Form

Figure 26 shows the complex number $x + yi$ that corresponds to a vector **OP** with direction θ and magnitude r. The following relationships among r, θ, x, and y can be verified from Figure 26. Notice the similarities between these and the ones given in Section 5.1 for vectors.

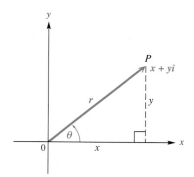

FIGURE 26

> **RELATIONSHIPS AMONG x, y, r, AND θ**
>
> $x = r \cos \theta \qquad r = \sqrt{x^2 + y^2}$
>
> $y = r \sin \theta \qquad \tan \theta = \dfrac{y}{x}, \quad \text{if } x \neq 0$

Substituting $x = r \cos \theta$ and $y = r \sin \theta$ from the results above into $x + yi$ gives

$$x + yi = r \cos \theta + (r \sin \theta)i$$
$$= r(\cos \theta + i \sin \theta).$$

> **TRIGONOMETRIC OR POLAR FORM OF A COMPLEX NUMBER**
>
> The expression
>
> $$r(\cos \theta + i \sin \theta)$$
>
> is called the **trigonometric form** or **polar form*** of the complex number $x + yi$. The expression $\cos \theta + i \sin \theta$ is sometimes abbreviated cis θ. Using this notation,
>
> $$r(\cos \theta + i \sin \theta) \text{ is written as } r \text{ cis } \theta.$$

The number r is called the **modulus** or **absolute value** of $x + yi$, while θ is the **argument** of $x + yi$. In this section we will choose the value of θ in the interval $[0°, 360°)$. However, keep in mind that the angle is not unique, since any angle coterminal with it is also acceptable.

---

* The terms *trigonometric form* and *polar form* are synonymous, and will be used interchangeably in the rest of this chapter.

EXAMPLE 2

Converting from
Polar Form to
Rectangular Form

Express $2(\cos 300° + i \sin 300°)$ in rectangular form.

SOLUTION   Since $\cos 300° = \frac{1}{2}$ and $\sin 300° = -\frac{\sqrt{3}}{2}$,

$$2(\cos 300° + i \sin 300°) = 2\left(\frac{1}{2} - i\frac{\sqrt{3}}{2}\right) = 1 - i\sqrt{3}. \qquad \blacksquare$$

TECHNOLOGICAL NOTE
The built-in functions described in the Technological Note found in Sec-
tion 5.1 can be adapted
to converting between po-
lar and rectangular forms
of complex numbers.

NOTE   In most of the examples of this section, we will write arguments using degree measure. Arguments may also be written with radian measure.

In order to convert from rectangular form to polar form, the following proce-dure is used.

STEPS FOR CONVERTING FROM RECTANGULAR TO POLAR FORM
1.   Sketch a graph of the number in the complex plane.
2.   Find r by using the equation $r = \sqrt{x^2 + y^2}$.
3.   Find θ by using the equation $\tan \theta = \frac{y}{x}$, $x \neq 0$. If $x = 0$, determine θ by inspection.

CAUTION   Errors often occur in Step 3 described above. Be sure that the correct quadrant of θ is chosen by referring to the graph sketched in Step 1.

EXAMPLE 3

Converting from
Rectangular Form to
Polar Form

Write the following complex numbers in polar form.

(a) $-\sqrt{3} + i$

SOLUTION   Start by sketching the graph of $-\sqrt{3} + i$ in the complex plane, as shown in Figure 27.

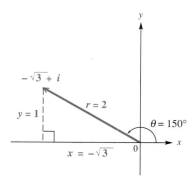

FIGURE 27

Next, find r. Since $x = -\sqrt{3}$ and $y = 1$,
$$r = \sqrt{x^2 + y^2} = \sqrt{(-\sqrt{3})^2 + 1^2} = \sqrt{3 + 1} = 2.$$

Then find θ.

$$\tan \theta = \frac{y}{x} = \frac{1}{-\sqrt{3}} = -\frac{\sqrt{3}}{3}$$

Since $\tan \theta = -\frac{\sqrt{3}}{3}$, the reference angle for θ is $30°$. From the sketch we see that θ is in quadrant II, so $\theta = 180° - 30° = 150°$. Therefore, in polar form,

$$-\sqrt{3} + i = 2(\cos 150° + i \sin 150°)$$
$$= 2 \text{ cis } 150°.$$

(b) $-3i$

SOLUTION The sketch of $-3i$ is shown in Figure 28. (We use the fact that $-3i = 0 - 3i$.)

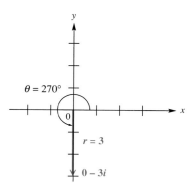

$\theta = 270°$

$r = 3$

$0 - 3i$

FIGURE 28

Since $-3i = 0 - 3i$, we have $x = 0$ and $y = -3$. Find r as follows.

$$r = \sqrt{0^2 + (-3)^2} = \sqrt{0 + 9} = \sqrt{9} = 3$$

We cannot find θ by using $\tan \theta = \frac{y}{x}$, since $x = 0$. In a case like this, refer to the graph and determine the argument directly from the sketch. A value for θ here is $270°$. In polar form,

$$-3i = 3(\cos 270° + i \sin 270°)$$
$$= 3 \text{ cis } 270°. \quad \blacksquare$$

NOTE In Example 3 we gave answers in both forms: $r(\cos \theta + i \sin \theta)$ and r cis θ. These forms will be used interchangeably throughout the rest of this chapter.

FOR **GROUP DISCUSSION**

Use the vector conversion techniques of your calculator (described in the previous section) in degree mode to verify the following results of Examples 2 and 3.

1. When $r = 2$ and $\theta = 300°$, we obtain $x = 1$ and $y = -\sqrt{3} \approx -1.732050808$ (from Example 2).
2. When $x = -\sqrt{3}$ and $y = 1$, we obtain $r = 2$ and $\theta = 150°$ (from Example 3a).
3. When $x = 0$ and $y = -3$, we obtain $r = 3$. The display for θ will depend on how your particular model is programmed. It will either equal $270°$ or an angle coterminal with $270°$, such as $-90°$ (from Example 3b).

(a) Write the complex number $6(\cos 115° + i \sin 115°)$ in rectangular form.

> [!example]
> **EXAMPLE 4**
>
> Converting Between Polar and Rectangular Forms

SOLUTION Since $115°$ does not have a special angle as a reference angle, we cannot find exact values for $\cos 115°$ and $\sin 115°$. Use a calculator set in the degree mode to find $\cos 115° \approx -.42261826$ and $\sin 115° \approx .90630779$. Therefore, in rectangular form,

$$6(\cos 115° + i \sin 115°) \approx 6(-.42261826 + .90630779i)$$
$$\approx -2.5357096 + 5.4378467i$$

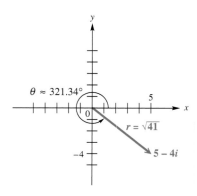

FIGURE 29

(b) Write $5 - 4i$ in polar form.

SOLUTION A sketch of $5 - 4i$ shows that θ must be in quadrant IV. See Figure 29. Here $r = \sqrt{5^2 + (-4)^2} = \sqrt{41}$ and $\tan \theta = \frac{-4}{5}$. Use a calculator to find that one measure of θ is approximately $-38.66°$. In order to express θ in the interval $[0, 360°)$, we find that $\theta \approx 360° - 38.66° \approx 321.34°$. Use these results to get

$$5 - 4i \approx \sqrt{41}(\cos 321.34° + i \sin 321.34°). \quad ■$$

> **FOR GROUP DISCUSSION**
>
> Support the results of Example 4 using the rectangular/polar conversion feature of your calculator.

Products of Complex Numbers in Polar Form

The product of two complex numbers $a + bi$ and $c + di$ in rectangular form is found by multiplying them as we would multiply binomials, and replacing i^2 with -1.

$$(a + bi)(c + di) = ac + adi + bci + bdi^2$$
$$= ac + (ad + bc)i + bd(-1)$$
$$= (ac - bd) + (ad + bc)i$$

We will apply this method to find the product of $1 + i\sqrt{3}$ and $-2\sqrt{3} + 2i$.

$$(1 + i\sqrt{3})(-2\sqrt{3} + 2i) = -2\sqrt{3} + 2i - 2i(3) + 2i^2\sqrt{3}$$
$$= -2\sqrt{3} + 2i - 6i - 2\sqrt{3}$$
$$= -4\sqrt{3} - 4i$$

This same product also can be found by first converting the complex numbers $1 + i\sqrt{3}$ and $-2\sqrt{3} + 2i$ to polar form. Using the method explained earlier in this section,

$$1 + i\sqrt{3} = 2(\cos 60° + i \sin 60°)$$

and

$$-2\sqrt{3} + 2i = 4(\cos 150° + i \sin 150°).$$

If the polar forms are now multiplied together and if the identities for the cosine and the sine of the sum of two angles are used, the result is

$$[2(\cos 60° + i \sin 60°)][4(\cos 150° + i \sin 150°)]$$
$$= 2 \cdot 4(\cos 60° \cdot \cos 150° + i \sin 60° \cdot \cos 150°$$
$$+ i \cos 60° \cdot \sin 150° + i^2 \sin 60° \cdot \sin 150°)$$
$$= 8[(\cos 60° \cdot \cos 150° - \sin 60° \cdot \sin 150°)$$
$$+ i(\sin 60° \cdot \cos 150° + \cos 60° \cdot \sin 150°)]$$
$$= 8[\cos (60° + 150°) + i \sin(60° + 150°)]$$
$$= 8(\cos 210° + i \sin 210°).$$

The modulus of the product, 8, is equal to the product of the moduli of the factors, $2 \cdot 4$, while the argument of the product, $210°$, is the sum of the arguments of the factors, $60° + 150°$.

As we would expect, the product obtained upon multiplying by the first method is the rectangular form of the product obtained upon multiplying by the second method.

$$8(\cos 210° + i \sin 210°) = 8\left(-\frac{\sqrt{3}}{2} - \frac{1}{2}i\right)$$
$$= -4\sqrt{3} - 4i$$

The work shown above is generalized in the following *product theorem*.

PRODUCT THEOREM

If $r_1(\cos \theta_1 + i \sin \theta_1)$ and $r_2(\cos \theta_2 + i \sin \theta_2)$ are any two complex numbers, then

$$[r_1(\cos \theta_1 + i \sin \theta_1)] \cdot [r_2(\cos \theta_2 + i \sin \theta_2)]$$
$$= r_1 r_2[\cos(\theta_1 + \theta_2) + i \sin(\theta_1 + \theta_2)].$$

In compact form, this is written

$$(r_1 \text{ cis } \theta_1)(r_2 \text{ cis } \theta_2) = r_1 r_2 \text{ cis}(\theta_1 + \theta_2).$$

EXAMPLE 5

Using the Product Theorem

Find the product of $3(\cos 45° + i \sin 45°)$ and $2(\cos 135° + i \sin 135°)$.

SOLUTION   Using the product theorem,

$$[3(\cos 45° + i \sin 45°)][2(\cos 135° + i \sin 135°)]$$
$$= 3 \cdot 2[\cos(45° + 135°) + i \sin(45° + 135°)]$$
$$= 6(\cos 180° + i \sin 180°),$$

which can be expressed as $6(-1 + i \cdot 0) = 6(-1) = -6$. The two complex numbers in this example are complex factors of -6.   ∎

Quotients of Complex Numbers in Polar Form

The quotient of two complex numbers $a + bi$ and $c + di$ in rectangular form is found by multiplying by a form of 1; we multiply by $\frac{c - di}{c - di}$. (The complex number $c - di$ is called the **complex conjugate** (or simply **conjugate**) of $c + di$.)

$$\frac{a + bi}{c + di} = \frac{a + bi}{c + di} \cdot \frac{c - di}{c - di}$$

$$= \frac{ac - adi + bci - bdi^2}{c^2 - d^2i^2}$$

$$= \frac{(ac + bd) + (bc - ad)i}{c^2 + d^2}$$

$$= \frac{ac + bd}{c^2 + d^2} + \frac{bc - ad}{c^2 + d^2}i$$

We will apply this method to find the quotient of $1 + i\sqrt{3}$ and $-2\sqrt{3} + 2i$.

$$\frac{1 + i\sqrt{3}}{-2\sqrt{3} + 2i} = \frac{(1 + i\sqrt{3})(-2\sqrt{3} - 2i)}{(-2\sqrt{3} + 2i)(-2\sqrt{3} - 2i)}$$

$$= \frac{-2\sqrt{3} - 2i - 6i - 2i^2\sqrt{3}}{12 - 4i^2}$$

$$= \frac{-8i}{16} = -\frac{1}{2}i.$$

Writing $1 + i\sqrt{3}$, $-2\sqrt{3} + 2i$, and $-\frac{1}{2}i$ in polar form gives

$$1 + i\sqrt{3} = 2(\cos 60° + i \sin 60°)$$

$$-2\sqrt{3} + 2i = 4(\cos 150° + i \sin 150°)$$

$$-\frac{1}{2}i = \frac{1}{2}[\cos(-90°) + i \sin(-90°)].$$

The modulus of the quotient, $\frac{1}{2}$, is the quotient of the two moduli, 2 and 4. The argument of the quotient, $-90°$, is the difference of the two arguments, $60° - 150° = -90°$. It would be easier to find the quotient of these two complex numbers in trigonometric form than in rectangular form. Generalizing from this example leads to another theorem, the *quotient theorem.*

QUOTIENT THEOREM

If $r_1(\cos \theta_1 + i \sin \theta_1)$ and $r_2(\cos \theta_2 + i \sin \theta_2)$ are complex numbers, where $r_2(\cos \theta_2 + i \sin \theta_2) \neq 0$, then

$$\frac{r_1(\cos \theta_1 + i \sin \theta_1)}{r_2(\cos \theta_2 + i \sin \theta_2)} = \frac{r_1}{r_2}[\cos(\theta_1 - \theta_2) + i \sin(\theta_1 - \theta_2)].$$

In compact form, this is written

$$\frac{r_1 \operatorname{cis} \theta_1}{r_2 \operatorname{cis} \theta_2} = \frac{r_1}{r_2} \operatorname{cis}(\theta_1 - \theta_2).$$

EXAMPLE 6

Using the Quotient Theorem

Find the quotient

$$\frac{10 \text{ cis}(-60°)}{5 \text{ cis } 150°}.$$

Write the result in rectangular form.

SOLUTION By the quotient theorem,

$$\frac{10 \text{ cis}(-60°)}{5 \text{ cis } 150°} = \frac{10}{5} \text{ cis}(-60° - 150°) \qquad \text{Quotient theorem}$$

$$= 2 \text{ cis}(-210°) \qquad \text{Subtract.}$$

$$= 2[\cos(-210°) + i \sin(-210°)]$$

$$= 2\left[-\frac{\sqrt{3}}{2} + i\left(\frac{1}{2}\right) \right] \qquad \begin{array}{l} \cos(-210°) = -\frac{\sqrt{3}}{2}; \\ \sin(-210°) = \frac{1}{2} \end{array}$$

$$= -\sqrt{3} + i \qquad \text{Rectangular form} \qquad ◗$$

EXAMPLE 7

Using the Product and Quotient Theorems with a Calculator

Use a calculator to find the following. Write the results in rectangular form.

(a) $(9.3 \text{ cis } 125.2°)(2.7 \text{ cis } 49.8°)$

SOLUTION By the product theorem,

$$(9.3 \text{ cis } 125.2°)(2.7 \text{ cis } 49.8°) = (9.3)(2.7) \text{ cis}(125.2° + 49.8°)$$

$$= 25.11 \text{ cis } 175°$$

$$= 25.11(\cos 175° + i \sin 175°)$$

$$\approx 25.11(-.99619470$$
$$+ i(.08715574))$$

$$\approx -25.014449 + 2.1884806i.$$

(b) $$\frac{10.42\left(\cos \dfrac{3\pi}{4} + i \sin \dfrac{3\pi}{4}\right)}{5.21\left(\cos \dfrac{\pi}{5} + i \sin \dfrac{\pi}{5}\right)}$$

SOLUTION Use the quotient theorem.

$$\frac{10.42\left(\cos \dfrac{3\pi}{4} + i \sin \dfrac{3\pi}{4}\right)}{5.21\left(\cos \dfrac{\pi}{5} + i \sin \dfrac{\pi}{5}\right)} = \frac{10.42}{5.21}\left[\cos\left(\frac{3\pi}{4} - \frac{\pi}{5}\right) + i \sin\left(\frac{3\pi}{4} - \frac{\pi}{5}\right) \right]$$

$$= 2\left(\cos \frac{11\pi}{20} + i \sin \frac{11\pi}{20} \right)$$

$$\approx -.31286893 + 1.9753767i \qquad ◗$$

Applications

One of the earliest encounters with the square root of a negative number was in A.D. 50 by Heron of Alexandria when he derived the expression $\sqrt{81 - 144}$. Taking square roots of negative numbers led to the need for the complex numbers. As one might imagine, if sixteenth- and seventeenth-century mathematicians felt uneasy about negative numbers, they felt even more uneasy about taking square roots of negative numbers. The famous mathematician René Descartes rejected complex numbers and coined the term "imaginary" numbers.

Today, complex numbers are readily accepted and play an important role in many new and exciting fields of applied mathematics and technology. Complex numbers are no more imaginary than are negative numbers and zero. Their development has been necessary in order to solve new problems like the design of airplane wings, ships, electrical circuits, noise control, and fractals.

During the past 20 years, computer graphics and complex numbers have made it possible to produce many beautiful fractals. In the 1970s, Benoit B. Mandelbrot first used the term **fractal.** Largely because of his efforts, fractal geometry has become a new field of study. At its basic level, a fractal is a unique and enchanting geometric figure with an endless self-similarity property. A fractal image repeats itself infinitely with ever decreasing dimensions. If you look at smaller and smaller portions of the figure, you will continue to see the whole—much like looking into two parallel mirrors that are facing each other. Although most current applications for fractals are related to creating fascinating images and pictures, they have a tremendous potential for applied science in the future. An example of a fractal is shown in Figure 30. It is an amazing graphical solution to a difficult problem first presented by Sir Arthur Cayley in 1879. The fractal is called *Newton's basins of attraction for the cube roots of unity*. Like many other fractals, it was created with the aid of high-resolution computer graphics and complex numbers.*

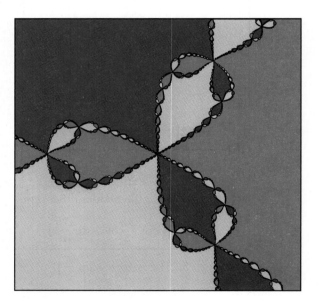

FIGURE 30

---

*Crownover R., *Introduction to Fractals and Chaos* (Boston: Jones and Bartlett Publishers, 1995).

Kline, M., *MATHEMATICS The Loss of Certainty* (New York: Oxford University Press, 1980).

Lauwerier, H., *Fractals* (Princeton, New Jersey: Princeton University Press, 1991).

National Council of Teachers of Mathematics, *Historical Topics for the Mathematics Classroom* (Washington, D.C: Thirty-first Yearbook, 1969).

<table>
<tr><td>

EXAMPLE **8**

Deciding Whether a
Complex Number Is
in the Julia Set

</td><td>

The fractal called the **Julia set** is shown in Figure 31. It can be created by graphing a special set of complex numbers. To determine if a complex number $z = a + bi$ is in this Julia set, perform the following sequence of calculations. Repeatedly compute the values of $z^2 - 1$, $(z^2 - 1)^2 - 1$, $[(z^2 - 1)^2 - 1]^2 - 1, \ldots$. If the modulus of any of the resulting complex numbers exceeds 2, then the complex number z is not in the Julia set. Otherwise z is part of this set and the point (a, b) should be shaded in the graph.*

</td></tr>
</table>

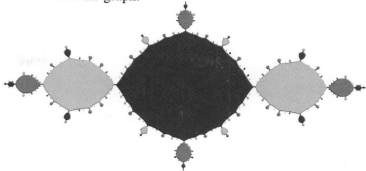

FIGURE 31

Determine whether or not the following numbers belong to the Julia set:
(a) $z = 0 + 0i$ **(b)** $z = 1 + 1i$.

SOLUTION
(a) $z = 0 + 0i$

Since $z = 0 + 0i = 0$, $z^2 - 1 = 0^2 - 1 = -1$,
$$(z^2 - 1)^2 - 1 = (-1)^2 - 1 = 0,$$
$$[(z^2 - 1)^2 - 1]^2 - 1 = 0^2 - 1 = -1,$$

and so on. We see that the calculations repeat as $0, -1, 0, -1$, and so on. The moduli are either 0 or 1, which do not exceed 2, so $0 + 0i$ is in the Julia set and the point $(0, 0)$ is part of the graph.

(b) $z = 1 + 1i$

We have $z^2 - 1 = (1 + i)^2 - 1 = (1 + 2i + i^2) - 1 = -1 + 2i$. The modulus is $\sqrt{(-1)^2 + 2^2} = \sqrt{5}$. Since $\sqrt{5}$ is greater than 2, $1 + 1i$ is not in the Julia set and $(1, 1)$ is not part of the graph.   ∎

5.2 EXERCISES

Graph each of the following complex numbers as a vector in the complex plane.

1. $-2 + 3i$ **2.** $-4 + 5i$ **3.** $8 - 5i$ **4.** $6 - 5i$ **5.** $2 - 2i\sqrt{3}$

6. $4\sqrt{2} + 4i\sqrt{2}$ **7.** $-4i$ **8.** $3i$ **9.** -8 **10.** 2

11. What must be true in order for a complex number to also be a real number?

12. If a real number is graphed in the complex plane, on what axis does the vector lie?

13. A complex number of the form $a + bi$ will have its corresponding vector lying on the y-axis provided $a =$ _____ .

14. The modulus of a complex number represents the _____ of the vector representing it in the complex plane.

---

*Crownover R., *Introduction to Fractals and Chaos* (Boston: Jones and Bartlett Publishers, 1995).

Find the resultant of each of the following pairs of complex numbers. Express in rectangular form a + bi.

15. $4 - 3i, -1 + 2i$

16. $2 + 3i, -4 - i$

17. $-3, 3i$

18. $6, -2i$

19. $2 + 6i, -2i$

20. $-5 - 8i, -1$

Write each of the following complex numbers in rectangular form. Give exact values for the real and imaginary parts. You may wish to use the capability of your calculator to support your answers.

21. $2(\cos 45° + i \sin 45°)$

22. $4(\cos 60° + i \sin 60°)$

23. $10 \text{ cis } 90°$

24. $8 \text{ cis } 270°$

25. $4(\cos 240° + i \sin 240°)$

26. $2(\cos 330° + i \sin 330°)$

27. $\cos \dfrac{\pi}{6} + i \sin \dfrac{\pi}{6}$

28. $3\left(\cos \dfrac{5\pi}{6} + i \sin \dfrac{5\pi}{6} \right)$

29. $5 \text{ cis } 300°$

30. $6 \text{ cis } 135°$

31. $\sqrt{2} \text{ cis } 180°$

32. $\sqrt{3} \text{ cis } 270°$

Write each of the following complex numbers in polar form r(cos θ + i sin θ), where r is exact, and −180° < θ ≤ 180°. You may wish to use the capability of your calculator to support your result.

33. $3 - 3i$

34. $-2 + 2i\sqrt{3}$

35. $-3 - 3i\sqrt{3}$

36. $1 + i\sqrt{3}$

37. $4\sqrt{3} + 4i$

38. $\sqrt{3} - i$

39. $-\sqrt{2} + i\sqrt{2}$

40. $-5 - 5i$

41. $2 + 2i$

42. $-\sqrt{3} + i$

43. -4

44. $5i$

45. $-2i$

46. 7

47. Use your calculator to find the polar form of $3 + 5i$. Give as many decimal places as the calculator displays.

48. Give the smallest possible *positive* degree measure of θ if $r > 0$ and **(a)** $r \text{ cis } \theta$ has real part equal to 0, **(b)** $r \text{ cis } \theta$ has imaginary part equal to 0.

Find each of the following products, and write the products in rectangular form, using exact values.

49. $[3(\cos 60° + i \sin 60°)][2(\cos 90° + i \sin 90°)]$

50. $[4(\cos 30° + i \sin 30°)][5(\cos 120° + i \sin 120°)]$

51. $[2(\cos 45° + i \sin 45°)][2(\cos 225° + i \sin 225°)]$

52. $[8(\cos 300° + i \sin 300°)][5(\cos 120° + i \sin 120°)]$

53. $[4(\cos 60° + i \sin 60°)][6(\cos 330° + i \sin 330°)]$

54. $[8(\cos 210° + i \sin 210°)][2(\cos 330° + i \sin 330°)]$

55. $[5 \text{ cis } 90°][3 \text{ cis } 45°]$

56. $[6 \text{ cis } 120°][5 \text{ cis}(-30°)]$

57. $[\sqrt{3} \text{ cis } 45°][\sqrt{3} \text{ cis } 225°]$

58. $[\sqrt{2} \text{ cis } 300°][\sqrt{2} \text{ cis } 270°]$

Find each of the following quotients, and write the quotients in rectangular form, using exact values.

59. $\dfrac{4(\cos 120° + i \sin 120°)}{2(\cos 150° + i \sin 150°)}$

60. $\dfrac{10(\cos 225° + i \sin 225°)}{5(\cos 45° + i \sin 45°)}$

61. $\dfrac{16(\cos 300° + i \sin 300°)}{8(\cos 60° + i \sin 60°)}$

62. $\dfrac{24(\cos 150° + i \sin 150°)}{2(\cos 30° + i \sin 30°)}$

63. $\dfrac{3 \text{ cis } 305°}{9 \text{ cis } 65°}$

64. $\dfrac{12 \text{ cis } 293°}{6 \text{ cis } 23°}$

*R*elating Concepts

Consider the complex numbers

$$w = -1 + i \quad and \quad z = -1 - i.$$

65. Multiply w and z using their rectangular forms and the "FOIL" method from algebra. Leave the product in rectangular form.

66. Find the polar forms of w and z.

67. Multiply w and z using their polar forms and the method described in this section.

68. Use the result of Exercise 67 to find the rectangular form of wz. How does this compare to your result in Exercise 65?

69. Find the quotient $\frac{w}{z}$ using their rectangular forms and multiplying both the numerator and the denominator by the conjugate of the denominator. Leave the quotient in rectangular form.

70. Use the polar forms of w and z, found in Exercise 66, to divide w by z using the method described in this section.

71. Use the result of Exercise 70 to find the rectangular form of $\frac{w}{z}$. How does this compare to your result in Exercise 69?

---

72. Notice that $(r \operatorname{cis} \theta)^2 = (r \operatorname{cis} \theta)(r \operatorname{cis} \theta) = r^2 \operatorname{cis}(\theta + \theta) = r^2 \operatorname{cis} 2\theta$. State in your own words how we can square a complex number in polar form. (In the next section, we will develop this idea.)

Solve each applied problem.

73. Refer to Example 8. Is $z = -.2i$ in the Julia set?

74. Refer to Example 8. The graph of the Julia set in Figure 31 appears to be symmetric with respect to both the x-axis and y-axis. Complete the following to show that this is true.
   (a) Show that complex conjugates have the same modulus.
   (b) Compute $z_1^2 - 1$ and $z_2^2 - 1$ where $z_1 = a + bi$ and $z_2 = a - bi$.
   (c) Discuss why if (a, b) is in the Julia set then so is $(a, -b)$.
   (d) Conclude that the graph of the Julia set must be symmetric with respect to the x-axis.
   (e) Using a similar argument, show that the Julia set must also be symmetric with respect to the y-axis.

*The fractal called the **Mandelbrot set** is shown in the figure. To determine if a complex number $z = a + bi$ is in this set, perform the following sequence of calculations. Repeatedly compute $z, z^2 + z, (z^2 + z)^2 + z, [(z^2 + z)^2 + z]^2 + z, \ldots$. In a manner analogous to the Julia set, the complex number z does not belong to the Mandelbrot set if any of the resulting moduli exceed 2. Otherwise z is in the set and the point (a, b) should be shaded in the graph. (Source: Lauwerier, H., Fractals (Princeton: Princeton University Press, 1991).)*

Determine whether or not the following numbers belong to the Mandelbrot set.

75. $1 - 1i$

76. $-.5i$

Impedance *is a measure of the opposition to the flow of alternating electrical current found in common electrical outlets. It consists of two parts called the resistance and reactance. Resistance occurs when a lightbulb is turned on, while reactance is produced when electricity passes through a coil of wire like that found in electric motors. Impedance Z in ohms (Ω) can be expressed as a complex number where the real part represents the resistance and the imaginary part represents the reactance. For example, if the resistive part is 3 ohms and the reactive part is 4 ohms, then the impedance could be described by the complex number Z = 3 + 4i. In the series circuit shown in the figure, the total impedance will be the sum of the individual impedances. (Source: Wilcox, G. and C. Hesselberth,* Electricity For Engineering Technology *(Boston: Allyn and Bacon, Inc., 1970).)*

77. The circuit contains two lightbulbs and two electric motors. If it is assumed that the lightbulbs are pure resistive and the motors are pure reactive, find the total impedance in this circuit and express it in the form $Z = a + bi$.

78. The phase angle θ measures the phase difference between the voltage and the current in an electrical circuit. θ can be determined by the equation $\tan \theta = \frac{b}{a}$. Find θ for this circuit.

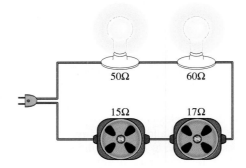

In work with alternating current, complex numbers are used to describe current, I, voltage, E, and impedance, Z (the opposition to current). These three quantities are related by the equation E = IZ. Thus, if any two of these quantities are known, the third can be found. In each of the following problems, solve the equation E = IZ for the missing variable.

79. $I = 8 + 6i, Z = 6 + 3i$

80. $I = 10 + 6i, Z = 8 + 5i$

81. $I = 7 + 5i, E = 28 + 54i$

82. $E = 35 + 55i, Z = 6 + 4i$

In the parallel electrical circuit shown in the figure below, the impedance Z can be

calculated using the equation $Z = \dfrac{1}{\frac{1}{Z_1} + \frac{1}{Z_2}}$ *where Z_1 and Z_2 are the impedances for each*

branch of the circuit.

83. If $Z_1 = 50 + 25i$ and $Z_2 = 60 + 20i$, calculate Z.

84. Determine the phase angle θ for Z as found in Exercise 83.

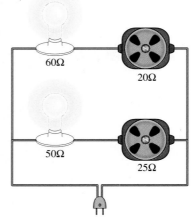

85. The alternating current in an electric inductor is

$$I = \frac{E}{Z}$$

amperes, where E is the voltage and $Z = R + X_L i$ is the impedance. If $E = 8(\cos 20° + i \sin 20°)$, $R = 6$, and $X_L = 3$, find the current. Give the answer in rectangular form.

86. The current I in a circuit with voltage E, resistance R,

capacitive reactance X_c, and inductive reactance X_L is

$$I = \frac{E}{R + (X_L - X_c)i}.$$

Find I if $E = 12(\cos 25° + i \sin 25°)$, $R = 3$, $X_L = 4$, and $X_c = 6$. Give the answer in rectangular form.

87. Prove the product theorem for complex numbers.

88. Prove the quotient theorem for complex numbers.

5.3 POWERS AND ROOTS OF COMPLEX NUMBERS

Powers of Complex Numbers (De Moivre's Theorem) ▌ Roots of Complex Numbers

Powers of Complex Numbers (De Moivre's Theorem)

In the previous section we studied the product and quotient theorems for complex numbers in polar form. Because raising a number to a positive integer power is a repeated application of the product rule, it would seem likely that a theorem for finding powers of complex numbers exists. This is indeed the case. For example, the square of the complex number $r(\cos \theta + i \sin \theta)$ is

$$[r(\cos \theta + i \sin \theta)]^2 = [r(\cos \theta + i \sin \theta)][r(\cos \theta + i \sin \theta)]$$
$$= r \cdot r[\cos(\theta + \theta) + i \sin(\theta + \theta)]$$
$$= r^2(\cos 2\theta + i \sin 2\theta).$$

In the same way,

$$[r(\cos \theta + i \sin \theta)]^3 = r^3(\cos 3\theta + i \sin 3\theta).$$

These results suggest the plausibility of the following theorem for positive integer values of n. Although the following theorem is stated and can be proved for all n, we will use it only for positive integer values of n and their reciprocals.

DE MOIVRE'S THEOREM

If $r(\cos \theta + i \sin \theta)$ is a complex number and if n is any real number, then

$$[r(\cos \theta + i \sin \theta)]^n = r^n(\cos n\theta + i \sin n\theta).$$

In compact form, this is written

$$[r \operatorname{cis} \theta]^n = r^n(\operatorname{cis} n\theta).$$

This theorem is named after the French expatriate friend of Isaac Newton, Abraham De Moivre (1667–1754), although he never explicitly stated it.

EXAMPLE **1**

Finding a Power of
a Complex Number

Find $(1 + i\sqrt{3})^8$ and express the result in rectangular form.

SOLUTION To use De Moivre's theorem, first convert $1 + i\sqrt{3}$ into polar form using the methods of Section 5.2.

$$1 + i\sqrt{3} = 2(\cos 60° + i \sin 60°)$$

Now apply De Moivre's theorem.

$$\begin{aligned}
(1 + i\sqrt{3})^8 &= [2(\cos 60° + i \sin 60°)]^8 \\
&= 2^8[\cos(8 \cdot 60°) + i \sin(8 \cdot 60°)] \\
&= 256(\cos 480° + i \sin 480°) \\
&= 256(\cos 120° + i \sin 120°) \\
&= 256\left(-\frac{1}{2} + i\frac{\sqrt{3}}{2}\right) \\
&= -128 + 128i\sqrt{3}
\end{aligned}$$

480° and 120°
are coterminal.

$\cos 120° = -\frac{1}{2}$;
$\sin 120° = \frac{\sqrt{3}}{2}$

Rectangular
form

Roots of Complex Numbers

In algebra it is shown that every nonzero complex number has exactly n distinct complex nth roots. De Moivre's theorem can be extended to find all nth roots of a complex number. An nth root of a complex number is defined as follows.

> ### nTH ROOT
>
> For a positive integer n, the complex number $a + bi$ is an **nth root** of the complex number $x + yi$ if
>
> $$(a + bi)^n = x + yi.$$

EXAMPLE **2**

Finding Complex
Roots

Find the three complex cube roots of $8(\cos 135° + i \sin 135°)$.

SOLUTION To find these three complex cube roots, we must look for a complex number, say $r(\cos \alpha + i \sin \alpha)$, that will satisfy

$$[r(\cos \alpha + i \sin \alpha)]^3 = 8(\cos 135° + i \sin 135°).$$

By De Moivre's theorem, this equation becomes

$$r^3(\cos 3\alpha + i \sin 3\alpha) = 8(\cos 135° + i \sin 135°).$$

One way to satisfy this equation is to set $r^3 = 8$ and also $\cos 3\alpha + i \sin 3\alpha = \cos 135° + i \sin 135°$. The first of these conditions implies that $r = 2$, and the second implies that

$$\cos 3\alpha = \cos 135° \qquad \text{and} \qquad \sin 3\alpha = \sin 135°.$$

For these equations to be satisfied, 3α must represent an angle that is coterminal with 135°. Therefore, we must have

$$3\alpha = 135° + 360° \cdot k, \quad k \text{ any integer,}$$

or

$$\alpha = \frac{135° + 360° \cdot k}{3}, \quad k \text{ any integer.}$$

Now let k take on the integer values 0, 1, and 2.

$$\text{If } k = 0, \quad \alpha = \frac{135° + 0°}{3} = 45°.$$

$$\text{If } k = 1, \quad \alpha = \frac{135° + 360°}{3} = \frac{495°}{3} = 165°.$$

$$\text{If } k = 2, \quad \alpha = \frac{135° + 720°}{3} = \frac{855°}{3} = 285°.$$

In the same way, $\alpha = 405°$ when $k = 3$. But note that $\sin 405° = \sin 45°$ and $\cos 405° = \cos 45°$. If $k = 4$, $\alpha = 525°$ which has the same sine and cosine values as $165°$. To continue with larger values of k would just be repeating solutions already found. Therefore, all of the cube roots (three of them) can be found by letting $k = 0$, 1, and 2.

When $k = 0$, the root is $2(\cos 45° + i \sin 45°)$.

When $k = 1$, the root is $2(\cos 165° + i \sin 165°)$.

When $k = 2$, the root is $2(\cos 285° + i \sin 285°)$.

In summary, $2(\cos 45° + i \sin 45°)$, $2(\cos 165° + i \sin 165°)$, and $2(\cos 285° + i \sin 285°)$ are the three cube roots of $8(\cos 135° + i \sin 135°)$. ∎

Generalizing from Example 2, we state the following theorem that can be applied to find nth roots.

---

nTH ROOT THEOREM

If n is any positive integer and r is a positive real number, then the nonzero complex number $r(\cos \theta + i \sin \theta)$ has exactly n distinct nth roots, given by

$$\sqrt[n]{r}(\cos \alpha + i \sin \alpha),$$

where

$$\alpha = \frac{\theta + 360° \cdot k}{n} \quad \text{or} \quad \alpha = \frac{\theta}{n} + \frac{360° \cdot k}{n},$$

$k = 0, 1, 2, \ldots, n - 1.$

EXAMPLE 3

Finding Complex Roots

Find all fourth roots of $-8 + 8i\sqrt{3}$. Write the roots in rectangular form.

SOLUTION First write $-8 + 8i\sqrt{3}$ in polar form as

$$-8 + 8i\sqrt{3} = 16(\cos 120° + i \sin 120°).$$

Here $r = 16$ and $\theta = 120°$. The fourth roots of this number have modulus $\sqrt[4]{16} = 2$ and arguments given as follows. Using the alternative formula for α,

$$\alpha = \frac{120°}{4} + \frac{360° \cdot k}{4} = 30° + 90° \cdot k.$$

If $k = 0$, $\quad \alpha = 30° + 90° \cdot 0 = 30°$.

If $k = 1$, $\quad \alpha = 30° + 90° \cdot 1 = 120°$.

If $k = 2$, $\quad \alpha = 30° + 90° \cdot 2 = 210°$.

If $k = 3$, $\quad \alpha = 30° + 90° \cdot 3 = 300°$.

Using these angles, the fourth roots are

$$2(\cos 30° + i \sin 30°),$$
$$2(\cos 120° + i \sin 120°),$$
$$2(\cos 210° + i \sin 210°),$$

and $\qquad 2(\cos 300° + i \sin 300°).$

These four roots can be written in rectangular form as $\sqrt{3} + i$, $-1 + i\sqrt{3}$, $-\sqrt{3} - i$, and $1 - i\sqrt{3}$. ∎

An interesting geometric interpretation involving vectors can be applied to nth roots of a complex number. If we use the values of r and α found in Example 3, the four fourth roots of $-8 + 8i\sqrt{3}$ can be illustrated as position vectors whose terminal points lie on a circle of radius $r = 2$. The vectors are spaced 90° apart; the 90° comes from the fact that since $n = 4$, $\frac{360°}{n} = \frac{360°}{4} = 90°$. See Figure 32.

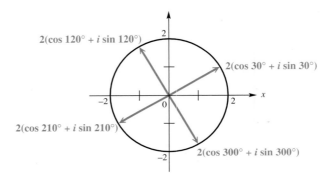

FIGURE 32

EXAMPLE 4

Solving an Equation by Finding Complex Roots

Find all complex number solutions of $x^5 - 1 = 0$. Graph them as vectors in the complex plane.

SOLUTION Write the equation as

$$x^5 - 1 = 0$$
$$x^5 = 1.$$

While there is only one real number solution, 1, there are five complex number solutions. To find these solutions, first write 1 in polar form as

$$1 = 1 + 0i = 1(\cos 0° + i \sin 0°).$$

The modulus of the fifth roots is $\sqrt[5]{1} = 1$, and the arguments are given by

$$0° + 72° \cdot k, \qquad k = 0, 1, 2, 3, \text{ or } 4.$$

By using these arguments, the fifth roots are

$$1(\cos 0° + i \sin 0°), \qquad k = 0$$
$$1(\cos 72° + i \sin 72°), \qquad k = 1$$
$$1(\cos 144° + i \sin 144°), \qquad k = 2$$
$$1(\cos 216° + i \sin 216°), \qquad k = 3$$

and
$$1(\cos 288° + i \sin 288°). \qquad k = 4$$

The solution set of the equation may be written as {cis 0°, cis 72°, cis 144°, cis 216°, cis 288°}. The first of these roots equals 1; the others cannot easily be expressed in rectangular form. The five fifth roots all lie on a unit circle and are equally spaced around it every 72°, as shown in Figure 33. ◘

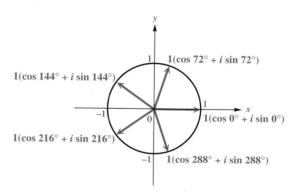

FIGURE 33

5.3 EXERCISES

Find each of the following powers. Write the answers in rectangular form.

1. $[3(\cos 30° + i \sin 30°)]^3$
2. $[2(\cos 135° + i \sin 135°)]^4$
3. $(\cos 45° + i \sin 45°)^8$
4. $[2(\cos 120° + i \sin 120°)]^3$
5. $[3 \text{ cis } 100°]^3$
6. $[3 \text{ cis } 40°]^3$
7. $(\sqrt{3} + i)^5$
8. $(2\sqrt{2} - 2i\sqrt{2})^6$
9. $(2 - 2i\sqrt{3})^4$
10. $\left(\dfrac{\sqrt{2}}{2} - \dfrac{\sqrt{2}}{2}i\right)^8$
11. $(-2 - 2i)^5$
12. $(-1 + i)^7$

Find the cube roots of each of the following complex numbers. Leave the answers in polar form. Then graph each cube root as a vector in the complex plane.

13. 1
14. i
15. $8(\cos 60° + i \sin 60°)$
16. $27(\cos 300° + i \sin 300°)$
17. $-8i$
18. $27i$
19. -64
20. 27
21. $1 + i\sqrt{3}$
22. $2 - 2i\sqrt{3}$
23. $-2\sqrt{3} + 2i$
24. $\sqrt{3} - i$

25. For the real number 1, find and graph all indicated roots. Give answers in rectangular form.
  (a) fourth **(b)** sixth **(c)** eighth

26. For the complex number i, find and graph all indicated roots. Give answers in polar form.
   (a) square   **(b)** fourth

27. Explain why a positive real number must have a positive real nth root.

28. True or false: **(a)** Every real number must have two real square roots.
                           (b) Some real numbers have three real cube roots.

Relating Concepts

We will examine how the three complex cube roots of -8 can be found in two different ways. Work Exercises 29–36 in order.

29. All complex roots of the equation $x^3 + 8 = 0$ are cube roots of -8. Factor $x^3 + 8$ as the sum of two cubes.

30. One of the factors found in Exercise 29 is linear. Set it equal to 0, solve and determine the real cube root of -8.

31. One of the factors found in Exercise 29 is quadratic. Set it equal to 0, solve, and determine the rectangular forms of the other two cube roots of -8.

32. Use the method described in this section to find the three complex cube roots of -8. Give them in polar form.

33. Convert the polar forms found in Exercise 32 to rectangular form.

34. Compare your results in Exercises 30 and 31, and Exercise 33. What do you notice?

35. Graph the function $f(x) = x^3 + 8$ in the standard window of your calculator, and find the x-intercept. How does this relate to the cube roots you found earlier?

36. The graph of $f(x) = x^6 - 1$ is shown here, along with the two x-intercepts as determined by a calculator. How do they relate to the answers in Exercise 25(b)?

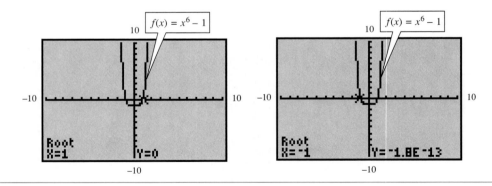

Find all complex solutions for each equation. Leave your answers in polar form.

37. $x^4 + 1 = 0$ **38.** $x^4 + 16 = 0$

39. $x^5 - i = 0$ **40.** $x^4 - i = 0$

Refer to the discussion of fractals in Section 5.2.

The fractal shown in the figure is the solution to Cayley's problem of determining the basins of attraction for the cube roots of unity. The three cube roots of unity are
$$w_1 = 1, \quad w_2 = -\frac{1}{2} + \frac{\sqrt{3}}{2}i, \quad \text{and} \quad w_3 = -\frac{1}{2} - \frac{\sqrt{3}}{2}i. \text{ This fractal can be generated by}$$

repeatedly evaluating the function $f(z) = \dfrac{2z^3 + 1}{3z^2}$ *where z is a complex number. One*
begins by picking $z_1 = a + bi$ *and then successively computing* $z_2 = f(z_1)$, $z_3 = f(z_2)$,
$z_4 = f(z_3)$, ... *If the resulting values of* $f(z)$ *approach* w_1, *color the pixel at* (a, b) *red.*
If it approaches w_2, *color it blue and if it approaches* w_3, *color it yellow. If this process*
continues for a large number of different z_1, *the fractal in the figure will appear.*
(Source: Crownover, R., Introduction to Fractals and Chaos *(Boston: Jones and Bartlett*
Publishers, 1995).)

Determine the appropriate color of the pixel for each value of z_1.

41. $z_1 = i$ **42.** $z_1 = 2 + i$ **43.** $z_1 = -1 - i$

Further Explorations

A calculator with TABLE capability can be used to generate the various values of the
arguments of the nth roots of a complex number. To illustrate, consider the fourth roots of
$-8 + 8i\sqrt{3}$ (see Example 3). The argument of $-8 + 8i\sqrt{3}$ is 120°, and $n = 4$. If we
define Y_1 to be $\dfrac{120 + 360\,X}{4}$ and let X take on the values 0, 1, 2, and 3, the TABLE will
display the four arguments for the fourth roots. See the figure. They are 30°, 120°, 210°,
and 300°.

X	Y1	
0	30	
1	120	
2	210	
3	300	

Y1冒(120+360X)/4

Compare these results with the ones found in Example 3. Knowing that for each root, $r = 2$, we can write the four fourth roots of $-8 + 8i\sqrt{3}$: 2 cis 30°, 2 cis 120°, 2 cis 210°, 2 cis 300°.

1. Use a TABLE to support the results of Example 2.

2. Find the arguments of the ninth roots of 1 using a TABLE. Express them in degree form.

3. Find the fourth roots of i using a TABLE.

4. If we want the arguments of roots to be in degree form, and we use a calculator with a TABLE feature to generate these arguments as described above, must the calculator be in degree mode? Why or why not?

Chapter 5   SUMMARY

Vectors are represented by directed line segments. Equal vectors have the same magnitude and the same direction. The parallelogram rule is used to find the sum (or resultant) of two vectors. The vector is called the resultant of its components. The horizontal and vertical components are most important in applications of vectors as forces.

A vector may also be determined by a direction angle (an angle with the horizontal) and its magnitude. The relationships of the horizontal and vertical components of a position vector to a direction angle and magnitude are basic rules for vectors. The vector operations of addition and scalar multiplication may be performed both geometrically and analytically. Vectors are particularly useful in the solution of navigation problems.

Complex numbers are graphed in a complex plane that has a real axis and an imaginary axis. Each complex number corresponds to a vector. Graphically, the sum of two complex numbers is the resultant of the corresponding vectors. Complex numbers are expressed in trigonometric (or polar) form by replacing the real and imaginary parts by $r \cos \theta$ and $r \sin \theta$ respectively.

Complex numbers in trigonometric (or polar) form can be multiplied and divided to give the product or quotient directly in the same form. Powers and roots of complex numbers in trigonometric (or polar) form also can be found directly from numbers expressed in that form using De Moivre's theorem. This is the only way to find *all* roots of many complex numbers.

Key Terms

SECTION 5.1	SECTION 5.2
scalar	rectangular form
vector	real axis
vertical component	imaginary axis
horizontal component	complex plane
zero vector	trigonometric (polar) form
scalar product	modulus (absolute value)
direction angle	argument
position vector (radius vector)	fractal
dot product	
angle θ between **u** and **v**	
airspeed	
groundspeed	

Chapter 5   REVIEW EXERCISES

In Exercises 1–3, use the vectors shown here. Sketch each of the following.

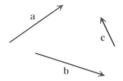

1. a + b   **2. a − b**   **3. a + 3c**

Vector **v** *has the given magnitude and makes an angle θ with the horizontal. Find the magnitudes of the horizontal and vertical components of* **v**, *first using the relationships* $x = r \cos \theta$ *and* $y = r \sin \theta$. *Then use the capabilities of your calculator to support your answers.*

4. $|\mathbf{v}| = 50$, $\theta = 45°$
(give exact values)

5. $|\mathbf{v}| = 69.2$, $\theta = 75°$

6. $|\mathbf{v}| = 964$, $\theta = 154°20'$

Given two forces and the angle between them, find the magnitude of the resultant force.

7. Forces of 142 and 215 newtons, forming an angle of 112°

8. Forces of 85.2 and 69.4 newtons, forming an angle of 58°20′

*Find (**a**) the dot product and (**b**) the angle between the pairs of vectors.*

9. $\mathbf{u} = \langle 6, 2 \rangle$, $\mathbf{v} = \langle 3, -2 \rangle$

10. $\mathbf{u} = \langle 2\sqrt{3}, 2 \rangle$, $\mathbf{v} = \langle 5, 5\sqrt{3} \rangle$

Solve each of the following problems.

11. A plane has an airspeed of 520 mph. The pilot wishes to fly on a bearing of 310°. A wind of 37 mph is blowing from a bearing of 212°. What direction should the pilot fly, and what will be her actual speed?

12. A boat travels 15 km per hr in still water. The boat is traveling across a large river, on a bearing of 130°.

The current in the river, coming from the west, has a speed of 7 km per hr. Find the resulting speed of the boat and its resulting direction of travel.

13. A long-distance swimmer starts out swimming a steady 3.2 mph due north. A 5.1-mph current is flowing on a bearing of 12°. What is the swimmer's resulting bearing and speed?

Graph each complex number as a vector in the complex plane.

14. $5i$ **15.** $-4 + 2i$ **16.** $3 - 3i\sqrt{3}$

Find and graph the resultant of each pair of complex numbers.

17. $7 + 3i$ and $-2 + i$ **18.** $2 - 4i$ and $5 + i$

19. The vector representing a complex number of the form bi will lie on the _____-axis in the complex plane.

20. Explain the geometric similarity between the absolute value of a real number and the absolute value (or modulus) of a complex number.

Perform the indicated operations. Give the answer in rectangular form.

21. $[5(\cos 90° + i \sin 90°)][6(\cos 180° + i \sin 180°)]$ **22.** $[3 \text{ cis } 135°][2 \text{ cis } 105°]$

23. $\dfrac{2(\cos 60° + i \sin 60°)}{8(\cos 300° + i \sin 300°)}$

24. $\dfrac{4 \text{ cis } 270°}{2 \text{ cis } 90°}$

25. $(\sqrt{3} + i)^3$

26. $(2 - 2i)^5$ **27.** $(\cos 100° + i \sin 100°)^6$ **28.** $(\text{cis } 20°)^3$

Complete the chart in Exercises 29–36.

Rectangular Form	Trigonometric Form
29. $-2 + 2i$	_____
30. _____	$3(\cos 90° + i \sin 90°)$
31. _____	$2(\cos 225° + i \sin 225°)$
32. $-4 + 4i\sqrt{3}$	_____
33. $1 - i$	_____
34. _____	$4(\cos 240° + i \sin 240°)$
35. $-4i$	_____
36. _____	$2 \text{ cis } 180°$

Find the indicated roots and graph as vectors in the complex plane. Leave your answers in polar form.

37. the cube roots of $-27i$ **38.** the fourth roots of $16i$ **39.** the fifth roots of 32

40. Solve the equation $x^4 + i = 0$. Leave solutions in polar form.

41. Without actually performing the operations, state why the products $[2(\cos 45° + i \sin 45°)] \cdot [5(\cos 90° + i \sin 90°)]$ and $[2(\cos(-315°) + i \sin(-315°))][5(\cos(-270°) + i \sin(-270°))]$ are the same.

42. Under what conditions is the difference between two nonreal complex numbers $a + bi$ and $c + di$ a real number?

*P*olar and Parametric Equations

The Polar Coordinate System ▮ Converting From Polar to Rectangular Coordinates ▮ Converting From Rectangular to Polar Coordinates

The Polar Coordinate System

Throughout this text we have been using the rectangular coordinate system to graph equations. Another coordinate system that is particularly useful for graphing many relations is the **polar coordinate system.** The system is based on a point, called the **pole,** and a ray, called the **polar axis.** The polar axis is usually drawn in the direction of the positive *x*-axis, as shown in Figure 1.

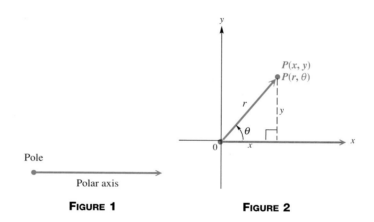

FIGURE 1

FIGURE 2

In Figure 2, the pole has been placed at the origin of a rectangular coordinate system, so that the polar axis coincides with the positive x-axis. Point P has coordinates (x, y) in the rectangular coordinate system. Point P can also be located by giving the directed angle θ from the positive x-axis to ray OP and the directed distance r from the pole to point P. The ordered pair (r, θ) gives the **polar coordinates** of point P.

In Chapter 5 we saw how the values of x, y, r, and θ are related, as they pertained to vectors and complex numbers in polar form. These same relationships hold here.

TECHNOLOGICAL NOTE
The built-in functions described in the Technological Note found in Section 5.1 can be adapted to converting between polar and rectangular coordinates.

CONVERTING BETWEEN POLAR AND RECTANGULAR COORDINATES

The following relationships hold between the point (x, y) in the rectangular coordinate plane and the point (r, θ) in the polar coordinate plane.

$$x = r \cos \theta \qquad r = \sqrt{x^2 + y^2}$$

$$y = r \sin \theta \qquad \tan \theta = \frac{y}{x}, \quad \text{if } x \neq 0$$

Converting From Polar to Rectangular Coordinates

The first example illustrates how points are plotted in the polar coordinate system and how polar coordinates can be converted to rectangular coordinates.

EXAMPLE 1

Plotting Points with Polar Coordinates and Determining Rectangular Coordinates

For each point, plot by hand in the polar coordinate system. Then determine the rectangular coordinates of the point:

(a) $P(2, 30°)$ **(b)** $Q(-4, 120°)$ **(c)** $R(5, -45°)$.

SOLUTION

(a)   In this case, $r = 2$ and $\theta = 30°$, so the point P is located 2 units from the origin in the positive direction on a ray making a $30°$ angle with the polar axis, as shown in Figure 3.

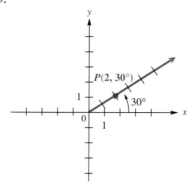

FIGURE 3

Using the conversion equations, we find the rectangular coordinates as follows.

$$\begin{aligned}
x &= r \cos \theta & y &= r \sin \theta \\
&= 2 \cos 30° & &= 2 \sin 30° \\
&= 2\left(\frac{\sqrt{3}}{2}\right) & &= 2\left(\frac{1}{2}\right) \\
&= \sqrt{3} & &= 1
\end{aligned}$$

The rectangular coordinates are $(\sqrt{3}, 1)$.

(b) Since r is negative, Q is 4 units in the negative direction from the pole on an extension of the 120° ray. See Figure 4. The rectangular coordinates are $x = -4 \cos 120° = -4(-\frac{1}{2}) = 2$ and $y = -4 \sin 120° = -4(\frac{\sqrt{3}}{2}) = -2\sqrt{3}$.

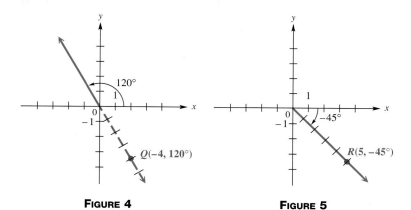

| **FIGURE 4** | **FIGURE 5** |

(c) Point R is shown in Figure 5. Since θ is negative, the angle is measured in the clockwise direction. Furthermore, we have

$$x = 5 \cos(-45°)$$
$$= \frac{5\sqrt{2}}{2}$$

and

$$y = 5 \sin(-45°)$$
$$= \frac{-5\sqrt{2}}{2}. \quad \blacksquare$$

FOR **GROUP DISCUSSION**

Use the polar/rectangular conversion feature of your graphing calculator to support the results in Example 1. Be aware of decimal approximations of exact values. Also, be sure your calculator is in degree mode to correspond to the example.

One important difference between rectangular coordinates and polar coordinates is that while a given point in the plane can have only one pair of rectangular coordinates, this same point has an infinite number of pairs of polar coordinates. For example, $(2, 30°)$ locates the same point as $(2, 390°)$ or $(2, -330°)$ or $(-2, 210°)$.

EXAMPLE 2

Giving Alternate Forms of a Pair of Polar Coordinates

Give three other pairs of polar coordinates for the point $P(3, 140°)$.

SOLUTION Three pairs that could be used for the point are $(3, -220°)$, $(-3, 320°)$, and $(-3, -40°)$. See Figure 6. $\quad \blacksquare$

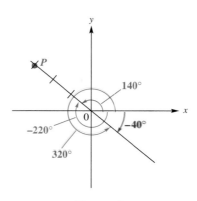

FIGURE 6

Converting from Rectangular to Polar Coordinates

So far in this section, we have plotted points determined by polar coordinates and determined rectangular coordinates for these points. We can also go the other way. That is, given a point in rectangular coordinates, we can determine a pair of polar coordinates for this point. The next example illustrates this process.

EXAMPLE 3

Plotting Points with Rectangular Coordinates and Determining Polar Coordinates

For each point, plot by hand in the rectangular coordinate system. Then determine *two* pairs of polar coordinates for the point.

(a) $(x, y) = (-1, 1)$

(b) $(x, y) = \left(\frac{3}{2}, \frac{\sqrt{3}}{2}\right)$

SOLUTION
(a)   As shown in Figure 7, the point $(-1, 1)$ lies in the second quadrant. Using the conversion equations we have,

$$\tan \theta = \frac{y}{x}$$

$$\tan \theta = \frac{1}{-1}$$

$$\tan \theta = -1.$$

This tells us that *one* choice for θ is $\theta = 135°$. Since this choice of θ lies in the second quadrant, the same quadrant as the point $(-1, 1)$, we can now use a positive value for r. To determine r, we use the conversion equation $r = \sqrt{x^2 + y^2}$ to get

$$r = \sqrt{x^2 + y^2}$$
$$r = \sqrt{(-1)^2 + (1)^2}$$
$$r = \sqrt{2}.$$

Hence, *one* pair of polar coordinates for the rectangular point $(x, y) = (-1, 1)$ is $(r, \theta) = (\sqrt{2}, 135°)$. A second pair is $(\sqrt{2}, 495°)$.

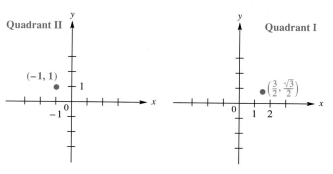

FIGURE 7   **FIGURE 8**

(b)   As shown in Figure 8, the point $(\frac{3}{2}, \frac{\sqrt{3}}{2})$ lies in the first quadrant. Using the conversion equations we have

$$\tan \theta = \frac{y}{x}$$

$$\tan \theta = \frac{\sqrt{3}/2}{3/2}$$

$$\tan \theta = \frac{\sqrt{3}}{3}.$$

So one possible choice for θ is $\theta = 30°$. As in part (a), since this choice for θ lies in the same quadrant as the point $(\frac{3}{2}, \frac{\sqrt{3}}{2})$, we can use a positive value for r.

$$r = \sqrt{x^2 + y^2}$$

$$r = \sqrt{\left(\frac{3}{2}\right)^2 + \left(\frac{\sqrt{3}}{2}\right)^2}$$

$$r = \sqrt{\frac{9}{4} + \frac{3}{4}}$$

$$r = \sqrt{3}$$

So *one* pair of polar coordinates for the rectangular point $(x, y) = (\frac{3}{2}, \frac{\sqrt{3}}{2})$ is $(\sqrt{3}, 30°)$. A second pair is $(\sqrt{3}, 390°)$.   ◻

FOR **GROUP DISCUSSION**

In Example 3, it was stated that we determined *one* pair of polar coordinates for the given rectangular coordinates. Why do we need to state that this is only *one* pair of polar coordinates?

6.1 EXERCISES

Plot by hand the graph of the point whose polar coordinates are given. Then give two other pairs of polar coordinates of the point. Finally, give the rectangular coordinates of the point.

1. $(1, 45°)$ **2.** $(3, 120°)$ **3.** $(-2, 135°)$ **4.** $(-4, 30°)$ **5.** $(5, -60°)$

6. $(2, -45°)$ **7.** $(-3, -210°)$ **8.** $(-1, -120°)$ **9.** $(3, 360°)$ **10.** $(4, 270°)$

11. If a point lies on an axis in the rectangular plane, then what kind of angle must θ be if (r, θ) represents the point in polar coordinates?

Plot the point whose rectangular coordinates are given. Then determine two pairs of polar coordinates for the point with $0° \leq \theta < 360°$.

12. $(-1, 1)$ **13.** $(1, 1)$ **14.** $(0, 3)$ **15.** $(0, -3)$ **16.** $(\sqrt{2}, \sqrt{2})$

17. $(-\sqrt{2}, \sqrt{2})$ **18.** $\left(\dfrac{\sqrt{3}}{2}, \dfrac{3}{2}\right)$ **19.** $\left(-\dfrac{\sqrt{3}}{2}, -\dfrac{1}{2}\right)$ **20.** $(3, 0)$ **21.** $(-3, 0)$

22. Explain in your own words why the equation $\tan \theta = \frac{y}{x}$ cannot be used when working Exercises 14 and 15.

6.2 POLAR EQUATIONS AND THEIR GRAPHS

Polar Equations ▌ Graphs of Polar Equations ▌ Conversion of Equations

Polar Equations

So far in this text, the graphs that we have studied have been sketched or generated by a calculator using rectangular coordinates. We will now examine how equations in polar coordinates are graphed. An equation like

$$r = 3 \sin \theta, \qquad r = 2 + \cos \theta, \qquad \text{or} \qquad r = \theta,$$

where r and θ are the variables, is a *polar equation*. The simplest equation for many types of curves turns out to be a polar equation.

The traditional method of graphing polar equations is much the same as the traditional method of graphing rectangular equations. We evaluate r for various values of θ until a pattern appears, and then we join the points with a smooth curve. However, this can be a very time-consuming task, and modern graphing calculators can perform the job in a matter of seconds. In this section we will present some examples of polar curves by including a table of representative points, a traditional graph, and a calculator-generated graph.

NOTE Refer to your owner's manual at this time to see how your particular model handles polar graphs. After reading the examples that follow, see if you can generate the curves shown. As always, pay close attention to the viewing window and the angle mode (degree or radian). Furthermore, you will need to decide on maximum and minimum values of θ. Keep in mind the periods of the functions, so that when you decide on these, the entire set of function values are generated.

Graphs of Polar Equations

EXAMPLE **1**
Graphing a Polar Equation (Cardioid)

Graph $r = 1 + \cos \theta$.

SOLUTION To graph this equation in a traditional manner, we find some ordered pairs (as in the table) and then connect the points in order—from $(2, 0°)$ to $(1.9, 30°)$ to $(1.7, 45°)$ and so on.

θ	0°	30°	45°	60°	90°	120°	135°	150°	180°	270°	315°	330°	360°
$\cos \theta$	1	.9	.7	.5	0	−.5	−.7	−.9	−1	0	.7	.9	1
$r = 1 + \cos \theta$	2	1.9	1.7	1.5	1	.5	.3	.1	0	1	1.7	1.9	2

It is not necessary to choose values greater than 360° or less than 0°, since one period of the cosine function has been completely covered. Joining the points with a smooth curve yields the graph shown in Figure 9(a). The calculator-generated graph is shown in Figure 9(b). Notice that the point (1.5, 60°) is indicated. This curve is called a **cardioid** because of its heart shape. ▮

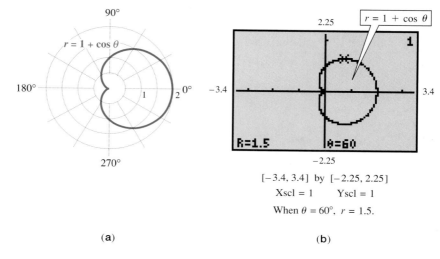

(a)

(b)

[−3.4, 3.4] by [−2.25, 2.25]
Xscl = 1   Yscl = 1
When $\theta = 60°$, $r = 1.5$.

FIGURE 9

Graph $r^2 = \cos 2\theta$.

SOLUTION First complete a table of ordered pairs as shown. The point $(-1, 0°)$, with r negative, may be plotted as $(1, 180°)$. Also, $(-.7, 30°)$ may be plotted as $(.7, 210°)$, and so on. This curve is called a **lemniscate.**

θ	0°	30°	45°	135°	150°	180°
2θ	0°	60°	90°	270°	300°	360°
$\cos 2\theta$	1	.5	0	0	.5	1
$r = \pm\sqrt{\cos 2\theta}$	± 1	$\pm.7$	0	0	$\pm.7$	± 1

Values of θ for $45° < \theta < 135°$ are not included in the table because the corresponding values of $\cos 2\theta$ are negative (quadrants II and III) and so do not have real square roots. Values of θ larger than 180° give 2θ larger than 360°, and would repeat the points already found.

A traditional graph of this lemniscate is shown in Figure 10(a). The calculator-generated graph shown in Figure 10(b) was obtained by letting θ take on values between $0°$ and $180°$. Again, we show a point on the calculator graph (this time, $(0, 45°)$). It was necessary to enter two equations into the calculator: $r_1 = \sqrt{\cos 2\theta}$ and $r_2 = -\sqrt{\cos 2\theta}$. ◨

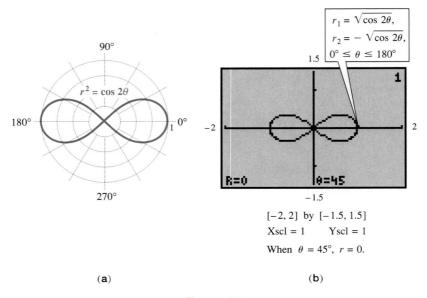

$[-2, 2]$ by $[-1.5, 1.5]$
Xscl = 1   Yscl = 1
When $\theta = 45°$, $r = 0$.

(a)   (b)

FIGURE 10

EXAMPLE **3**
Graphing a Polar Equation (Rose)

Graph $r = 3 \cos 2\theta$.

SOLUTION   Because of the 2θ, the graph requires a large number of points. A few ordered pairs are given below. You should complete the table similarly through the first $360°$ if you are graphing in a traditional manner.

θ	$0°$	$15°$	$30°$	$45°$	$60°$	$75°$	$90°$
2θ	$0°$	$30°$	$60°$	$90°$	$120°$	$150°$	$180°$
$\cos 2\theta$	1	.9	.5	0	$-.5$	$-.9$	-1
r	3	2.6	1.5	0	-1.5	-2.6	-3

Plotting these points in order gives the graph, called a **four-leaved rose.** Notice in Figure 11(a) how the graph is developed with a continuous curve, beginning with the upper half of the right horizontal leaf and ending with the lower half of that leaf. As the graph is traced, the curve goes through the pole four times.

This pattern is easily seen if a calculator is used to graph the rose. See Figure 11(b) for such a graph. ◨

Figure 11(a) uses arrowheads and circled numbers to indicate the pattern in which this four-leaved rose is traced. Use your calculator to duplicate the calculator graph seen in Figure 11(b), and watch closely to confirm this pattern. (Remember that the smaller the step, the longer it will take to trace the graph.)

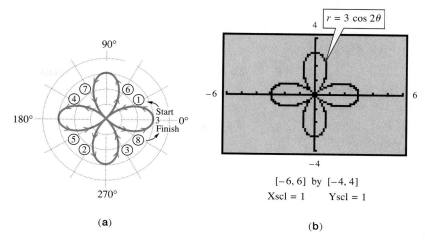

(a)

(b)

FIGURE 11

The graph in Figure 11 is one of a family of curves called **roses.** The graphs of $r = \sin n\theta$ and $r = \cos n\theta$ are roses, with n petals if n is odd, and $2n$ petals if n is even.

FOR **GROUP DISCUSSION**

Keeping in mind that the minimum and maximum values of the sine and cosine function are -1 and 1, discuss how the minimum and maximum values of r in a polar equation can be found. This will help you in determining the appropriate window when using your calculator to graph a polar equation.

EXAMPLE 4

Graphing a Polar Equation (Spiral of Archimedes)

Graph $r = 2\theta$, where θ is measured in radians, $-2\pi \le \theta \le 2\pi$. Use a calculator to generate the graph.

SOLUTION This graph is an example of a **spiral of Archimedes.** It is quite tedious to graph using traditional methods, so we show only a calculator graph. See Figure 12. ▮

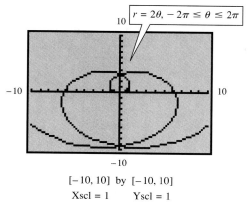

$[-10, 10]$ by $[-10, 10]$
Xscl = 1   Yscl = 1

FIGURE 12

TECHNOLOGICAL NOTE
To see how the size of the step can greatly affect the accuracy of your graph, try various values such as .1, .5, 1, and 1.5 with the polar equation in Example 4. The graph shown in Figure 12 was generated with θ-step = .1.

FOR **GROUP DISCUSSION**

Experiment with various minimum and maximum values of θ for the spiral of Archimedes in Example 4, and discuss the various behaviors you observe.

Conversion of Equations

Using the relationships stated earlier in Section 6.1 (see the box headed "Converting Between Polar and Rectangular Coordinates"), we can transform polar equations to rectangular ones and vice versa.

EXAMPLE 5

Converting a Polar Equation to a Rectangular Equation and Graphing

For the polar equation $r = \frac{4}{1 + \sin \theta}$:

(a) convert to a rectangular equation,

(b) use a graphing calculator to graph the polar equation for $0° \leq \theta \leq 360°$, and

(c) use a graphing calculator to graph the rectangular equation.

SOLUTION

(a) Multiply both sides of the equation by the denominator on the right, to clear the fraction.

$$r = \frac{4}{1 + \sin \theta}$$

$$r + r \sin \theta = 4$$

Now substitute $\sqrt{x^2 + y^2}$ for r and y for $r \sin \theta$.

$$\sqrt{x^2 + y^2} + y = 4$$
$$\sqrt{x^2 + y^2} = 4 - y$$

Square both sides to eliminate the radical.

$$x^2 + y^2 = (4 - y)^2$$
$$x^2 + y^2 = 16 - 8y + y^2$$
$$x^2 = -8y + 16$$
$$x^2 = -8(y - 2)$$

(b) Figure 13(a) shows a calculator-generated graph using polar coordinates.

(c) Solving $x^2 = -8(y - 2)$ for y, we obtain

$$y = 2 - \frac{1}{8}x^2.$$

The graph of this rectangular equation is shown in Figure 13(b). Notice that the two graphs are the same. ∎

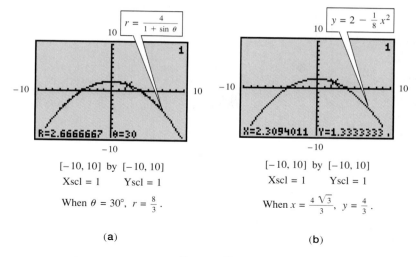

$[-10, 10]$ by $[-10, 10]$
Xscl = 1   Yscl = 1

When $\theta = 30°$, $r = \frac{8}{3}$.

(a)

$[-10, 10]$ by $[-10, 10]$
Xscl = 1   Yscl = 1

When $x = \frac{4\sqrt{3}}{3}$, $y = \frac{4}{3}$.

(b)

FIGURE 13

FOR GROUP DISCUSSION

The displays at the bottom of the screens in Figures 13(a) and 13(b) indicate the same point. Discuss how you would go about verifying analytically that the polar coordinates shown in (a) and the rectangular coordinates shown in (b) are equivalent.

EXAMPLE 6

Converting a Rectangular Equation to a Polar Equation and Graphing

For the rectangular equation $3x + 2y = 4$:

(a) convert to a polar equation,

(b) use a graphing calculator to graph the rectangular equation, and

(c) use a graphing calculator to graph the polar equation for $0° \le \theta \le 360°$.

SOLUTION

(a) Use $x = r \cos \theta$ and $y = r \sin \theta$ to get

$$3x + 2y = 4$$
$$3r \cos \theta + 2r \sin \theta = 4.$$

Now solve for r. First factor out r on the left.

$$r(3 \cos \theta + 2 \sin \theta) = 4$$

$$r = \frac{4}{3 \cos \theta + 2 \sin \theta}$$

The polar equation of the line $3x + 2y = 4$ is

$$r = \frac{4}{3 \cos \theta + 2 \sin \theta}.$$

(b)   We solve the given rectangular equation for *y* to get

$$y = -\frac{3}{2}x + 2.$$

Its graph is shown in Figure 14(a).

(c)   Using the polar graphing capability of the calculator and the equation found in part (a), we obtain the graph shown in Figure 14(b).   ∎

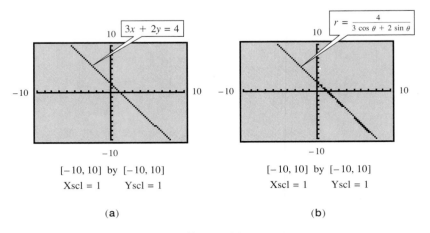

(a)   (b)

FIGURE 14

6.2   EXERCISES

Use a graphing calculator to graph each of the following equations for $0° \le \theta \le 360°$.
Use a square viewing window.

1. $r = 2 + 2 \cos \theta$ (cardioid)

2. $r = 1 + 3 \cos \theta$ (limaçon with a loop)

3. $r = 8 + 6 \cos \theta$ (dimpled limaçon)

4. $r = 4 - 4 \sin \theta$ (cardioid)

5. $r = 4 \cos 2\theta$ (four-leaved rose)

6. $r = 3 \sin 4\theta$ (eight-leaved rose)

7. $r = 4 \cos 3\theta$ (three-leaved rose)

8. $r = 3 \sin 5\theta$ (five-leaved rose)

9. $r^2 = 4 \cos 2\theta$ (lemniscate)

10. $r^2 = 3 \sin 2\theta$ (lemniscate)

11. $r = 2 \sin \theta \tan \theta$ (cissoid)

12. $r = \dfrac{\cos 2\theta}{\cos \theta}$ (cissoid with a loop)

13. Explain the method you would use to graph (r, θ) by hand if $r < 0$.

14. Explain why, if $r > 0$, the points (r, θ) and $(-r, \theta + 180°)$ have the same graph.

For each of the following equations, find an equivalent equation in rectangular coordi-
nates. Then use the polar graphing capability of your calculator to graph the equation
in its original form. (It may be necessary to first solve for r.) Use a square window.

15. $r = 2 \sin \theta$

16. $r = 2 \cos \theta$

17. $r = \dfrac{2}{1 - \cos \theta}$

18. $r = \dfrac{3}{1 - \sin \theta}$

19. $r + 2 \cos \theta = -2 \sin \theta$

20. $r = \dfrac{3}{4 \cos \theta - \sin \theta}$

21. $r = 2 \sec \theta$

22. $r = -5 \csc \theta$

23. $r(\cos \theta + \sin \theta) = 2$

24. $r(2 \cos \theta + \sin \theta) = 2$

For each of the following equations, find an equivalent equation in polar coordinates.

25. $x + y = 4$ **26.** $2x - y = 5$ **27.** $x^2 + y^2 = 16$

28. $x^2 + y^2 = 9$ **29.** $y = 2$ **30.** $x = 4$

The graph of $r = a\theta$ in polar coordinates is an example of the spiral of Archimedes. With your calculator set to radian mode and polar graphing capability, use the given value of a and interval of θ to graph the spiral in the window specified.

31. $a = 1, 0 \le \theta \le 4\pi, [-15, 15]$ by $[-15, 15]$

32. $a = 2, -4\pi \le \theta \le 4\pi, [-30, 30]$ by $[-30, 30]$

33. $a = 1.5, -4\pi \le \theta \le 4\pi, [-20, 20]$ by $[-20, 20]$

34. $a = -1, 0 \le \theta \le 12\pi, [-40, 40]$ by $[-40, 40]$

35. Refer to Example 6. Would you find it easier to graph the equation using the rectangular or the polar form? Why?

36. Show that the distance between (r_1, θ_1) and (r_2, θ_2) is $\sqrt{r_1^2 + r_2^2 - 2r_1 r_2 \cos(\theta_1 - \theta_2)}$.

*F*urther Explorations

If your graphing calculator has a TABLE feature, it can be used to help you understand how polar equations are graphed, and also how graphing calculators can sometimes display very misleading results. Your calculator should be in POLAR and DEGREE modes. Use the following WINDOW and TABLE settings:

$$\theta\min = 0, \quad \theta\max = 360, \quad \theta\text{step} = 7.5,$$
$$\text{Xmin} = -15, \quad \text{Xmax} = 15, \quad \text{Ymin} = -10, \quad \text{Ymax} = 10$$
$$\text{Tblmin} = 0, \quad \Delta\text{Tbl} = 7.5$$

1. Enter $r1 = 8$. Recall that the graph of $y = 8$ in standard rectangular coordinates is a horizontal line. Examine the graph and the TABLE for $r1$. Use the values for $r1$ in the TABLE and the definition of a circle to explain why this equation is the graph of a circle with a radius of 8.

2. Enter $r1 = 8 \cos 4\theta$. Examine (follow the path of the cursor while this rose is being graphed) the graph and the TABLE for $r1$.
As θ goes from $0°$ to $22.5°$, $r1$ goes from 8 to 0.
(a) As θ goes from $22.5°$ to $45°$, $r1$ goes from _____ .
(b) As θ goes from $45°$ to $67.5°$, $r1$ goes from _____ .
(c) As θ goes from $67.5°$ to $112.5°$, $r1$ goes from _____ .
(d) As θ goes from $112.5°$ to $157.5°$, $r1$ goes from _____ .
(e) How many petals does this rose have? Does this match the statement in this section that $r = \cos n\theta$ has $2n$ petals if n is even?
(f) Points are plotted on the graph only for those values of θ that are evaluated (as determined by θ step and ΔTbl). From the TABLE, how many points are plotted to draw each petal of this graph? You may also want to watch the graph drawn in DOT mode.

3. Change $r1$ so that $r1 = 8 \cos 6\theta$. Examine the graph and TABLE for $r1$.
(a) How many petals does this rose have? Does this match the statement in this section that $r = \cos n\theta$ has $2n$ petals if n is even?
(b) How many points are plotted to draw each petal of this graph?

4. Now things start to get strange!
Change $r1$ so that $r1 = 8 \cos 12\theta$. Examine the graph and TABLE for $r1$.
(a) From the TABLE, how many points are plotted to draw each petal?
(b) Does the graph look like a rose? Why or why not? What has happened to the petals?

5. Change $r1$ so that $r1 = 8 \cos 48\theta$. Examine the graph and TABLE for $r1$. Be sure you still have θ step $= 7.5$ and ΔTbl $= 7.5$.
   (a) What does the graph look like? Surprised?
   (b) From what you've learned about polar roses, what should this graph look like? Use the TABLE to help you explain why the graph has been distorted.

NOTE This is an excellent example of why it is so important to not always believe your graphing calculator and why the graphing calculator will not replace the need to understand algebraic and geometric concepts.

6. What can you do so that the calculator will draw more realistic graphs of $r = \cos 12\theta$ and $r = \cos 48\theta$? Try your conjecture on your graphing calculator.

6.3 PARAMETRIC EQUATIONS AND THEIR GRAPHS

Basic Concepts ▌ Graphs of Parametric Equations and Their Rectangular Equivalents ▌ Alternative Forms of Parametric Equations ▌ Applications and the Cycloid

TECHNOLOGICAL NOTE
Refer to the Technological Note in Section 6.2 regarding *step* or *increment*. When graphing in parametric mode, it will be necessary to decide on what step to use for the parameter t when making the window settings. You should be aware of the *default* values built into the calculator. The calculator-generated graphs in this section were created by using t-step $= .1$.

Basic Concepts

In Chapter 2 we saw how the graph of the unit circle can be found on a graphing calculator using parametric equations. We will now investigate parametric equations in more detail. Throughout this text, we have concentrated on graphing sets of ordered pairs of real numbers that corresponded to a function of the form $y = f(x)$ or $r = f(\theta)$. Another way to determine a set of ordered pairs involves two functions f and g defined by $x = f(t)$ and $y = g(t)$, where t is a real number in some interval I. Each value of t leads to a corresponding x-value and a corresponding y-value, and thus to an ordered pair (x, y).

> **PARAMETRIC EQUATIONS OF A PLANE CURVE**
>
> A **plane curve** is a set of points (x, y) such that $x = f(t)$, $y = g(t)$, and f and g are both defined on an interval I. The equations $x = f(t)$ and $y = g(t)$ are **parametric equations** with **parameter t.**

NOTE Just as graphing calculators are capable of graphing rectangular and polar equations, they are also capable of graphing plane curves defined by parametric equations. You should familiarize yourself with how your particular model handles them.

Graphs of Parametric Equations and Their Rectangular Equivalents

EXAMPLE 1

Graphing a Plane Curve Defined Parametrically

For the plane curve defined by the parametric equations

$$x = t^2, \ y = 2t + 3, \quad \text{for } t \text{ in the interval } [-3, 3],$$

graph the curve using a graphing calculator, and then find an equivalent rectangular equation.

SOLUTION Figure 15 shows the graph of the curve, which appears to be a horizontal parabola.

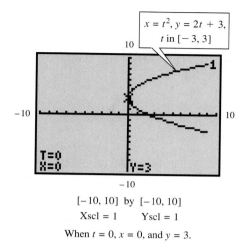

[−10, 10] by [−10, 10]

Xscl = 1   Yscl = 1

When $t = 0$, $x = 0$, and $y = 3$.

FIGURE 15

To find an equivalent rectangular equation, we analytically eliminate the parameter t. For this curve, we will solve for t in the second equation, $y = 2t + 3$, to begin.

$$y = 2t + 3$$
$$2t = y - 3$$
$$t = \frac{y - 3}{2}$$

Now substitute the result in the first equation to get

$$x = t^2 = \left(\frac{y - 3}{2}\right)^2 = \frac{(y - 3)^2}{4}$$

or

$$x = \frac{1}{4}(y - 3)^2.$$

This is indeed an equation of a parabola. It has a horizontal axis and opens to the right. Because t is in $[-3, 3]$, x is in $[0, 9]$ and y is in $[-3, 9]$. The rectangular equation must be given with its restricted domain as

$$x = \frac{1}{4}(y - 3)^2, \quad \text{for } x \text{ in } [0, 9]. \quad \blacksquare$$

FOR GROUP DISCUSSION

The display at the bottom of Figure 15 indicates particular values of t, x, and y. Discuss how these relate to both the parametric form and the rectangular form of the equation of this parabola.

EXAMPLE **2**

Graphing a Plane
Curve Defined
Parametrically

Repeat Example 1 for the plane curve defined by the parametric equations

$$x = 2 \sin t, \, y = 3 \cos t, \quad \text{for } t \text{ in the interval } [0, 2\pi].$$

Find a rectangular equation for the curve.

SOLUTION   Taking care to see that the calculator is in radian mode and that the minimum and maximum t values are 0 and 2π, respectively, we obtain the graph shown in Figure 16. It appears to be an ellipse.* (A square window is necessary to gain the correct perspective.)

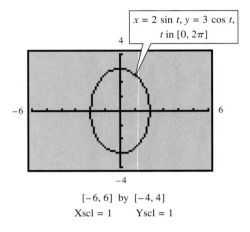

$[-6, 6]$ by $[-4, 4]$
Xscl $= 1$ Yscl $= 1$

FIGURE 16

Because it is awkward to solve either equation for t, we can use another approach. Because $\sin^2 t + \cos^2 t = 1$, we can square both sides of each equation, and then solve one for $\sin^2 t$ and the other for $\cos^2 t$.

$$x = 2 \sin t \qquad\qquad y = 3 \cos t$$
$$x^2 = 4 \sin^2 t \qquad\qquad y^2 = 9 \cos^2 t$$
$$\frac{x^2}{4} = \sin^2 t \qquad\qquad \frac{y^2}{9} = \cos^2 t$$

Now add corresponding sides of the two equations to get

$$\frac{x^2}{4} + \frac{y^2}{9} = \sin^2 t + \cos^2 t$$
$$\frac{x^2}{4} + \frac{y^2}{9} = 1.$$

In algebra it is shown that this is the equation of an ellipse with x-intercepts ± 2 and y-intercepts ± 3.   ◧

Alternative Forms of Parametric Equations

Parametric representations of a curve are not unique. In fact, there are infinitely many parametric representations of a given curve. If the curve can be described by

---

*An ellipse is an example of a *conic section.* Other conic sections are the circle, the parabola, and the hyperbola. See a standard college algebra text for more information on conic sections.

a rectangular equation $y = f(x)$, with domain I, then one simple parametric representation is

$$x = t, \, y = f(t), \quad \text{for } t \text{ in } I.$$

The next example shows how one plane curve has alternative parametric equation forms.

EXAMPLE 3

Finding Alternative Parametric Equation Forms

Give three parametric representations for the parabola

$$y = (x - 2)^2 + 1.$$

SOLUTION   The simplest choice is to let

$$x = t, \quad y = (t - 2)^2 + 1, \quad \text{for } t \text{ in } (-\infty, \infty).$$

Another choice, that leads to a simpler equation for y is

$$x = t + 2, \quad y = t^2 + 1, \quad \text{for } t \text{ in } (-\infty, \infty).$$

Sometimes trigonometric functions are desirable; one choice here might be

$$x = 2 + \tan t, \quad y = \sec^2 t, \quad \text{for } t \text{ in } \left(-\frac{\pi}{2}, \frac{\pi}{2}\right). \quad \blacksquare$$

Applications and the Cycloid

Of the many applications of parametric equations, one of the most useful allows us to determine the path of a moving object whose position is given by the function $x = f(t)$, $y = g(t)$, where t represents time. The parametric equations give the position of the object at any time t.

EXAMPLE 4

Examining Parametric Equations Defining the Position of an Object in Motion

The motion of a projectile (neglecting air resistance) is given by

$$x = (v_0 \cos \theta)t, \quad y = -16t^2 + (v_0 \sin \theta)t, \quad \text{for } t \text{ in } [0, k],$$

where t is time in seconds, v_0 is the initial speed of the projectile in the direction θ with the horizontal, x and y are in feet, and k is a positive real number. (See Figure 17.) Find the rectangular form of the equation.

SOLUTION   Solving the first equation for t and substituting the result into the second equation gives (after simplification)

$$y = (\tan \theta)x - \frac{16}{v_0^2 \cos^2 \theta}x^2,$$

the equation of a vertical parabola opening downward, as shown in Figure 17. $\quad \blacksquare$

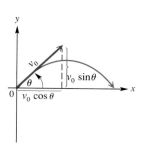

FIGURE 17

The path traced by a fixed point on the circumference of a circle rolling along a line is called graphing a **cycloid.** See Figure 18.

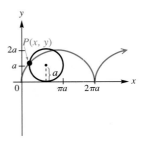

FIGURE 18

The cycloid is defined by the parametric equations

$$x = at - a \sin t, \quad y = a - a \cos t, \quad \text{for } t \text{ in } (-\infty, \infty).$$

A graphing calculator provides an excellent means of graphing a cycloid.

EXAMPLE 5

Graphing a Cycloid Using a Graphing Calculator

Graph the cycloid for $a = 3$ and t in the interval $[-2\pi, 2\pi]$.

SOLUTION   The equations

$$x = 3t - 3 \sin t, \quad y = 3 - 3 \cos t$$

are used. The calculator should be set in radian mode, with minimum and maximum t values -2π and 2π, respectively. Notice that the graph never falls below the x-axis, and the maximum y-value is $2(3) = 6$. See Figure 19, where a square window is used.   ▯

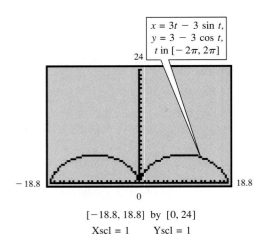

$x = 3t - 3 \sin t,$
$y = 3 - 3 \cos t,$
t in $[-2\pi, 2\pi]$

$[-18.8, 18.8]$ by $[0, 24]$
Xscl = 1   Yscl = 1

FIGURE 19

The cycloid has an interesting physical property. If a flexible cord or wire goes through points P and Q as in Figure 20 and a bead is allowed to slide without friction along this path from P to Q, the path that requires the shortest time takes the shape of the graph of an inverted cycloid.

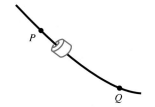

FIGURE 20

6.3 EXERCISES

For each plane curve, use a graphing calculator to generate the curve over the interval for the parameter t, using the window specified. Then find a rectangular equation for the curve.

1. $x = 2t, y = t + 1,$ for t in $[-2, 3]$
   Window: $[-8, 8]$ by $[-8, 8]$

2. $x = t + 2, y = t^2,$ for t in $[-1, 1]$
   Window: $[0, 4]$ by $[-2, 2]$

3. $x = \sqrt{t}, y = 3t - 4,$ for t in $[0, 4]$
   Window: $[-6, 6]$ by $[-6, 10]$

4. $x = t^2, y = \sqrt{t},$ for t in $[0, 4]$
   Window: $[-2, 20]$ by $[0, 4]$

5. $x = t^3 + 1, y = t^3 - 1,$ for t in $[-3, 3]$
   Window: $[-30, 30]$ by $[-30, 30]$

6. $x = 2t - 1, y = t^2 + 2,$ for t in $[-10, 10]$
   Window: $[-20, 20]$ by $[0, 120]$

7. $x = 2^t, y = \sqrt{3t - 1},$ for t in $\left[\dfrac{1}{3}, 4\right]$

   Window: $[-2, 30]$ by $[-2, 10]$

8. $x = \ln(t - 1), y = 2t - 1,$ for t in $(1, 10]$
   Window: $[-5, 5]$ by $[-2, 20]$

9. $x = 2 \sin t, y = 2 \cos t,$ for t in $[0, 2\pi]$
   Window: $[-6, 6]$ by $[-4, 4]$

10. $x = \sqrt{5} \sin t, y = \sqrt{3} \cos t,$ for t in $[0, 2\pi]$
   Window: $[-6, 6]$ by $[-4, 4]$

11. $x = 3 \tan t, y = 2 \sec t,$ for t in $[0, 2\pi]$
   Window: $[-6, 6]$ by $[-4, 4]$ (Use dot mode.)

12. $x = \cot t, y = \csc t,$ for t in $(0, 2\pi)$
   Window: $[-6, 6]$ by $[-4, 4]$ (Use dot mode.)

For each plane curve, find a rectangular equation.

13. $x = \sin \theta, y = \csc \theta,$ for θ in $(0, \pi)$

14. $x = \tan \theta, y = \cot \theta,$ for θ in $\left(0, \dfrac{\pi}{2}\right)$

15. $x = t, y = \sqrt{t^2 + 2},$ for t in $(-\infty, \infty)$

16. $x = \sqrt{t}, y = t^2 - 1,$ for t in $[0, \infty)$

17. $x = e^t, y = e^{-t},$ for t in $(-\infty, \infty)$

18. $x = e^{2t}, y = e^t,$ for t in $(-\infty, \infty)$

19. $x = 2 + \sin \theta, y = 1 + \cos \theta,$ for θ in $[0, 2\pi]$

20. $x = 1 + 2 \sin \theta, y = 2 + 3 \cos \theta,$ for θ in $[0, 2\pi]$

21. $x = t + 2, y = \dfrac{1}{t + 2}, \; t \neq -2$

22. $x = t - 3, y = \dfrac{2}{t - 3}, \; t \neq 3$

23. $x = t^2, y = 2 \ln t, \; t$ in $(0, \infty)$

24. $x = \ln t, y = 3 \ln t, \; t$ in $(0, \infty)$

Use a graphing calculator to graph each cycloid for t in $[0, 4\pi]$. Use the window specified.

25. $x = t - \sin t, y = 1 - \cos t$
   Window: $[0, 4\pi]$ by $[0, 6]$

26. $x = 2t - 2 \sin t, y = 2 - 2 \cos t$
   Window: $[0, 4\pi]$ by $[0, 4]$

27. A projectile is fired with an initial velocity of 400 ft per sec. at an angle of 45° with the horizontal. Find each of the following. **(a)** the time when it strikes the ground **(b)** the range (horizontal distance covered) **(c)** the maximum altitude

28. Repeat Exercise 27 if the projectile is fired at 800 ft per sec. at an angle of 30° with the horizontal.

29. Show that the rectangular equation for the curve describing the motion of a projectile defined by

$$x = (v_0 \cos \theta)t, \ y = (v_0 \sin \theta)t - 16t^2,$$

for t in $[0, k]$, is

$$y = (\tan \theta)x - \frac{16}{v_0^2 \cos^2 \theta}x^2.$$

30. Find the vertex of the parabola given by the rectangular equation of Exercise 29.

31. Give two parametric representations of the line through the point (x_1, y_1) with slope m.

32. Give two parametric representations of the parabola $y = a(x - h)^2 + k$.

33. Give a parametric representation of the hyperbola

$$\frac{x^2}{a^2} - \frac{y^2}{b^2} = 1.$$

34. Give a parametric representation of the ellipse

$$\frac{x^2}{a^2} + \frac{y^2}{b^2} = 1.$$

35. The spiral of Archimedes has polar equation $r = a\theta$, where $r^2 = x^2 + y^2$. Show that a parametric representation of the spiral is

$$x = at \cos t, \ y = at \sin t, \text{ for } t \text{ in } (-\infty, \infty).$$

(The parameter t must be in radians.)

36. Show that the *hyperbolic spiral* $r\theta = a$, where $r^2 = x^2 + y^2$, is given parametrically by

$$x = \frac{a \cos t}{t}, y = \frac{a \sin t}{t},$$

for t in $(-\infty, 0) \cup (0, \infty)$.

(The parameter t must be in radians.)

6.4 APPLICATIONS OF PARAMETRIC EQUATIONS

Path of a Projectile ▌ Airplane Ascent

In the previous section we introduced the concepts of a parameter and parametric equations. Now we explore several applications of parametric equations that illustrate their power and usefulness.

Path of a Projectile

Refer back to Example 4 in Section 6.3. If the projectile in this example is fired with initial velocity $v_0 = 48$ feet per second and at an angle $\theta = 45°$ with respect to the horizontal, then the rectangular equation

$$y = (\tan \theta)x - \frac{16}{v_0^2 \cos^2 \theta}x^2$$

becomes

$$y = (\tan 45°)x - \frac{16}{48^2 \cos^2 45°}x^2.$$

Simplifying this last equation, we get

$$y = x - \frac{1}{72}x^2. \qquad \text{(Recall that } x \text{ and } y \text{ are in feet.)}$$

Figure 21 shows the graph of this rectangular equation. Here we see the parabolic path of the projectile. In other words, we see *where* the projectile has been. However, the rectangular equation does not tell us *when* the projectile was at a

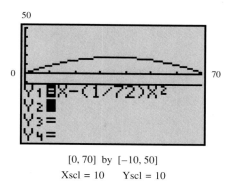

[0, 70] by [−10, 50]
Xscl = 10   Yscl = 10

FIGURE 21

particular point (x, y). The first example illustrates how we can use the parameter t, representing time, along with the graphing capabilities of a calculator, to determine when the projectile was at any point (x, y). (*Note:* In this section we will assume that the only force acting on any projectile is gravity.)

EXAMPLE 1

Analyzing the Path of a Projectile Using Parametric Equations

Recall from Example 4 in Section 6.3 that the parametric equations for a projectile fired with an initial velocity v_0 and an angle θ in degrees with respect to the horizontal are given by

$$x = (v_0 \cos \theta)t \quad \text{and} \quad y = -16t^2 + (v_0 \sin \theta)t.$$

Find the coordinates of the point at which the projectile is located at times $t = 0$ and $t = 1$. Solve analytically and support with a calculator.

SOLUTION   If $\theta = 45°$ and $v_0 = 48$ feet per second, then the equations become

$$x = (48 \cos 45°)t \quad \text{and} \quad y = -16t^2 + (48 \sin 45°)t$$
$$x = 24\sqrt{2}\, t \qquad\qquad\qquad y = -16t^2 + 24\sqrt{2}\, t$$

Letting $t = 0$, we get $x = 0$ and $y = 0$, so at time $= 0$, the coordinates of the location of the projectile are $(0, 0)$. Letting $t = 1$, we find that the coordinates are $(24\sqrt{2}, -16 + 24\sqrt{2}) \approx (33.94, 17.94)$. Figure 22 supports both of these results. The first is supported with the display at the bottom of the upper half of the screen, while the second is supported in the bottom entry of the table in the lower half of the screen.

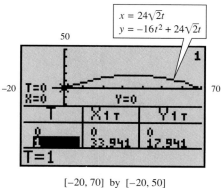

[−20, 70] by [−20, 50]
Xscl = 10   Yscl = 10

FIGURE 22

Notice that the graphs in Figures 21 and 22 are the same. They were, however, generated differently: in Figure 21, the function mode was used while in Figure 22, the parametric mode was used.   ∎

So far we have analyzed the path of a projectile that has been launched from ground level. In general, the path of a projectile fired with an initial velocity v_0 feet per second, from a height h feet, and with an angle θ degrees from the horizontal has parametric equations

$$x = (v_0 \cos \theta)t \qquad \text{and} \qquad y = h - 16t^2 + (v_0 \sin \theta)t$$

where x and y are in feet and t is in seconds. The next example utilizes these equations.

EXAMPLE 2

Using the Graph of Parametric Equations to Analyze a Projectile in Motion

---

A small rocket is launched from a table that is 3.36 feet above the ground. Its initial velocity is 64 feet per second and it is launched at an angle of 30° with respect to the ground.

(a) Determine parametric equations for the flight of the rocket and graph the flight of the rocket with a calculator in parametric mode.

(b) Determine graphically the maximum height of the rocket and the time it reaches its maximum height.

(c) Determine graphically the total time of the flight and the horizontal distance it travels.

(d) Find a rectangular equation for the path of the rocket.

SOLUTION

(a) Since the table is 3.36 feet above the ground, we know that $h = 3.36$. We also know that $\theta = 30°$ and $v_0 = 64$. So our parametric equations are

$$x = (64 \cos 30°)t \qquad \text{and} \qquad y = 3.36 - 16t^2 + (64 \sin 30°)t$$
$$x = 32\sqrt{3}\, t \qquad \qquad \text{and} \qquad y = -16t^2 + 32t + 3.36.$$

Figure 23 shows the path of the rocket.

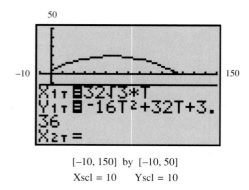

[−10, 150] by [−10, 50]
Xscl = 10   Yscl = 10

FIGURE 23

(b) The maximum height of the rocket will occur at the highest point on the graph. Tracing to this point, we see that at the highest point, $y = 19.36$ and $t = 1$. This means that after one second, the rocket reaches its maximum height, 19.36 feet. See Figure 24.

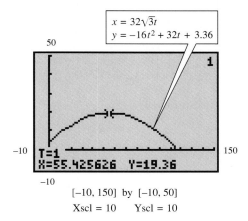

$$x = 32\sqrt{3}t$$
$$y = -16t^2 + 32t + 3.36$$

[−10, 150] by [−10, 50]
Xscl = 10   Yscl = 10

FIGURE 24

(c) Tracing to the point at which the graph intersects the *x*-axis, we find that when $t = 2.1, x \approx 116.4$. See Figure 25. This means that after 2.1 seconds of flight time, the rocket has traveled a distance of 116.4 feet.

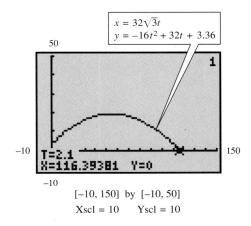

$$x = 32\sqrt{3}t$$
$$y = -16t^2 + 32t + 3.36$$

[−10, 150] by [−10, 50]
Xscl = 10   Yscl = 10

FIGURE 25

(d) We know that the parametric equations are

$$x = 32\sqrt{3}t \quad \text{and} \quad y = -16t^2 + 32t + 3.36.$$

From $x = 32\sqrt{3}t$ we get

$$t = \frac{x}{32\sqrt{3}}.$$

Substituting into the other parametric equation for *t* yields

$$y = -16\left(\frac{x}{32\sqrt{3}}\right)^2 + 32\left(\frac{x}{32\sqrt{3}}\right) + 3.36.$$

Upon simplification, we determine the rectangular equation to be

$$y = -\frac{1}{192}x^2 + \frac{\sqrt{3}}{3}x + 3.36. \quad \blacksquare$$

FOR GROUP DISCUSSION

Graph the parametric equations in Example 2. Trace to where $t = 2$ and observe the y-value. In your own words, describe the relationship between this y-value and the height from which the rocket was launched. Make a conjecture as to the total time of the flight if the rocket had been launched from ground level.

In the exercises for this section, you will be asked to model a physical situation parametrically, graph the parametric equations and answer questions from the graph. We should also remember that we have analytic means to support our graphical answers. In our next example we illustrate how to support our answers from Example 2 analytically.

EXAMPLE 3

Using Analytic Methods to Verify Graphical Solutions

Support the answers to Example 2(c) analytically.

SOLUTION   To determine the total time the rocket is in the air, we utilize the equation

$$y = -16t^2 + 32t + 3.36$$

since it tells us the vertical position of the rocket for any time t. We need to determine those values of t for which $y = 0$, since this corresponds to the rocket at ground level. This yields

$$0 = -16t^2 + 32t + 3.36$$

Using the quadratic formula to solve for t, we determine that $t = -.1$ or $t = 2.1$. Since t represents time, $t = -.1$ is an unacceptable answer. However, the solution $t = 2.1$ does support our graphical answer from Example 2.

Since we know that the rocket was in the air for 2.1 seconds, we can use $t = 2.1$ and the parametric equation that describes the horizontal position, $x = 32\sqrt{3}t$, to get

$$x = 32\sqrt{3}\,(2.1)$$
$$x = 67.2\sqrt{3} \text{ feet.}$$

Using a calculator to approximate $67.2\sqrt{3}$ indicates that this does indeed support our graphical solution found in Example 2(c).   ◼

FOR GROUP DISCUSSION

How would you support analytically the result of Example 2(b)?

The next two examples illustrate problems that have become more popular to explore and analyze with the advent of graphing calculators.

EXAMPLE 4

Using the Graph of Parametric Equations to Analyze the Flight of a Baseball

A major-league baseball player hits a baseball when it is 3 feet above the ground. The ball leaves his bat at an angle of 31° from the horizontal with a velocity of 121 feet per second.

(a) Determine parametric equations for the flight of the baseball and graph the flight of the ball on a calculator in parametric mode.

(b) Determine graphically the maximum height of the ball and when it reaches the maximum height.

(c) Determine graphically how far the ball travels and how long it is in the air.

SOLUTION

(a) The parametric equations for the flight of the baseball are

$$x = (121 \cos 31°)t \qquad \text{and} \qquad y = 3 - 16t^2 + (121 \sin 31°)t.$$

Figure 26 shows the graph of these parametric equations which models the flight of the baseball.

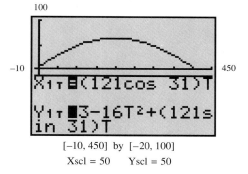

$[-10, 450]$ by $[-20, 100]$
Xscl = 50   Yscl = 50

FIGURE 26

(b) In Figure 27 we have traced the flight of the ball to the highest point and determined that the maximum height of the ball is approximately 63.6 feet (y-value) and occurs about 1.9 seconds (t-value) after being hit.

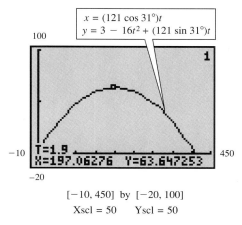

$[-10, 450]$ by $[-20, 100]$
Xscl = 50   Yscl = 50

FIGURE 27

(c)   Figure 28 indicates that the ball will travel a little more than 404 feet (*x*-value) and it is in the air slightly more than 3.9 seconds (*t*-value). We determine these answers by looking for the point where *y is close to* 0 since this corresponds to being on the ground. ▯

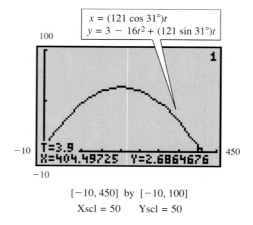

$x = (121 \cos 31°)t$
$y = 3 - 16t^2 + (121 \sin 31°)t$

[−10, 450] by [−10, 100]
Xscl = 50   Yscl = 50

FIGURE 28

In Example 4(c) we said the ball will travel more than 404 feet and it would be in the air more than 3.9 seconds. This is due to the *y*-value being 2.69. Since *y* is the vertical component, after 3.9 seconds the ball has not quite reached the ground. In Figure 29, we see that after 4 seconds the baseball has traveled approximately 415 feet but the *y*-value is now −3.72. This means that the ball would have hit the ground and gone 3.72 feet below the surface! Since a *t*-value of 3.9 puts the ball closer to the ground, we answered part (c) with this value (and its corresponding *x*-value), being careful to use the proper language.

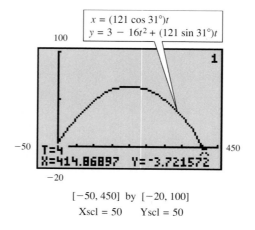

$x = (121 \cos 31°)t$
$y = 3 - 16t^2 + (121 \sin 31°)t$

[−50, 450] by [−20, 100]
Xscl = 50   Yscl = 50

FIGURE 29

EXAMPLE 5

Determining Whether a Batted Ball Will be a Home Run

Assume the ball in Example 4 is heading toward the 360-foot sign on the outfield fence. (This indicates the distance from home plate, where the batter is, to the fence.) Will the ball clear the fence and be a home run if the fence is 10 feet high?

SOLUTION To answer this, we can trace to a point near $x = 360$ and see if the y-value is greater than 10. Figure 30 shows that at $x \approx 363$, $y \approx 25.1$. We conclude that the ball does clear the fence for a home run. ◧

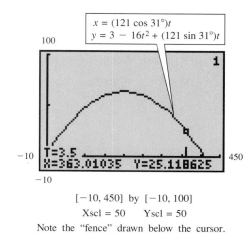

$x = (121 \cos 31°)t$
$y = 3 - 16t^2 + (121 \sin 31°)t$
100
−10
450
−10
T=3.5
X=363.01035 Y=25.118625

[−10, 450] by [−10, 100]
Xscl = 50 Yscl = 50
Note the "fence" drawn below the cursor.

FIGURE 30

An alternate way to look at Example 5 is to use the LINE command on your calculator to simulate a 10-foot-high fence that is 360 feet from the batter. Again, see Figure 30. Here, we clearly see that the path of the ball goes over the "fence."

Airplane Ascent

Suppose that a small plane leaves Cleveland with Pittsburgh as its destination. Furthermore, suppose that the plane has a groundspeed of 100 feet per second (horizontally) and that it is rising 25 feet per second (vertically). Figure 31 depicts its flight path. From the moment of liftoff, we can create a table that shows the position of the plane for different values of t. See Figure 32.

10
0
10
−2

[0, 10] by [−2, 10]
Xscl = 1 Yscl = 1

FIGURE 31

T	X1T	Y1T
0	0	0
1	100	25
2	200	50
3	300	75
4	400	100
5	500	125
6	600	150

T=0

FIGURE 32

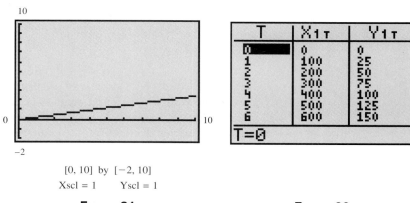

Figure 32 indicates that the horizontal position, x, and the vertical position, y, can be represented with the parametric equations $x = 100t$ and $y = 25t$. Now that we have a model, we can simulate the flight path graphically.

EXAMPLE **6**

Simulating a Flight Path and Determining a Rectangular Equation for the Path

(a) Use a calculator to simulate the first 40 seconds of flight of the aforementioned plane.

(b) Determine a rectangular equation for the plane's flight path.

SOLUTION

(a) Figure 33 illustrates the graph of the parametric equations $x = 100t$ and $y = 25t$, as t takes on values from 0 to 40.

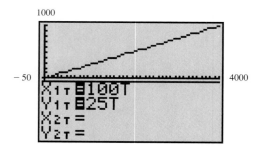

$[-50, 4000]$ by $[-50, 1000]$
Xscl = 100   Yscl = 100

FIGURE 33

(b) From the equation $x = 100t$, we get

$$\frac{x}{100} = t.$$

Substituting this into the other parametric equation yields

$$y = 25\left(\frac{x}{100}\right)$$

$$y = \frac{1}{4}x. \quad \blacksquare$$

The result in Example 6(b) should not be surprising. The flight path for the first 40 seconds, as shown in Figure 33, appears to be linear. Also looking back at Figure 32, we see that each y-value is $\frac{1}{4}$ of the corresponding x-value. From algebra, you may recall that this average rate of change is called *slope*. In the airplane problem, we see that the slope of the line determining the flight path is $\frac{1}{4}$. (Recognizing that the answer to Example 6(b) is in slope-intercept form reaffirms this.)

6.4 EERCISES

1. A model rocket is launched from the ground with a velocity of 48 feet per second and at an angle of 60° with respect to the ground.
   (a) Determine parametric equations describing the path of the rocket. (Give exact numerical values for the coefficients.)
   (b) Draw the path of the rocket on your calculator.
   (c) Use the graph to approximate the total time of the flight.
   (d) Use the graph to approximate the maximum height of the rocket.
   (e) Determine an *xy*-equation for the path of the rocket. (Give exact numerical values for the coefficients.)

2. Use an analytic method to determine the maximum height of the rocket in Exercise 1.

3. Sally hits a softball when it is 2 feet above the ground. The ball leaves her bat at an angle of 20° with respect to the ground at a velocity of 88 feet per second.
   (a) Use parametric equations to draw the flight of the ball.
   (b) Graphically approximate the maximum height of the ball and when it reaches the maximum height.
   (c) Graphically approximate how far the ball travels and how long it is in the air.

4. Brian is playing golf. He hit a golf ball, from the ground, at an angle of 60° with respect to the ground at a velocity of 150 feet per second.
   (a) Use parametric equations to draw the flight of the ball.
   (b) Graphically approximate the maximum height of the ball and when it reaches the maximum height.
   (c) Graphically approximate how far the ball travels and how long it is in the air.

5. In Exercise 4, there was an oak tree between Brian and where he was aiming his golf shot. The tree was 100 feet from where he was hitting the ball and was 90 feet tall. Did Brian's ball make it over the tree?

6. Carlos hits a baseball when it is 2.5 feet above the ground. The ball leaves his bat at an angle of 29° from the horizontal with a velocity of 136 feet per second.
   (a) Use parametric equations to draw the flight of the ball.
   (b) Graphically approximate how long the ball is in the air and how far it travels.
   (c) For what values of *t* is the ball at least 50 feet above the ground?

7. Assume in Exercise 6 that the ball Carlos hit was heading toward the 414-foot sign on the center field fence. If the fence is 12 feet high, would Carlos' ball clear the fence for a home run?

8. Grover hits a baseball when it is 3.5 feet above the ground. The ball leaves his bat at an angle of 20° from the horizontal with a velocity of 120 feet per second.
   (a) If the ball is heading toward the 390-foot sign on the left-center field fence and if the fence is 10 feet high, would his ball clear the fence for a home run?
   (b) If you answered no to part (a), experiment with θ in 5° increments to determine the smallest θ that would produce a home run. Use the TABLE feature of your calculator.
   (c) Working in 5° increments, experiment to determine what value of θ produces the farthest distance traveled by the baseball.

9. Lisa is playing golf. She is 150 yards from the green and there is a 250-foot-tall oak tree exactly halfway between her ball and the green. Assuming her ball leaves the golf club with a velocity of 150 feet per second, determine θ, to the nearest 5°, that will propel her ball over the tree and have it land nearest to the green. (*Hint:* Start with θ = 30° and work from there.)

10. A National Football League (NFL) punter kicks a football from his 15-yard line. The ball leaves his foot approximately 1 foot above the ground with a velocity of 85 feet per second and at an angle of 60° with respect to the ground.
   (a) Use parametric equations to graph the flight of the ball.
   (b) Graphically determine how far the ball travels, to the nearest yard.
   (c) Graphically determine the maximum height of the football and when it reaches the maximum height.

11. Most NFL coaches are interested in knowing the hang time of a punted football. *Hang time* is simply the total time the football is in the air. Determine the hang time for the football kicked in Exercise 10 if
   (a) it hits the ground.
   (b) a player catches the football approximately 4.5 feet above the ground.

12. An antiaircraft missile is launched from the ground with a velocity of 1224 feet per second and at an angle of 80° with respect to the ground.
   (a) Use parametric equations to graph the flight of the missile.
   (b) Graphically determine how far the missile travels and how long it is in the air.

13. If you are the pilot of a plane that is being fired at by the missile described in Exercise 12, what is the lowest altitude (to the nearest thousand) that you would fly the plane to ensure that it would not hit your plane?

14. For what values of t is the missile in Exercise 12 at least 15,000 feet above the ground?

15. Assume that the missile in Exercise 12 is now used to hit targets on land. Assume that the missile is launched from the ground with a velocity of 1224 feet per second.

(a) Determine the maximum range for the missile, i.e., determine the maximum distance the missile can travel. To do this, you will need to experiment with different values of θ. (*Hint:* Does Exercise 8 help?)

(b) Determine θ to the nearest 5° that is required to hit a target 35,000 feet away.

(c) Is it possible to hit a target 10 miles away?

In Exercises 16–19, a projectile has been launched from the ground with an initial velocity of 88 feet per second. You are supplied with the parametric equations modeling the path of the projectile.

(a) *Graph the parametric equations.*

(b) *Approximate θ, the angle the projectile makes with the horizontal at launch, to the nearest tenth of a degree.*

(c) *Based on your answer to (b), write parametric equations for the projectile and graph your equations, along with the ones supplied to you, to see how accurate you are.*

16. $x = 82.69265063t$ and $y = -16t^2 + 30.09777261t$

17. $x = 56.56530965t$ and $y = -16t^2 + 67.41191099t$

18. $x = 62.22539674t$ and $y = -16t^2 + 62.22539674t$

19. $x = 44t$ and $y = -16t^2 + 76.21023553t$

If a ball is hit into a wind, the parametric equation for its horizontal component is given by $x = (v_0 \cos \theta - wind)t$, while if it is hit with the wind it would be given by $x = (v_0 \cos \theta + wind)t$, where wind is given in feet per second. The parametric equation for the vertical component is the same as discussed throughout this section. Use this information to work Exercises 20–23.

20. Rework Exercise 3 except have Sally hit the softball into a 7.2-feet-per-second wind.

21. Rework Exercise 3 except have Sally hit the softball with a 7.2-feet-per-second wind.

22. Rework Exercise 4 except have Brian hit the golf ball into a 10.2-feet-per-second wind.

23. Rework Exercise 4 except have Brian hit the golf ball with a 10.2-feet-per-second wind.

24. If a major-league baseball player hits a ball 3 feet above the ground that leaves his bat at 124 feet per second and at an angle of 90° with respect to the ground, describe the path of the ball.

25. Suppose a major-leaguer hits a ball with the same velocity as in Exercise 24. Describe the path of the ball if $90° < \theta < 180°$. What if $-90° < \theta < 0°$?

26. While commuting on a small propeller-driven airplane from Phoenix to Tucson, the pilot supplied data modeling her flight path from takeoff, where

T	X1T	Y1T
0	0	0
1	90	30
2	180	60
3	270	90
4	360	120
5	450	150
6	540	180

T=0

$t = 0$ corresponds to takeoff. (See the accompanying table.) Recall that x is the horizontal position and y is the vertical position.

(a) Determine parametric equations modeling the flight path.

(b) Use a calculator to simulate the airplane's first 30 seconds of flight.

(c) Determine how long the plane was on this flight path to reach the cruising altitude of 4000 feet.

(d) Utilize right triangle trigonometry to determine θ, the angle of ascent (elevation), to the nearest tenth of a degree.

27. While approaching the airport in Tucson (see Exercise 26), the pilot stated that she wanted to approach the runway on a descent that was similar to the initial flight path. Assuming the plane is cruising at an altitude of 4000 feet and the change in the horizontal

T	X1T	Y1T
0	0	4000
1	90	3970
2	180	3940
3	270	3910
4	360	3880
5	450	3850
6	540	3820

T=0

and vertical components are the same as in Exercise 26, the accompanying table models the descent. Here $t = 0$ corresponds to the moment the descent begins, x is the horizontal distance traveled once the descent has begun, and y is feet above ground.
   (a) Determine parametric equations modeling the flight path toward landing.
   (b) At what distance from the airport should the pilot begin the descent assuming she wants to stay on this flight path?
   (c) How long is it before the plane lands?
   (d) What is the angle of descent (depression) to the nearest tenth of a degree?
   (e) Review your answers to Exercise 26. Do your answers to parts (c) and (d) here surprise you?

On the moon, the force due to gravity is significantly less than it is on Earth. As a result, parametric equations modeling the path of a projectile would be slightly different for a projectile on the moon than they would be for a projectile on Earth. Assume that the parametric equations for the path of a projectile on the moon, launched from the lunar surface, are given by

$$x = (v_0 \cos \theta)t \qquad \text{and} \qquad y = -2.65t^2 + (v_0 \sin \theta)t.$$

Notice that the only difference between these equations and the ones for Earth is the coefficient on t^2 in the equation that describes the vertical position. Use these equations in the following problem.

28. On a moon landing, an astronaut hit a golf ball while on the moon. Assume that the ball was on the lunar surface, and it left the golf club at an angle of 38° with a velocity of 104 feet per second.
   (a) Graph the path of the golf ball on a calculator.
   (b) How far did the ball travel, to the nearest yard, and how long was it in the air?
   (c) Assume the same conditions except now have the astronaut on Earth. Graph the path of the golf ball on a calculator.
   (d) On Earth, how far did the golf ball travel, to the nearest yard, and how long was it in the air?

*C*hapter 6   SUMMARY

A point in the plane may be graphed in the polar coordinate system by finding a direction angle θ and the magnitude r of the corresponding vector. The ordered pair (r, θ) gives the polar coordinates of the point. There are many ways to express a specific point with polar coordinates. Polar equations are expressed with the variables r and θ, and may be graphed by plotting many ordered pairs (r, θ) or with a graphing calculator. Polar equations and rectangular equations may be converted from one form to the other using the relationships among x, y, r, and θ.

A third way to graph a set of ordered pairs in the plane is with parametric equations, which express x and y in terms of a third variable t. The graphs can be found by using the equations $x = f(t)$ and $y = g(t)$ to find enough ordered pairs (x, y) to determine the graph. A graphing calculator is also capable of graphing parametric equations. Alternative pairs of parametric equations can lead to the same set of ordered pairs. Parametric equations are often the simplest way to describe the position of an object in motion. A cycloid, the path traced by a fixed point on the circumference of a circle rolling along a line, is defined by parametric equations.

The path of a projectile can be modeled on a calculator by utilizing parametric equations. The maximum height, time at which maximum height occurs, total distance traveled and total time of travel are some of the items that can be analyzed from the graph. Besides baseballs, footballs, golf balls, and rockets in flight, parametric equations can also model the initial flight path for an airplane.

Key Terms

SECTION 6.1

rectangular coordinate system
polar coordinate system
pole
polar axis
polar coordinates

SECTION 6.2

polar equation
cardioid
lemniscate
rose
spiral of Archimedes

SECTION 6.3

parametric equation
parameter
cycloid

SECTION 6.4

projectile motion
horizontal component
vertical component
hang time
ascent
descent
groundspeed

Chapter 6　　REVIEW EXERCISES

Convert to rectangular coordinates. Give exact values.

1. $(6, 30°)$　　　　　　**2.** $(12, 225°)$　　　　　　**3.** $(-8, -60°)$

Convert to polar coordinates, with $-180° < \theta \le 180°$, *with* $r > 0$. *Give exact values.*

4. $(-6, 6)$　　　　　　**5.** $(0, -5)$　　　　　　**6.** $(-\sqrt{5}, -\sqrt{5})$

Use a graphing calculator to graph each polar equation for $0° \le \theta \le 360°$. *Use a square window.*

7. $r = 4 \cos \theta$ (circle)

8. $r = -1 + \cos \theta$ (cardioid)

9. $r = 1 - 2 \sin \theta$ (limaçon with a loop)

10. $r = 2 \sin 4\theta$ (eight-leaved rose)

11. $r = 3 \cos 3\theta$ (three-leaved rose)

Find an equivalent equation in rectangular coordinates.

12. $r = \dfrac{3}{1 + \cos \theta}$　　　**13.** $r = \dfrac{4}{2 \sin \theta - \cos \theta}$　　　**14.** $r = \sin \theta + \cos \theta$　　　**15.** $r = 2$

Find an equivalent equation in polar coordinates.

16. $x = -3$　　　　　**17.** $y = x$　　　　　**18.** $y = x^2$　　　　　**19.** $x = y^2$

Use a graphing calculator to graph the plane curve defined by the parametric equations. Use the window specified.

20. $x = 4t - 3$, $y = t^2$,   for t in $[-3, 4]$
Window: $[-20, 20]$ by $[-20, 20]$

21. $x = t^2$, $y = t^3$,   for t in $[-2, 2]$
Window: $[-15, 15]$ by $[-10, 10]$

22. $x = t + \ln t$, $y = t + e^t$,   for t in $(0, 2]$
Window: $[-10, 5]$ by $[0, 10]$

23. $x = 3t - 3 \sin t$, $y = 3 - 3 \cos t$,   for t in $[0, 2\pi]$
Window: $[0, 6\pi]$ by $[0, 6]$

Find a rectangular equation for each plane curve with the given parametric equations.

24. $x = 3t + 2$, $y = t - 1$,   for t in $[-5, 5]$

25. $x = \sqrt{t - 1}$, $y = \sqrt{t}$,   for t in $[1, \infty)$

26. $x = 5 \tan t$, $y = 3 \sec t$,   for t in $\left(-\dfrac{\pi}{2}, \dfrac{\pi}{2}\right)$

Solve each of the following problems.

27. A baseball is hit when it is 3.2 feet above the ground. It leaves the bat with a velocity of 118 feet per second at an angle of 27° with respect to the ground.
  (a) How far does it travel and how long is it in the air?
  (b) What is the maximum height it attains?

28. A baseball is hit when it is 2.3 feet above the ground. It leaves the bat with a velocity of 127 feet per second at an angle of 51° with respect to the ground. If the ball is heading toward the 410-foot sign in center field and the fence is 12 feet high, will the ball clear the fence for a home run?

29. At liftoff, an airplane had a groundspeed of 100 feet per second and was climbing vertically at a rate of 20 feet per second.

  (a) Determine parametric equations to model the flight path.
  (b) What angle does the flight path make with the ground?

30. A golf ball that left the golf club with a velocity of 154 feet per second has parametric equations

$$x = 122.9898685t \quad \text{and}$$
$$y = -16t^2 + 92.67951357t.$$

Use the method of Section 6.4, Exercises 16–19, to estimate θ, the angle the ball makes with respect to the ground, to the nearest degree.

31. Rework Exercise 28 except have the ball hit into a 7.9-feet-per-second wind.

32. Rework Exercise 28 except have the ball hit with a 5.1-feet-per-second wind.

Answers to Selected Exercises

In this section we provide the answers that we think most students will obtain when they work the exercises using the methods explained in the text. If your answer does not look exactly like the one given here, it is not necessarily wrong. In many cases there are equivalent forms of the answer that are correct. For example, if the answer section shows $\frac{3}{4}$ and your answer is .75, you have obtained the correct answer but written it in a different (yet equivalent) form. Unless the directions specify otherwise, .75 is just as valid an answer as $\frac{3}{4}$. In general, if your answer does not agree with the one given, see whether it can be transformed into the other form. If it can, then it is the correct answer. If you still have doubts, talk with your instructor.

CHAPTER 1

Section 1.1 (page 8)

1. (a) 10 **(b)** 0, 10 **(c)** $-6, -\frac{12}{4}$ (or -3), 0, 10 **(d)** $-6, -\frac{12}{4}$ (or -3), $-\frac{5}{8}$, 0, .31, $.\overline{3}$, 10 **(e)** $-\sqrt{3}, 2\pi, \sqrt{17}$
(f) All are real numbers. **3. (a)** None are natural numbers. **(b)** None are whole numbers. **(c)** $-\sqrt{100}, -1$
(d) $-\sqrt{100}, -\frac{13}{6}, -1, 5.23, 9.\overline{14}, 3.14, \frac{22}{7}$ **(e)** None are irrational numbers. **(f)** All are real numbers.
5.
7.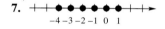
9.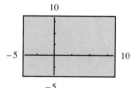

11. D **13.** B **15.** C **17.** A **19.** J **21.** $.8\overline{3}$ **23.** $-4.\overline{3}$ **25.** $.\overline{2}$ **27.** $.0\overline{81}$
29. The rational number .87 represents $\frac{87}{100}$. The repeating decimal $.\overline{87}$ is larger than $\frac{87}{100}$, since it contains alternating 8 and 7 in all decimal places after the second. .87 contains zeros in all other decimal places.
31. 7.615773106 **33.** 3.20753433 **35.** 3.045261646 **37.** 2.236067977
39. 2.620741394 **41.** 3.66643574 **43.** $<, \leq$ **45.** $>, \geq$ **47.** $\leq, \geq$
49. I **51.** III **53.** no quadrant **55.** II **57.** no quadrant
59. I or III **61.** II or IV **63.** It must lie on the y-axis.
65. Answers will vary. For example, for the TI-82, use ZOOM 6.

67.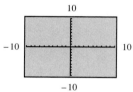
69.
71.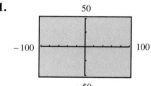

73. There are no tick marks. To set a screen with no tick marks on the axes, use Xscl $= 0$ and Yscl $= 0$.

Section 1.2 (page 19)

1. $(-1, 4)$

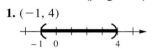

3. $(-\infty, 0)$

5. $[1, 2)$

7. $(-\infty, -9)$

9. $\{x \mid -4 < x < 3\}$

11. $\{x \mid x \le -1\}$ **13.** $\{x \mid -2 \le x < 6\}$ **15.** $\{x \mid x \le -4\}$
17. Use a parenthesis if the symbol is $<$ or $>$, and use a square bracket if the symbol is $\le$ or $\ge$.
19. domain: $\{5, 3, 4, 7\}$; range: $\{1, 2, 9, 6\}$; function   **21.** domain: $\{2, 0\}$; range: $\{4, 2, 5\}$; not a function
23. domain: $\{-3, 4, -2\}$; range: $\{1, 7\}$; function   **25.** domain: $\{1, 4, 0, 7\}$; range: $\{3, 7, 6, 2\}$; function
27. domain: $(-\infty, \infty)$; range: $(-\infty, \infty)$; function   **29.** domain: $[3, \infty)$; range: $(-\infty, \infty)$; not a function
31. domain: $[-4, 4]$; range: $[-3, 3]$; not a function   **33.** domain: $[2, \infty)$; range: $[0, \infty)$; function
35. domain: $[-9, \infty)$; range: $(-\infty, \infty)$; not a function   **37.** domain: $\{2, 5, 11, 17, 3\}$; range: $\{1, 7, 20\}$; function
39. domain: $\{1, 2, 3, 5\}$; range: $\{10, 15, 19, 27\}$; not a function   **41.** 3   **43.** 8   **45.** 7   **47.** 0
49. -4 **51.** -3 **53.** $-4, 0, 4$ **55.** $3a - 1$ **57.** $2r^2 + r + 3$ **59.** $7p + 2s - 1$

Further Explorations

1. -15 **3.** 701

Section 1.3 (page 41)

1. $\{10\}$ **3.** $\{-1.3\}$ **5.** $\{2\}$ **7.** The y-value is the function value (range value) for both y_1 and y_2 when
the solution of the equation is substituted for x.   **9.** $\{-.8\}$ **11.** $\{-16\}$ **13.** $\{12\}$ **15.** $\{3\}$
17. $\{-2\}$ **19.** $\{7\}$ **21.** $\{3\}$ **23.** $(-\infty, 3)$ **25.** $(3, \infty)$ **27.** $[3, \infty)$ **29.** $\{3\}$
31. $\{3\}$ **33. (a)** $(20, \infty)$ **(b)** $(-\infty, 20)$ **(c)** $[20, \infty)$ **(d)** $(-\infty, 20]$ **35. (a)** $[.4, \infty)$ **(b)** $(-\infty, .4)$
(c) $(.4, \infty)$ **(d)** $(-\infty, .4]$ **37. (a)** $(-\infty, -3]$ **(b)** $(-3, \infty)$ **39. (a)** $\left(-\frac{3}{4}, \infty\right)$ **(b)** $\left(-\infty, -\frac{3}{4}\right]$
41. (a) $(-\infty, 15]$ **(b)** $(15, \infty)$ **43. (a)** $(-6, \infty)$ **(b)** $(-\infty, -6]$ **45.** $\{\pm 4\}$ **47.** $\{\pm 3\}$
49. $\{\pm 4i\}$ **51.** $\{\pm\sqrt{2}\}$ **53.** $\{3, 5\}$ **55.** $\{1 \pm \sqrt{5}\}$ **57.** $\left\{-\frac{1}{2} \pm \frac{1}{2}i\right\}$ **59.** $\left\{\frac{1 \pm \sqrt{5}}{2}\right\}$
61. $\{3 \pm \sqrt{2}\}$ **63.** $\left\{\frac{3}{2} \pm \frac{\sqrt{2}}{2}i\right\}$ **65.** $\{2, 4\}$ **67.** $(-\infty, 2) \cup (4, \infty)$
69. $(-\infty, 3) \cup (3, \infty)$ **71.** $(-\infty, \infty)$ **73. (a)** $(-\infty, -3] \cup [-1, \infty)$
(b) the interval $(-3, -1)$ **75. (a)** $\left(-\infty, \frac{1}{2}\right) \cup (4, \infty)$ **(b)** $\left[\frac{1}{2}, 4\right]$ **77. (a)** $[1 - \sqrt{2}, 1 + \sqrt{2}]$
(b) $(-\infty, 1 - \sqrt{2}) \cup (1 + \sqrt{2}, \infty)$ **79. (a)** $\emptyset$ **(b)** $(-\infty, \infty)$

Section 1.4 (page 59)

1. $(-\infty, \infty)$ **3.** $[0, \infty)$ **5.** $(-\infty, -3) \cup (-3, \infty)$ **7.** $(-\infty, -2) \cup (-2, \infty)$ **9.** While it seems to
be continuous at first glance, there is a discontinuity at $x = -3$, as indicated by the calculator not giving a
corresponding value of y when we trace to $x = -3$. This happens because -3 causes the denominator to equal
zero, giving an undefined expression.   **11. (a)** $(-\infty, 0)$ **(b)** $(0, \infty)$ **(c)** none   **13. (a)** $(-2, \infty)$
(b) $(-\infty, -2)$ **(c)** none   **15. (a)** $(-\infty, 1)$ **(b)** $(4, \infty)$ **(c)** $(1, 4)$ **17. (a)** none   **(b)** $(-\infty, \infty)$
(c) none   **19. (a)** none   **(b)** $(-\infty, -2) \cup (3, \infty)$ **(c)** $(-2, 3)$ **21.** increasing   **23.** decreasing
25. increasing   **27.** increasing   **29.** increasing   **31.** decreasing   **33. (a)** no   **(b)** yes
(c) no   **35. (a)** yes   **(b)** no   **(c)** no   **37. (a)** yes   **(b)** yes   **(c)** yes   **39. (a)** yes   **(b)** no   **(c)** no
41. (a) no   **(b)** no   **(c)** yes   **43. (a)** no   **(b)** no   **(c)** no   **45.** $(-1.625, 2.0352051)$
47. $(.5, -.84089642)$ **49.** $(-5.092687, .9285541)$
51. symmetric with respect to the origin   **53.** symmetric with respect to the y-axis

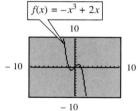

$f(x) = -x^3 + 2x$

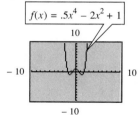

$f(x) = .5x^4 - 2x^2 + 1$

55. neither symmetric with respect to the origin nor the y-axis

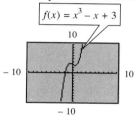

For the graphs preceding Exercises 57–66, see Figures 47, 48, 49, 52, 53 and 54 in Section 1.4.

57. true   **59.** false   **61.** true   **63.** true   **65.** true   **67.** odd   **69.** even   **71.** neither

73. (a)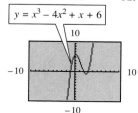

(b) $(2.54, -.88)$; not an absolute minimum
(c) $(.13, 6.06)$; not an absolute maximum
(d) $(-\infty, \infty)$
(e) x-intercepts: $-1, 2, 3$; y-intercept: 6

75. (a)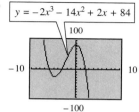

(b) $(-4.74, -27.03)$; not an absolute minimum
(c) $(.07, 84.07)$; not an absolute maximum
(d) $(-\infty, \infty)$
(e) x-intercepts: $-6, -3.19, 2.19$; y-intercept: 84

77. (a)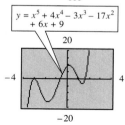

(b) $(-1.73, -16.39)$; $(1.35, -3.49)$; neither is an absolute minimum
(c) $(-3, 0)$; $(17, 9.52)$; neither is an absolute maximum
(d) $(-\infty, \infty)$
(e) x-intercepts: $-3, -.62, 1, 1.62$; y-intercept: 9

79. (a)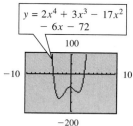

(b) $(-2.63, -132.69)$ is an absolute minimum; $(1.68, -99.90)$
(c) $(-.17, -71.48)$; no absolute maximum
(d) $[-132.69, \infty)$
(e) x-intercepts: $-4, 3$; y-intercept: -72

81. (a)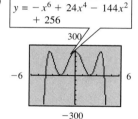

(b) $(-2, 0)$; $(2, 0)$; neither is an absolute minimum
(c) $(-3.46, 256)$, $(0, 256)$, and $(3.46, 256)$ all are absolute maximum points
(d) $(-\infty, 256]$
(e) x-intercepts: $-4, -2, 2, 4$; y-intercept: 256

Further Explorations

1. even   **3.** even

Section 1.5 (page 79)

1. D   **3.** C   **5.** B   **7.** A   **9.** C   **11.** B   **13.** C   **15.** A   **17.** E   **19.** B
21. D   **23.** E   **25.** A   **27. (a)** $(-\infty, \infty)$ **(b)** $[-3, \infty)$ **29. (a)** $[2, \infty)$ **(b)** $[-4, \infty)$
31. (a) $(-\infty, \infty)$ **(b)** $(-\infty, \infty)$ **33.** B   **35.** D   **37.**

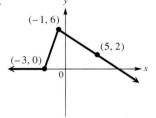

39.

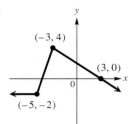

41. B   **43.** A   **45. (a)** $(-4, \infty)$ **(b)** $(-\infty, -4)$ **47. (a)** $(-\infty, \infty)$
(b) none   **49. (a)** $(-\infty, \infty)$ **(b)** none

51.

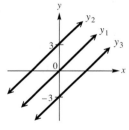

53.

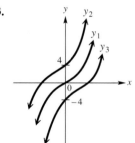

55.

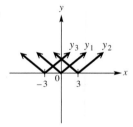

57.

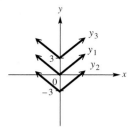

59.

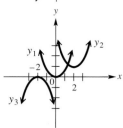

61.

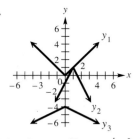

63. 4; x **65.** 2; left; $\frac{1}{4}$; x; 3; downward (or negative)   **67.** 3; right; 6   **69.** $y = \frac{1}{2}x^2 - 7$
71. $y = 4.5\sqrt{x - 3} - 6$
73. (a)   **(b)**   **(c)**   **(d)** $f(0) = 1$

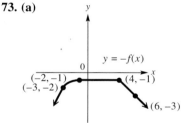

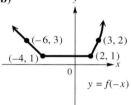

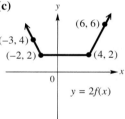

75. (a)

(b)

(c)

(d) −1 and 4

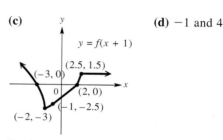

77. (a)

(b) The graph of $y = f(-x)$ is the same as that of $y = -f(x)$, shown in part (a).
(c)
(d) symmetry with respect to the origin

79. (a) r is an x-intercept.   **(b)** $-r$ is an x-intercept.   **(c)** $-r$ is an x-intercept.   **81.** decreases
83. increases   **85. (a)** It is symmetric with respect to the y-axis.   **(b)** It is symmetric with respect to the y-axis.

Chapter 1   Review Exercises (page 87)

1. I   **2.** K   **3.** B   **4.** A   **5.** I   **6.** M   **7.** O   **8.** K   **9.** $\{\frac{46}{7}\}$ **10.** $\{8\}$
11. $\{-3.81\}$ **12.** $(-\infty, -3.81)$ **13.** $[-3.81, \infty)$ **14.** $(2.5, \infty)$ **15.** $\{3\}$ **16.** $(-3, \infty)$
17. F   **18.** C   **19.** G   **20.** A   **21.** B   **22.** H   **23.** E   **24.** D
25. $(-\infty, -2), [-2, 1], (1, \infty)$ **26.** $(-2, 1)$ **27.** $(-\infty, -2)$ **28.** $(1, \infty)$ **29. (a)** $(-\infty, \infty)$
(b) $\{-2\} \cup [-1, 1] \cup (2, \infty)$ **30. (a)** neither   **(b)** even   **31.** x-axis symmetry, y-axis symmetry, origin
symmetry; not a function   **32.** none of these symmetries; neither even nor odd   **33.** y-axis symmetry;
even function   **34.** none of these symmetries; neither even nor odd   **35.** x-axis symmetry; not a
function   **36.** y-axis symmetry; even function   **37.** Start with the graph of $y = x^2$. Shift it 4 units to the
left, stretch vertically by a factor of 3, reflect across the x-axis, and shift 8 units downward.
38. $y = -\frac{2}{3}\sqrt{-x} + 4$ **39.** $\{0, 2\}$ **40.** $(-\infty, 0) \cup (2, \infty)$ **41.** $[0, 2]$ **42.** $[2, \infty)$
43. $(-\infty, 0]$ **44.** $\emptyset$ **45. (a)** $\{-1, 4\}$ **(b)** $(-\infty, -1) \cup (4, \infty)$ **(c)** $[-1, 4]$
46. The graph intersects the x-axis at -1 and 4, supporting the answer in (a). It lies above the x-axis when $x < -1$
or $x > 4$, supporting the answer in (b). It lies on or below the x-axis when x is between -1 and 4 (inclusive),
supporting the answer in (c).   **47.** The graph is concave up for all values in the domain.
48. The calculator gives $\{-.4342585, .76759188\}$. The solution set with exact values is $\{\frac{1 \pm \sqrt{13}}{6}\}$.
49. (a) $(-\infty, \frac{1-\sqrt{13}}{6}) \cup (\frac{1+\sqrt{13}}{6}, \infty)$ **(b)** $(\frac{1-\sqrt{13}}{6}, \frac{1+\sqrt{13}}{6})$ **50.** The calculator gives $(.16666851, -1.083333)$,
supporting the exact values —the point is $(\frac{1}{6}, -\frac{13}{12})$.   **51.** two   **52.** $(2, 0)$ **53.** $(-\infty, \infty)$
54. $(-.97, -54.15)$ **55. (a)** $\{-1, 2, 3\}$ (Note: -1 is of multiplicity two.)   **(b)** the open interval $(2, 3)$
(c) $(-\infty, -1) \cup (-1, 2) \cup (3, \infty)$
56. For Exercise 45: $y = (x + 4)^2 + 3$; For Exercise 46: $y = \sqrt{x + 5}$; For Exercise 47: $y = x^3 - 5$;
For Exercise 48: $y = |x + 10|$

CHAPTER 2

Section 2.1 (page 102)

1. (a) $60°$ **(b)** $150°$ **3. (a)** $45°$ **(b)** $135°$ **5. (a)** $\frac{\pi}{3}$ **(b)** $\frac{5\pi}{6}$ **7. (a)** $\frac{\pi}{4}$ **(b)** $\frac{3\pi}{4}$ **9.** $45°$
11. $320°$ **13.** $90°$ **15.** $\frac{7\pi}{4}$ **17.** $\frac{\pi}{2}$ **19.** $\frac{\pi}{6}$ **21.** $\frac{\pi}{4}$ **23.** $120°$ **25.** $-660°$
27. $20.900°$ **29.** $91.598°$ **31.** $-274.316°$ **33.** $31° 25' 47''$ **35.** $89° 54' 01''$
37. $-178° 35' 58''$ **39.** $30° + n \cdot 360°$; I   **41.** $230° + n \cdot 360°$; III
43. $270° + n \cdot 360°$; no quadrant

45. $\frac{\pi}{4} + 2n\pi$; I   **47.** $\frac{3\pi}{4} + 2n\pi$; II

Angles other than the ones listed are possible in Exercises 49–53.

49. 435°; −285°; I   **51.** 308°; −412°; IV   **53.** $\frac{11\pi}{3}$; −$\frac{\pi}{3}$; IV

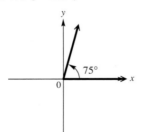

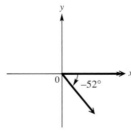

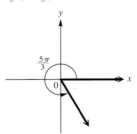

55. 25.8 cm   **57.** 318 m   **59.** 5.05 m   **61.** 1200 km   **63.** 3500 km   **65.** 5900 km

Further Explorations

1. $Y_1 = (180/\pi)x$; $Y_2 = 6400(Y_1*(\pi/180))$

Section 2.2 (page 113)

1.

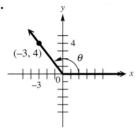

3.

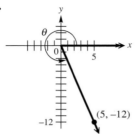

In Exercises 5–19 we give, in order: sine, cosine, tangent, cotangent, secant, and cosecant.

5. $\frac{4}{5}$; $-\frac{3}{5}$; $-\frac{4}{3}$; $-\frac{3}{4}$; $-\frac{5}{3}$; $\frac{5}{4}$ **7.** $-\frac{12}{13}$; $\frac{5}{13}$; $-\frac{12}{5}$; $-\frac{5}{12}$; $\frac{13}{5}$; $-\frac{13}{12}$ **9.** $\frac{24}{25}$; $-\frac{7}{25}$; $-\frac{24}{7}$; $-\frac{7}{24}$; $-\frac{25}{7}$; $\frac{25}{24}$ **11.** 1;

0; undefined; 0; undefined; 1   **13.** 0; −1; 0; undefined; −1; undefined   **15.** $-\frac{2}{3}$; $\frac{\sqrt{5}}{3}$; $-\frac{2\sqrt{5}}{5}$; $-\frac{\sqrt{5}}{2}$; $\frac{3\sqrt{5}}{5}$;

$-\frac{3}{2}$ **17.** $\frac{\sqrt{2}}{2}$; $\frac{\sqrt{2}}{2}$; 1; 1; $\sqrt{2}$; $\sqrt{2}$ **19.** $\frac{\sqrt{3}}{2}$; $\frac{1}{2}$; $\sqrt{3}$; $\frac{\sqrt{3}}{3}$; 2; $\frac{2\sqrt{3}}{3}$ **21.** Reciprocals must always have the

same sign.   **23.** It is the distance from the origin to the point (x, y) on the terminal side of the angle.

25. positive   **27.** negative   **29.** positive   **31.** positive   **33.** negative   **35.** negative

In Exercises 37–41, we give, in order: sine, cosine, tangent, cotangent, secant, and cosecant.

37. $-\frac{2\sqrt{5}}{5}$; $\frac{\sqrt{5}}{5}$; −2; $-\frac{1}{2}$; $\sqrt{5}$; $-\frac{\sqrt{5}}{2}$ **39.** $-\frac{4\sqrt{65}}{65}$; $-\frac{7\sqrt{65}}{65}$; $\frac{4}{7}$; $\frac{7}{4}$; $-\frac{\sqrt{65}}{7}$; $-\frac{\sqrt{65}}{4}$

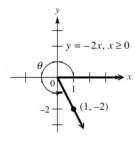

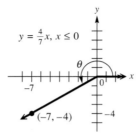

41. $\frac{5\sqrt{34}}{34}$; $-\frac{3\sqrt{34}}{34}$; $-\frac{5}{3}$; $-\frac{3}{5}$; $-\frac{\sqrt{34}}{3}$; $\frac{\sqrt{34}}{5}$ **43.** $\frac{1}{3}$ **45.** -5

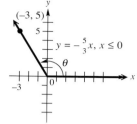

47. $2\sqrt{2}$ **49.** $-\frac{3\sqrt{5}}{5}$ **51.** $.70069071$ **53.** 2.2778902 **55.** No, because the sine and cosecant functions are reciprocals and they must have the same sign for a particular angle. **57.** Because $\cot 90° = 0$, $\frac{1}{\cot 90°}$ is undefined. The equality symbol would indicate that the expressions represent real numbers.
59. II **61.** III **63.** IV **65.** II or IV **67.** I or II **69.** I or III **71.** $-\frac{\sqrt{5}}{3}$
73. $-\frac{\sqrt{5}}{2}$ **75.** $-\frac{4}{3}$ **77.** $-\frac{\sqrt{3}}{2}$ **79.** 3.44701905 **81.** $-.56616682$

In Exercises 83–91, we give, in order: sine, cosine, tangent, cotangent, secant, and cosecant.

83. $-\frac{4}{5}$; $-\frac{3}{5}$; $\frac{4}{3}$; $\frac{3}{4}$; $-\frac{5}{3}$; $-\frac{5}{4}$ **85.** $\frac{7}{25}$; $-\frac{24}{25}$; $-\frac{7}{24}$; $-\frac{24}{7}$; $-\frac{25}{24}$; $\frac{25}{7}$ **87.** $\frac{1}{2}$; $-\frac{\sqrt{3}}{2}$; $-\frac{\sqrt{3}}{3}$; $-\sqrt{3}$; $-\frac{2\sqrt{3}}{3}$; 2
89. $\frac{8\sqrt{67}}{67}$; $\frac{\sqrt{201}}{67}$; $\frac{8\sqrt{3}}{3}$; $\frac{\sqrt{3}}{8}$; $\frac{\sqrt{201}}{3}$; $\frac{\sqrt{67}}{8}$ **91.** $.164215$; $-.986425$; $-.166475$; -6.00691; -1.01376; 6.08958

Further Explorations

1. $45°, 135°, 225°, 315°$ **3.** Whenever x and y equal ($y = x$) or whenever they are opposites ($y = -x$) of each other, $|x| = |y|$. Dividing both sides by $r = |r|$, we have $|\frac{x}{r}| = |\frac{y}{r}|$, and thus $|\cos \theta| = |\sin \theta|$.

Section 2.3 (page 125)

In Exercises 1–5, we give, in order: sine, cosine, tangent, cotangent, secant, and cosecant.

1. $\frac{\sqrt{2}}{2}$; $-\frac{\sqrt{2}}{2}$; -1; -1; $-\sqrt{2}$; $\sqrt{2}$ **3.** $-\frac{1}{2}$; $\frac{\sqrt{3}}{2}$; $-\frac{\sqrt{3}}{3}$; $-\sqrt{3}$; $\frac{2\sqrt{3}}{3}$; -2 **5.** $-\frac{\sqrt{3}}{2}$; $-\frac{1}{2}$; $\sqrt{3}$; $\frac{\sqrt{3}}{3}$; -2; $-\frac{2\sqrt{3}}{3}$
7. $\tan 17°$ **9.** $\cos 51° 31'$ **11.** $\sin \frac{3\pi}{10}$ **13.** $\cot (\frac{\pi}{2} - .5)$ **15.** The *exact* value of $\sin 45°$ is $\frac{\sqrt{2}}{2}$. The decimal value he gave was just an approximation.

The number of digits in part (c) of Exercises 17–33 will vary depending upon the model of calculator used.

17. (a) $\frac{\sqrt{3}}{3}$ **(b)** irrational **(c)** $.5773502692$ **19. (a)** $\frac{1}{2}$ **(b)** rational **21. (a)** $\frac{2\sqrt{3}}{3}$ **(b)** irrational
(c) 1.154700538 **23. (a)** $\sqrt{2}$ **(b)** irrational **(c)** 1.414213562 **25. (a)** $\frac{\sqrt{2}}{2}$ **(b)** irrational
(c) $.7071067812$ **27. (a)** 1 **(b)** rational **29. (a)** $\frac{\sqrt{3}}{2}$ **(b)** irrational **(c)** $.8660254038$
31. (a) $\sqrt{3}$ **(b)** irrational **(c)** 1.732050808 **33. (a)** 2 **(b)** rational

The number of digits displayed in Exercises 35–45 will vary.

35. $.62478851$ **37.** $-.32281638$ **39.** -3.1791978 **41.** $.48775041$ **43.** 1.0170372
45. $.95544269$ **47.** $\frac{\sqrt{3}}{3}$; $\sqrt{3}$ **49.** $\frac{\sqrt{3}}{2}$; $\frac{\sqrt{3}}{3}$; $\frac{2\sqrt{3}}{3}$ **51.** -1; -1 **53.** $-\frac{\sqrt{3}}{2}$; $-\frac{2\sqrt{3}}{3}$
55. (a) $-\sin \frac{\pi}{6}$ **(b)** $-\frac{1}{2}$ **(c)** $\sin \frac{7\pi}{6} = -\sin \frac{\pi}{6} = -.5$ **57. (a)** $-\tan \frac{\pi}{4}$ **(b)** -1
(c) $\tan \frac{3\pi}{4} = -\tan \frac{\pi}{4} = -1$ **59. (a)** $-\cos \frac{\pi}{6}$ **(b)** $-\frac{\sqrt{3}}{2}$ **(c)** $\cos \frac{7\pi}{6} = -\cos \frac{\pi}{6} \approx -.86602540$
61. $30°$; $150°$ **63.** $60°$; $240°$ **65.** $120°$; $240°$ **67.** $240°$; $300°$

The number of displayed digits will vary in Exercises 69–73.

69. $46.593881°$; $313.40612°$ **71.** $24.392576°$; $155.60742°$ **73.** $41.248183°$; $221.24818°$
75. $.20952066$; 3.3511133 **77.** 1.4429646; 1.6986280 **79.** 1.3631380; 1.7784546 **81.** It represents the distance from the point (x_1, y_1) to the origin. **83.** $60°$ **85.** It is a measure of the angle formed by the positive x-axis and the ray $y = \sqrt{3}\, x$, $x \geq 0$. **87.** slope; tangent **89.** They agree, as they are both approximately 1.154700538. **91.** $\cos 60° = .5$. This is the x-coordinate of the point found in Exercise 90. Because $r = 1$, here $\cos 60° = \frac{x}{1} = x = .5$.
93. $\sqrt{(x - 1)^2 + (y - 0)^2} = \sqrt{(\frac{\sqrt{3}}{2} - \frac{\sqrt{3}}{2})^2 + (\frac{1}{2} - (-\frac{1}{2}))^2}$; $(x - 1)^2 + y^2 = 1$
95. x is positive in the first quadrant. **97.** $x = \frac{1}{2}$, $y = \frac{\sqrt{3}}{2}$, $r = 1$

Section 2.4 (page 135)

In Exercises 1–13, calculator approximations of irrational numbers are shown for the purpose of calculator support.

1. $-\frac{1}{2}$ **3.** -1 **5.** $-\frac{\sqrt{3}}{2}$ ($\approx -.8660254038$)   **7.** -2 **9.** $-\sqrt{3}$ (≈ -1.732050808)
11. $-\frac{1}{2}$ **13.** $\frac{2\sqrt{3}}{3}$ (≈ 1.154700538)   **15.** $.5736049112$ **17.** $.4067524531$ **19.** 1.206484913
21. 14.33376901 **23.** -1.045975716 **25.** -3.866512664 **27.** $\cos 1.2 \approx .36235775$,
$\sin 1.2 \approx .93203909$ **29.** $\cos 3.5 \approx -.9364567$, $\sin 3.5 \approx -.3507832$ **31.** $.1865123694$
33. $-.9824526126$ **35.** positive; negative; IV   **37.** $\cos 4.38$ **39.** $-\sin .5$ **41.** $-\tan \frac{\pi}{7}$
43. $\sec 8$ **45.** $-\csc \frac{1}{4}$ **47.** $-\cot 10^5$

Further Explorations

1. (a) $45°, 225°$ **(b)** Cos x decreases from $0°$ to $180°$ and increases from $180°$ to $360°$.   **(c)** Sin x decreases from $90°$ to $270°$ and increases from $0°$ to $90°$ and from $270°$ to $360°$.   **(d)** In both cases, the maximum value is 1 and the minimum value is -1.   **(e)** One example is ΔTbl $= 360$ and Tblmin $= 0$ (degrees). Cos $x = 1$ for multiples of $360°$.   **(f)** One example is ΔTbl $= 180$ and Tblmin $= 90$ (degrees). Cos $x = 0$ for odd multiples of $90°$.   **(g)** One example is ΔTbl $= 360$ and Tblmin $= 90$ (degrees). Sin $x = 1$ for $x = (4k + 1) \cdot 90°$, where k is an integer.   **(h)** One example is ΔTbl $= 180$ and Tblmin $= 0$ (degrees). Sin $x = 0$ for multiples of $180°$.

Section 2.5 (page 147)

In Exercises 1–5, verify the equation by using the appropriate circular function key, or the parametrically-generated unit circle $x = \cos t$, $y = \sin t$.

1. $\sin(-2.75) \approx -.381661$ **3.** $\cos(-3.5) \approx -.9364567$ **5.** $\cos \frac{\pi}{2} = 0$ **7.** G   **9.** E
11. B   **13.** F   **15.** D   **17.** H   **19.** B   **21.** F   **23.** a **25.** $-c$
27. (a) 2   **(b)** 2π **(c)** π **(d)** none   **(e)** $[-2, 2]$ **29. (a)** 4   **(b)** 4π **(c)** $-\pi$ **(d)** none   **(e)** $[-4, 4]$
31. (a) 1   **(b)** $\frac{2\pi}{3}$ **(c)** $\frac{\pi}{15}$ **(d)** up 2 units   **(e)** $[1, 3]$ **33. (a)** 3   **(b)** 2   **(c)** none   **(d)** up 2 units   **(e)** $[-1, 5]$
35. Shift the graph of $y = \sin x$ $\frac{\pi}{2}$ units to the left, stretch vertically by a factor of 2, and shift 3 units downward.
37. Shift the graph of $y = \cos x$ $\frac{\pi}{6}$ units to the right, change the period to $\frac{2\pi}{3}$, and shift $\frac{5}{2}$ units downward.
39.

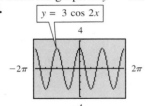

$y = 3 \cos 2x$

41.

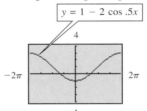

$y = 1 - 2 \cos .5x$

43. (a) $1; 1; 3; -2$ **(b)** $-1; -1; -3; -8$ **45.** The standard trig window as defined in the text has Ymin $= -4$ and Ymax $= 4$. Because the range of f includes values less than -4, the local minimum points will not appear in the standard trig window.   **47.** The period of f is $\frac{2\pi}{2} = \pi$, and the distance between -2π and 2π is 4π units. Since $\frac{4\pi}{\pi} = 4$, these values will show four periods of the graph.   **49.** Ymin $= -8$; Ymax $= -2$
51. $y = -4 \sin 3x$ **53.** $y = -2 \cos 3x$ **55.** $y = 3 \sin 6x$

Further Explorations

1. $y_2 = 2 \sin x$ **3.** $y_2 = \sin 2x$ **5.** $y_2 = \sin(x - \frac{\pi}{6})$ **7.** $y_2 = 1 + \sin x$

Section 2.6 (page 159)

1. $\sec 2.1 \approx -1.980802$ **3.** $\csc(-.75) \approx -1.467053$ **5.** $\tan 2 \approx -2.18504$ **7.** $\cot \frac{\pi}{4} = 1$
9. To be a parabola, the function must be of the form $f(x) = ax^2 + bx + c$ $(a \neq 0)$. Furthermore, the graph of a vertical parabola does not lie between vertical asymptotes.   **11.** B   **13.** E   **15.** D
17. (a)

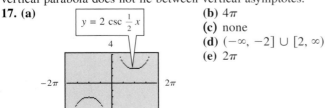

$y = 2 \csc \frac{1}{2}x$

(b) 4π
(c) none
(d) $(-\infty, -2] \cup [2, \infty)$
(e) 2π

19. (a)

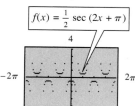

$y = -2 \sec\left(x + \frac{\pi}{2}\right)$

(b) 2π
(c) $-\frac{\pi}{2}$
(d) $(-\infty, -2] \cup [2, \infty)$
(e) π

21. (a)

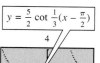
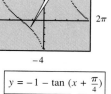

$y = \frac{5}{2} \cot \frac{1}{3}\left(x - \frac{\pi}{2}\right)$

(b) 3π
(c) $\frac{\pi}{2}$
(d) $(-\infty, \infty)$
(e) $\frac{\pi}{2}$

23. (a)

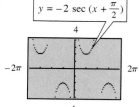

$f(x) = \frac{1}{2} \sec(2x + \pi)$

(b) π
(c) $-\frac{\pi}{2}$
(d) $\left(-\infty, -\frac{1}{2}\right] \cup \left[\frac{1}{2}, \infty\right)$
(e) $\frac{\pi}{4}$

25. (a)

$y = -1 - \tan\left(x + \frac{\pi}{4}\right)$

(b) π
(c) $-\frac{\pi}{4}$
(d) $(-\infty, \infty)$
(e) $\frac{\pi}{4}$

27. Start with the graph of $y = \csc x$. Shift π units to the right, change the period to π, stretch vertically by a factor of 3, and shift 2 units downward.   **29.** (The given function is equivalent to $y = 4 - 2 \tan 2(x - 1)$.) Start with the graph of $y = \tan x$. Shift 1 unit to the right, change the period to $\frac{\pi}{2}$, stretch vertically by a factor of 2, reflect across the x-axis, and shift 4 units upward.   **31.** π **33.** $\frac{5\pi}{4} + n\pi$
35. $\pi + .32175055 \approx 3.463343208$ **37.** origin; $-\tan x$ **39.** origin; $-\cot 1.75$; $.1811469526$

Further Explorations

1. The two x-values that give error messages are for $x = \frac{\pi}{2}$ and $x = \frac{3\pi}{2}$. When angles of these radian measures are sketched in standard position, a point on the terminal side has $a = 0$, leading to an undefined value of $\frac{b}{a}$. Thus, an error message occurs.

Chapter 2 Review Exercises (page 164)

1.

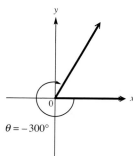

$\theta = -300°$

2. $60°$ **3.** $-660°$ (There are others.)
4. $-300° + n \cdot 360°$ (There are others.)   **5.** $-\frac{5\pi}{3}$ **6.** $s = \pi$
7. $\theta = 3$ **8.** $r = \frac{48}{\pi}$ **9.** 35.8 centimeters   **10.** 10π inches
11. 4500 kilometers   **12.** $-\frac{7\sqrt{53}}{53}$ **13.** $-\frac{2\sqrt{53}}{53}$ **14.** $\frac{7}{2}$
15. $-\frac{\sqrt{53}}{7}$ **16.** $-\frac{\sqrt{53}}{2}$ **17.** $\frac{2}{7}$ **18. (a)** $254.05°$ **(b)** $74.05°$

19.

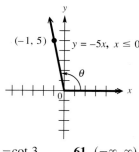

$(-1, 5)$
$y = -5x, \; x \le 0$
θ

20. $\sin \theta = \frac{5\sqrt{26}}{26}$; $\cos \theta = -\frac{\sqrt{26}}{26}$ **21.** $101° \, 19'$ **22.** III   **23.** II
24. $-\frac{\sqrt{22}}{5}$ **25.** $\frac{5\sqrt{3}}{3}$ **26.** $-\frac{\sqrt{66}}{3}$ **27.** $\frac{\sqrt{2}}{2}$ **28.** -1
29. undefined   **30.** $-\frac{2\sqrt{3}}{3}$ **31.** 2 **32.** 0 **33.** $\frac{\sqrt{2}}{2}$
34. $-\frac{\sqrt{3}}{2}$ **35.** $\sqrt{3}$ **36.** undefined   **37.** -1 **38.** $-\sqrt{3}$
39. $.5495089781$ **40.** 3.525320086 **41.** $-.726542528$
42. 2.076521397 **43.** $\sqrt{3}$ **44.** $-\frac{1}{2}$ **45.** $-\frac{1}{2}$ **46.** $-\sqrt{3}$
47. $-\sqrt{3}$ **48.** 2 **49.** 2 **50.** $\frac{\sqrt{3}}{3}$ **51.** $.8660266282$
52. 1.67553317 **53.** 1.131194347 **54.** -1.344734732
55. $\cos 3$ **56.** $-\sin 3$ **57.** $-\tan 3$ **58.** $\sec 3$ **59.** $-\csc 3$
60. $-\cot 3$ **61.** $(-\infty, \infty)$ **62.** $[3, 7]$ **63.** 2 **64.** $\frac{2\pi}{3}$ **65.** $-\frac{\pi}{12}$ **66.** 5 units upward

62. $[3, 7]$ **63.** 2   **64.** $\frac{2\pi}{3}$ **65.** $-\frac{\pi}{12}$ **66.** 5 units upward

67.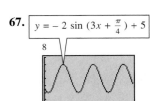

$y = -2 \sin(3x + \frac{\pi}{4}) + 5$

68. tangent   **69.** sine   **70.** cosine   **71.** cosecant   **72.** cotangent   **73.** secant
74. In both cases, the definition leads to the cosine function in the denominator. Thus when the cosine is zero, the function value is undefined.   **75.** $y = 3 \sin 2(x - \frac{\pi}{4})$ **76.** $y = 4 \sin \frac{1}{2}x$ **77.** $y = \frac{1}{3} \sin \frac{\pi}{2} x$
78. $y = \pi \sin \pi(x - \frac{1}{2})$ **79.** $y = 3 \cos 2(x + \frac{\pi}{2})$ **80.** $y = 4 \cos \frac{1}{2}(x - \pi)$

CHAPTER 3

Section 3.1 (page 175)

1. (b)   **3.** (e)   **5.** (a)   **7.** (a)   **9.** (d)   **11.** The student has neglected to write the argument of the functions (for example, θ or t).   **13.** $\frac{\pm\sqrt{1 + \cot^2 \theta}}{1 + \cot^2 \theta}$; $\frac{\pm\sqrt{\sec^2 \theta - 1}}{\sec \theta}$ **15.** $\frac{\pm\sin \theta \sqrt{1 - \sin^2 \theta}}{1 - \sin^2 \theta}$; $\frac{\pm\sqrt{1 - \cos^2 \theta}}{\cos \theta}$;

$\pm\sqrt{\sec^2 \theta - 1}$; $\frac{\pm\sqrt{\csc^2 \theta - 1}}{\csc^2 \theta - 1}$ **17.** $\frac{\pm\sqrt{1 - \sin^2 \theta}}{1 - \sin^2 \theta}$; $\pm\sqrt{\tan^2 \theta + 1}$; $\frac{\pm\sqrt{1 + \cot^2 \theta}}{\cot \theta}$; $\frac{\pm\csc \theta \sqrt{\csc^2 \theta - 1}}{\csc^2 \theta - 1}$;   **19.** $\sin \theta$
21. 1   **23.** $\tan^2 \beta$ **25.** $\tan^2 x$ **27.** $\sec^2 x$ **29.** $\sin x$ **31.** 1

In Exercises 33–37, we provide only the graphs in parts (b). The left side of the identity is graphed as y_1, and the right side as y_2.

33. (b)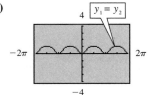

35. (b)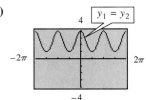

37. (b)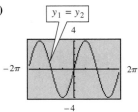

57. $\sin x$ **59.** 1   **61.** While the equation is true for the *particular* value $\theta = \frac{\pi}{2}$, it is not true in *general*. To be an identity, the equation must be true in *all* cases for which the functions involved are defined.   **65.** $y = \cos 2x$

Section 3.2 (page 184)

1. $\{\frac{\pi}{3}, \frac{5\pi}{3}\}$ **3.** $[0, \frac{\pi}{3}] \cup [\frac{5\pi}{3}, 2\pi)$ **5.** none   **7.** (a) $\{\frac{2\pi}{3}, \frac{4\pi}{3}\}$ (b)

$f(x) = 2 \cos x + 1$

(c) $[0, \frac{2\pi}{3}) \cup (\frac{4\pi}{3}, 2\pi)$
(d) $(\frac{2\pi}{3}, \frac{4\pi}{3})$

9. (a) $\{\frac{\pi}{3}, \frac{2\pi}{3}, \frac{4\pi}{3}, \frac{5\pi}{3}\}$ (b)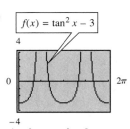

$f(x) = \tan^2 x - 3$

11. (a) $\{\frac{\pi}{6}, \frac{\pi}{2}, \frac{3\pi}{2}, \frac{11\pi}{6}\}$ (b)

$f(x) = 2 \cos^2 x - \sqrt{3} \cos x$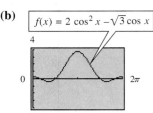

(c) $(\frac{\pi}{3}, \frac{\pi}{2}) \cup (\frac{\pi}{2}, \frac{2\pi}{3}) \cup (\frac{4\pi}{3}, \frac{3\pi}{2}) \cup (\frac{3\pi}{2}, \frac{5\pi}{3})$
(d) $[0, \frac{\pi}{3}) \cup (\frac{2\pi}{3}, \frac{4\pi}{3}) \cup (\frac{5\pi}{3}, 2\pi)$

(c) $[0, \frac{\pi}{6}) \cup (\frac{\pi}{2}, \frac{3\pi}{2}) \cup (\frac{11\pi}{6}, 2\pi)$
(d) $(\frac{\pi}{6}, \frac{\pi}{2}) \cup (\frac{3\pi}{2}, \frac{11\pi}{6})$

13. (a) $\left\{\frac{\pi}{4}, \frac{3\pi}{4}, \frac{5\pi}{4}, \frac{7\pi}{4}\right\}$ **(b)**   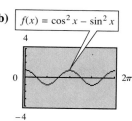   **(c)** $\left[0, \frac{\pi}{4}\right) \cup \left(\frac{3\pi}{4}, \frac{5\pi}{4}\right) \cup \left(\frac{7\pi}{4}, 2\pi\right)$
(d) $\left(\frac{\pi}{4}, \frac{3\pi}{4}\right) \cup \left(\frac{5\pi}{4}, \frac{7\pi}{4}\right)$

15. (a) $\left\{\frac{\pi}{4}, \frac{5\pi}{4}\right\}$ **(b)**   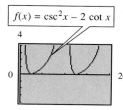   **(c)** $\left(0, \frac{\pi}{4}\right) \cup \left(\frac{\pi}{4}, \pi\right) \cup \left(\pi, \frac{5\pi}{4}\right) \cup \left(\frac{5\pi}{4}, 2\pi\right)$
(d) $\emptyset$

17. (a) $\left\{\frac{\pi}{2}, 3.87, 5.55\right\}$ **(b)**   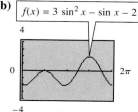   **(c)** $(3.87, 5.55)$
(d) $\left[0, \frac{\pi}{2}\right) \cup \left(\frac{\pi}{2}, 3.87\right) \cup (5.55, 2\pi)$

19. (a) $\{1.86, 2.61, 5.00, 5.75\}$ **(b)**   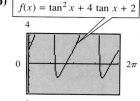   **(c)** $\left[0, \frac{\pi}{2}\right) \cup \left(\frac{\pi}{2}, 1.86\right) \cup \left(2.61, \frac{3\pi}{2}\right) \cup \left(\frac{3\pi}{2}, 5.00\right) \cup$
$(5.75, 2\pi)$
(d) $(1.86, 2.61) \cup (5.00, 5.75)$

21. (a) $\{1.20, 5.09\}$ **(b)**   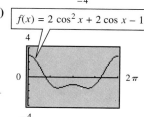   **(c)** $[0, 1.20) \cup (5.09, 2\pi)$
(d) $(1.20, 5.09)$

23. $\{1.01, 2.78\}$ **25.** $\{.75, 2.39\}$ **27.** $\{1.38\}$

Section 3.3 (page 198)

The answer to part (c) in Exercises 3–11 is found in a manner similar to the answer to part (c) in Exercise 1.
1. (a) $\frac{\sqrt{6} - \sqrt{2}}{4}$ **(b)** .2588190451   **(c)**

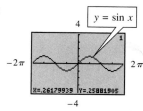

$$\frac{\pi}{12} \approx .26179939$$
$$\sin \frac{\pi}{12} \approx .25881905$$

3. (a) $\frac{-\sqrt{6}+\sqrt{2}}{4}$ **(b)** $-.2588190451$ **5. (a)** $\frac{\sqrt{6}+\sqrt{2}}{4}$ **(b)** $.9659258263$ **7. (a)** $\frac{\sqrt{6}-\sqrt{2}}{4}$

(b) $.2588190451$ **9. (a)** $\frac{-\sqrt{6}-\sqrt{2}}{4}$ **(b)** $-.9659258263$ **11. (a)** $\frac{-\sqrt{6}+\sqrt{2}}{4}$ **(b)** $-.2588190451$

13. (a) $\frac{\sqrt{6}-\sqrt{2}}{4}$ **(b)** $.2588190451$ **15. (a)** $2-\sqrt{3}$ **(b)** $.2679491924$ **17. (a)** $\frac{\sqrt{6}+\sqrt{2}}{4}$

(b) $.9659258263$ **19. (a)** $\frac{\sqrt{6}+\sqrt{2}}{4}$ **(b)** $.9659258263$ **21. (a)** $-2-\sqrt{3}$ **(b)** -3.732050808

23. (a) $\frac{\sqrt{6}+\sqrt{2}}{4}$ **(b)** $.9659258263$ **25.** $\sin x$ **27.** $\sin x$ **29.** $-\sin x$ **31.** $\tan x$ **33.** $-\cos x$

35.

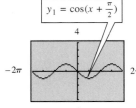

37. $-.841471$ **39.** It is the same as the graph of $y_1 = \cos(x + \frac{\pi}{2})$.

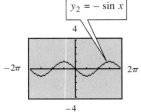

41. (a) $\frac{63}{65}$ **(b)** $\frac{33}{65}$ **(c)** $\frac{63}{56}$ **(d)** $\frac{33}{56}$ **(e)** I **(f)** I   **43. (a)** $\frac{77}{85}$ **(b)** $\frac{13}{85}$ **(c)** $-\frac{77}{36}$ **(d)** $\frac{13}{84}$ **(e)** II **(f)** I

45. $\cos 2x = \frac{119}{169}$; $\sin 2x = -\frac{120}{169}$; $\tan 2x = -\frac{120}{119}$; $\cot 2x = -\frac{119}{120}$; $\sec 2x = \frac{169}{119}$; $\csc 2x = -\frac{169}{120}$

47. $\sin 2x = \frac{15}{17}$; $\cos 2x = -\frac{8}{17}$; $\tan 2x = -\frac{15}{8}$; $\cot 2x = -\frac{8}{15}$; $\sec 2x = -\frac{17}{8}$; $\csc 2x = \frac{17}{15}$

49. (a) $\frac{\sqrt{2-\sqrt{3}}}{2}$ **(b)** $.2588190451$ **51. (a)** $1-\sqrt{2}$ (or $-\sqrt{3-2\sqrt{2}}$) **(b)** $-.4142135624$

53. (a) $\frac{\sqrt{2+\sqrt{2}}}{2}$ **(b)** $.9238795325$ **55.** $\frac{\sqrt{10}}{4}$ **57.** 3   **59.** $-\sqrt{7}$ **65.** $\cos 3x = 4\cos^3 x - 3\cos x$

67. $\tan 4x = \frac{4\tan x - 4\tan^3 x}{1 - 6\tan^2 x + \tan^4 x}$

In Exercises 75–81, we provide only the graphical support. By graphing y_1 as the left side and y_2 as the right side, we get the same graph.

75.

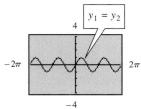

77.

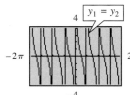

Connected Mode
(The vertical lines are
not part of the graph.)

79.

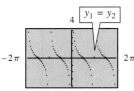

Dot Mode

81.

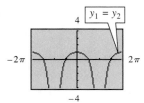

Further Explorations

Answers will vary. We give only one possibility.

1. ΔTbl $= \pi$, Tblmin $= \frac{\pi}{4}$ **3.** ΔTbl $= \frac{\pi}{2}$, Tblmin $= \frac{\pi}{4}$ **5.** ΔTbl $= \frac{\pi}{2}$, Tblmin $= \frac{\pi}{8}$

Section 3.4 (page 208)

1. (a) $\{\frac{\pi}{12}, \frac{11\pi}{12}, \frac{13\pi}{12}, \frac{23\pi}{12}\}$ **(b)** $[0, \frac{\pi}{12}) \cup (\frac{11\pi}{12}, \frac{13\pi}{12}) \cup (\frac{23\pi}{12}, 2\pi)$ **(c)** $(\frac{\pi}{12}, \frac{11\pi}{12}) \cup (\frac{13\pi}{12}, \frac{23\pi}{12})$ **3. (a)** $\{\frac{\pi}{2}, \frac{7\pi}{6}, \frac{11\pi}{6}\}$

(b) $[0, \frac{\pi}{2}) \cup (\frac{\pi}{2}, \frac{7\pi}{6}) \cup (\frac{7\pi}{6}, \frac{11\pi}{6}) \cup (\frac{11\pi}{6}, 2\pi)$ **(c)** $\emptyset$ **5. (a)** $\{\frac{3\pi}{8}, \frac{5\pi}{8}, \frac{11\pi}{8}, \frac{13\pi}{8}\}$

(b) $[0, \frac{3\pi}{8}] \cup [\frac{5\pi}{8}, \frac{11\pi}{8}] \cup [\frac{13\pi}{8}, 2\pi)$ **(c)** $[\frac{3\pi}{8}, \frac{5\pi}{8}] \cup [\frac{11\pi}{8}, \frac{13\pi}{8}]$ **7. (a)** $\{\frac{\pi}{2}, \frac{3\pi}{2}\}$ **(b)** $(\frac{\pi}{2}, \frac{3\pi}{2})$
(c) $[0, \frac{\pi}{2}) \cup (\frac{3\pi}{2}, 2\pi)$ **9. (a)** $\{0, \frac{2\pi}{3}, \frac{4\pi}{3}\}$ **(b)** $[0, \frac{2\pi}{3}] \cup [\frac{4\pi}{3}, 2\pi)$ **(c)** $[\frac{2\pi}{3}, \frac{4\pi}{3}]$ **11. (a)** $\{\frac{\pi}{2}\}$ **(b)** $(\frac{\pi}{2}, 2\pi)$
(c) $[0, \frac{\pi}{2})$ **13. (a)** $\{\frac{\pi}{3}, \pi, \frac{5\pi}{3}\}$ **(b)** $[0, \frac{\pi}{3}] \cup \{\pi\} \cup [\frac{5\pi}{3}, 2\pi)$ **(c)** $[\frac{\pi}{3}, \frac{5\pi}{3}]$ **15. (a)** $\{0, \frac{2\pi}{3}, \frac{4\pi}{3}\}$
(b) $[0, \frac{\pi}{6}) \cup [\frac{2\pi}{3}, \frac{5\pi}{6}) \cup [\frac{4\pi}{3}, \frac{3\pi}{2})$ **(c)** $(\frac{\pi}{6}, \frac{2\pi}{3}) \cup (\frac{5\pi}{6}, \frac{4\pi}{3}) \cup (\frac{11\pi}{6}, 2\pi)$, with $x \neq \frac{\pi}{2}, \frac{7\pi}{6}, \frac{11\pi}{6}$
17. (a) $\{\frac{\pi}{3}, \frac{\pi}{2}, \frac{3\pi}{2}, \frac{5\pi}{3}\}$ **(b)** $[0, \frac{\pi}{3}] \cup [\frac{\pi}{2}, \frac{3\pi}{2}] \cup [\frac{5\pi}{3}, 2\pi)$ **(c)** $[\frac{\pi}{3}, \frac{\pi}{2}] \cup [\frac{3\pi}{2}, \frac{5\pi}{3}]$
19. $\{.262, 1.309, 1.571, 3.403, 4.451, 4.712\}$ **21.** $\{1.047, 3.142, 5.236\}$
23. $\{.259, 1.372, 3.142, 4.911, 6.024\}$ **25.** We cannot say that $\frac{\tan 2\theta}{2} = \tan \theta$. The 2 in the numerator on the left is not a factor of the entire numerator, so it cannot cancel with the 2 in the denominator.
27.

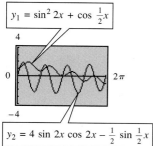

Section 3.5 (page 221)

1. $-\frac{\pi}{6}$ **3.** $\frac{\pi}{4}$ **5.** π **7.** $-\frac{\pi}{3}$ **9.** $-\frac{\pi}{4}$ **11.** $\frac{\pi}{4}$ **13.** $\frac{5\pi}{6}$ **15.** 0
17. (a) 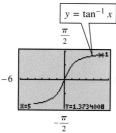 **(b)** $.3575711036$ **19. (a)** **(b)** 2.214297436

21. (a) 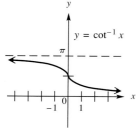 **(b)** 1.373400767

23. The expression $\cos^{-1} 1.5$ is undefined, because 1.5 is not in the domain of the inverse cosine function.
25. domain: $(-\infty, \infty)$; range: $(0, \pi)$ **27.** domain: $(-\infty, -1] \cup [1, \infty)$; range: $[0, \frac{\pi}{2}) \cup (\frac{\pi}{2}, \pi]$

29. $\frac{3\pi}{4}$ **31.** $\frac{5\pi}{6}$ **33.** $\frac{2\pi}{3}$ **35.** 1.003 is not in the domain of the inverse sine function.
37. true **39.** false; The statement is true only if x is in the interval $[-\frac{\pi}{2}, \frac{\pi}{2}]$. **41.** true **43.** $\frac{\sqrt{7}}{3}$

45. $\frac{\sqrt{5}}{5}$ **47.** $-\frac{\sqrt{5}}{2}$ **49.** $\frac{120}{169}$ **51.** $-\frac{7}{25}$ **53.** $\frac{4\sqrt{6}}{25}$ **55.** 2   **57.** $\sqrt{1-u^2}$

59. $\frac{\sqrt{1-u^2}}{u}$ **61.** $\frac{\sqrt{u^2-4}}{u}$ **63.** $\frac{u\sqrt{2}}{2}$ **65.** $\frac{2\sqrt{4-u^2}}{4-u^2}$ **67.** $\cos A = \frac{1}{2}$, $\sin A = \frac{\sqrt{3}}{2}$

69. $\cos (A + B) = \frac{1}{2} \cdot \frac{2\sqrt{2}}{3} - \frac{\sqrt{3}}{2} \cdot \frac{1}{3}$ **71.** $\frac{63}{65}$ **73.** $\frac{\sqrt{10} - 3\sqrt{30}}{20}$

Section 3.6 (page 234)

1. (a) approximately 1.998 w/m² **(b)** approximately −46.461 w/m² **(c)** 46.478 **(d)** $N = 82.5$ **3.** 8.6 hours;
15.4 hours   **5. (a)** amplitude: 17.5; period: 12; phase shift: 4; vertical translation: 67.5

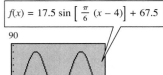

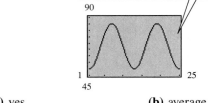

(b) approximately 52°
(c) 85° F in July; 50°F in January
(d) 67.5° F

7. (a) yes

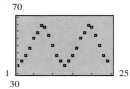

Data for Vancouver temperatures

(b) average yearly temperature **(c)** 14; 12; 4.2
(d) $f(x) = 14 \sin \left[\frac{\pi}{6} (x - 4.2)\right] + 50$

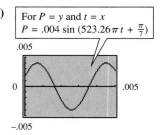

(e)

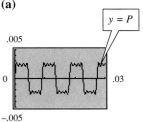

9. 3   **11. (a)** 5, $\frac{1}{40}$ **(b)** 40 **(c)** 5, 1.55, −4.05, −4.05, 1.55
13. (a) .25 second **(b)** .17 second **(c)** .21 second

15. (a)

For $V = y$ and $t = x$
$V = 30 \sin 120 \pi t + 40 \cos 120 \pi t$

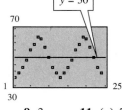

(b) $a = 50$, $\phi = -5.353$

17. (a)

For $P = y$ and $t = x$
$P = .004 \sin (523.26 \pi t + \frac{\pi}{7})$

(b) $.00164 \leq t \leq .00355$

19. (a)

$y = P$

(b) The graph can be described as a wave with "squared-off"
tops and bottoms.
(c) $.0045 < t < .909$, $.0136 < t < .0182$, $.0227 < t < .0273$

21. (a) The pressure P is oscillating. **(b)** The pressure oscillates and amplitude decreases as r increases.

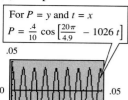

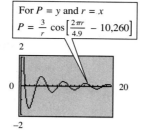

(c) $P = \frac{a}{n\lambda}\cos(ct)$ **23.** $113°$ **25.** $60°$ **27. (a)** 5000 feet **(b)** 1250 feet **29. (a)** $42.2°$
(b) $90°$ **(c)** $48.0°$ **31.** 980.799 cm per second^2 **33.** .50 second and .83 second
37. (a) $L(x) = .022x^2 + .55x + 316 + 3.5\sin(2\pi x)$ **(b)** maximums : $x = \frac{1}{4}, \frac{5}{4}, \frac{9}{4}, \ldots$; minimums : $x = \frac{3}{4}, \frac{7}{4}, \frac{11}{4}, \ldots$

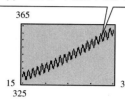

Chapter 3 Review Exercises (page 242)

1. (b) **2. (a)** **3. (c)** **4. (f)** **5. (h)** **6. (e)** **7. (d)** **8. (g)** **25.** $\{0, \pi\}$
26. $f(x) = \cos^2 x \sin x - \sin x$ **27. (a)** the open interval $(\pi, 2\pi)$ **(b)** the open interval $(0, \pi)$

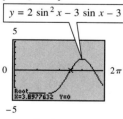

28. $\{3.90, 5.53\}$

$y = 2\sin^2 x - 3\sin x - 3$

$y = 2\sin^2 x - 3\sin x - 3$

29. $\{\frac{\pi}{4}, \frac{\pi}{2}, \frac{5\pi}{4}, \frac{3\pi}{2}\}$ **30.** $\{\frac{\pi}{12}, \frac{5\pi}{12}, \frac{13\pi}{12}, \frac{17\pi}{12}\}$ **31.** $\{0, \frac{\pi}{3}, \frac{\pi}{2}, \frac{2\pi}{3}, \pi, \frac{4\pi}{3}, \frac{3\pi}{2}, \frac{5\pi}{3}\}$ **32.** $\{\frac{\pi}{3}, \pi, \frac{5\pi}{3}\}$
33. $\{\frac{\pi}{6}, \frac{2\pi}{3}, \frac{7\pi}{6}, \frac{5\pi}{3}\}$ **34.** $\{0\}$ **35. (a)** $(\frac{\pi}{4}, \frac{\pi}{2}) \cup (\frac{5\pi}{4}, \frac{3\pi}{2})$ **(b)** $[0, \frac{\pi}{4}) \cup (\frac{\pi}{2}, \frac{5\pi}{4}) \cup (\frac{3\pi}{2}, 2\pi)$
36. $\tan x$ $f(x) = \frac{1 - \cos 2x}{\sin 2x}$ **37.** $-\frac{63}{65}$ **38.** $\frac{16}{65}$ **39.** $-\frac{63}{65}$ **40.** $\frac{33}{65}$ **41.** $-\frac{56}{65}$
42. $-\frac{33}{56}$ **43.** $\frac{24}{25}$ **44.** $-\frac{119}{169}$ **45.** $\frac{24}{7}$ **46.** $\frac{3\sqrt{10}}{10}$
47. $-\frac{\sqrt{10}}{10}$ **48.** -3 **49.** $\frac{\pi}{4}$ **50.** $\frac{2\pi}{3}$ **51.** $-\frac{\pi}{3}$
52. $-\frac{\pi}{2}$ **53.** $\frac{3\pi}{4}$ **54.** $\frac{\pi}{6}$ **55.** $\frac{2\pi}{3}$ **56.** $\frac{\pi}{3}$
57. $\frac{3\pi}{4}$ **58.** There is no real number whose sine value is -3.
59. $\frac{1}{2}$ **60.** $\frac{\sqrt{7}}{4}$ **61.** $\frac{\sqrt{10}}{10}$ **62.** $\frac{9}{7}$ **63.** $\frac{\pi}{2}$
64. $\frac{294 + 125\sqrt{6}}{92}$ **65.** $\frac{u\sqrt{1 + u^2}}{1 + u^2}$

Dot Mode

66. $\frac{1}{u}$ **67. (a)**

80

1   25
20

Data for Chicago temperatures

(b) $f(x) = 25 \sin\left[\frac{\pi}{6}(x - 4.2)\right] + 50$ **(c)**

$f(x) = 25 \sin\left[\frac{\pi}{6}(x - 4.2)\right] + 50$

80

1   25
20

68. (a) It indicates that 41 days after March 21 (May 1), the number of daylight hours is about 13.2.
(b) It indicates that the longest day of the year has 14 daylight hours, 91 days after March 21 (June 20).
69. $t = \frac{1}{2\pi f} \arcsin \frac{e}{E_{\max}}$ **(b)** .00068 second   **70.** about .8, 2.4, 7.0, and 8.7 months; never   **71. (a)** 100
(b) 258   **(c)** 122   **(d)** 296   **72. (a)** about 20 years   **(b)** from about 5000 to about 150,000
73. (a) $t = \frac{3}{4\pi} \arcsin 3y$ **(b)** .27 second   **74.** May   **75.** $\frac{1}{\omega}$

CHAPTER 4

Section 4.1 (page 257)

In Exercises 1 and 3, we give, in order: sine, cosine, tangent, cotangent, secant, and cosecant.

1. $\frac{3}{5}; \frac{4}{5}; \frac{3}{4}; \frac{4}{3}; \frac{5}{4}; \frac{5}{3}$ **3.** $\frac{n}{p}; \frac{m}{p}; \frac{n}{m}; \frac{m}{n}; \frac{p}{m}; \frac{p}{n}$ **5.** $B = 53°\ 40'$; $a = 571$ m; $b = 777$ m
7. $M = 38.8°$; $n = 154$ m; $p = 198$ m   **9.** $A = 60°\ 20'$; $B = 29°\ 40'$; $c = 8.06$ **11.** $B = 62°\ 00'$;
$a = 8.17$ ft; $b = 15.4$ ft   **13.** $A = 17°\ 00'$; $a = 39.1$ in; $c = 134$ in   **15.** $c = 85.9$ yd; $A = 62°\ 50'$;
$B = 27°\ 10'$ **17.** No, because there are infinitely many such right triangles.   **19.** The transversal AB
intersecting the parallel lines AD and BC forms equal alternate interior angles.   **21.** 9.35 m
23. $62°\ 50'$ **25.** 33.4 m   **27.** 26.92 in   **29.** 13.3 ft   **31.** $37°\ 40'$ **33.** 42,600 ft
35. $26°\ 20'$ **37.** $A = 35°\ 59'$; $B = 54°\ 1'$ **39.** 52.4 ft   **41.** 8.229 cm   **43.** 446   **45.** 147 m
47. 2.47 km   **49.** 150 km   **51.** 5856 m   **53.** 2.01 mi

Section 4.2 (page 270)

1. $C = 95°$, $b = 13$ m, $a = 11$ m   **3.** $C = 80°\ 40'$, $a = 79.5$ mm, $c = 108$ mm
5. $A = 36.54°$, $b = 44.17$ m, $a = 28.10$ m   **7.** $A = 49°\ 40'$, $b = 16.1$ cm, $c = 25.8$ cm
9. $C = 91.9°$, $BC = 490$ ft, $AB = 847$ ft   **11.** It is not a right triangle.   **13.** 118 m   **15.** 1.93 mi
17. 10.4 in   **19.** 111°   **21.** first location: 5.1 mi;   second location: 7.2 mi   **23.** 26.5 km
25. 2.18 km   **27.** 38.3 cm   **29.** approximately 419,171 km   **31.** approximately 6100 ft
33. $B = 20.6°$, $C = 116.9°$, $c = 20.6$ ft   **35.** There is no such triangle.
37. $B_1 = 49°\ 20'$, $C_1 = 92°\ 00'$, $c_1 = 15.5$ km; $B_2 = 130°\ 40'$, $C_2 = 10°\ 40'$, $c_2 = 2.88$ km
39. $A_1 = 52°\ 10'$, $C_1 = 95°\ 00'$, $c_1 = 9520$ cm; $A_2 = 127°\ 50'$, $C_2 = 19°\ 20'$, $c_2 = 3160$ cm
41. $B = 37.77°$, $C = 45.43°$, $c = 4.174$ ft   **43.** $\sin C = 1$; $c = 90°$; ABC is a right triangle.
45. If $A = 103°\ 20'$, a must be the longest side of the triangle, which is not the case here, since $14.6 < 20.4$.
47. Such a piece of property cannot exist.   **49.** 2×10^8 m per sec   **51.** 19°   **53.** 48.7°

Section 4.3 (page 279)

1. $c = 2.83$ in, $A = 44.9°$, $B = 106.8°$ **3.** $c = 6.46$ m, $A = 53.1°$, $B = 81.3°$
5. $a = 156$ cm, $B = 64°\ 50'$, $C = 34°\ 30'$ **7.** $A = 82°$, $B = 37°$, $C = 61°$
9. $A = 42°\ 00'$, $B = 35°\ 50'$, $C = 102°\ 10'$ **11.** $A = 50°\ 50'$; $B = 44°\ 40'$; $C = 84°\ 30'$
13. We get a value for the cosine which cannot possibly be, leading to an error message or a complex number if we
use the inverse cosine function on a calculator.   **15.** 257 m   **17.** 281 km   **19.** 10.8 mi
21. 115 km   **23.** 18 ft   **25.** 5500 m   **27.** 438.14 ft   **29.** 350°   **31.** 2000 km
33. 163.5°   **35.** 25.24983 mi   **37.** 22 ft   **39.** 551.59 ft

Section 4.4 (page 287)

1. approximately 1700 feet   **3.** 55 miles per hour   **5.** about 46°   **7.** .03°   **9.** approximately
78 miles per hour   **11. (a)** approximately 550 feet   **(b)** approximately 369 feet   **13.** 46.4 m²
15. 722.9 in² .   **17.** 78 m²   **19.** 3650 ft²   **21.** 228 yd²   **23.** 100 m²   **25.** 33 cans
33. $\sin A = \frac{h}{c}$ **35.** $\mathcal{A} = \frac{1}{2}b(c \sin A)$, or $\mathcal{A} = \frac{1}{2}bc \sin A$ **37.** 1800°   **39.** 12.5 rotations per hour

41. 12.7 cm   **43. (a)** 39,616 rotations   **(b)** 62.9 mi; yes   **45.** $\frac{16\pi}{3}$ m per second; $\frac{16\pi}{3}$ radians per second
47. (a) $\frac{2\pi}{365}$ radian   **(b)** $\frac{\pi}{4380}$ radian per hour   **(c)** 66,700 miles per hour
49. larger pulley: $\frac{25\pi}{18}$ radians per second; smaller pulley: $\frac{125\pi}{48}$ radians per second   **51.** 3.73 cm
53. 114 cm^2 **55. (a)** 13.85°   **(b)** 76 m^2

Chapter 4 Review Exercises (page 294)

In Exercises 1 and 2, we give, in order: sine, cosine, tangent, cotangent, secant, and cosecant.

1. $\frac{21}{29}$; $\frac{20}{29}$; $\frac{21}{20}$; $\frac{20}{21}$; $\frac{29}{20}$; $\frac{29}{21}$ **2.** $\frac{45}{53}$; $\frac{28}{53}$; $\frac{45}{28}$; $\frac{28}{45}$; $\frac{53}{28}$; $\frac{53}{45}$ **3.** $A = 47.9108°$; $c = 84.816$ cm; $a = 62.942$ cm
4. $A = 21.4858°$; $b = 3330.68$ m; $a = 1311.04$ m   **5.** $\frac{\sqrt{133}}{13}$ **6.** $\frac{13\sqrt{133}}{133}$ **7.** 62.5°
8. The sine and cosine functions are cofunctions and θ and $90° - \theta$ are complements. Cofunction values of
complementary angles are equal.   **9.** $B = 50.28°$; $a = 32.38$ m; $c = 50.66$ m
10. $A = 42°07'$; $B = 47°53'$; $c = 402.6$ m   **11.** 73.7 ft   **12.** 20.4 m   **13.** 18.75 cm
14. 50.24 m   **15.** 1200 m   **16.** 110 km   **17.** 140 mi   **18.** 110 ft
19. 419   **20. (a)** approximately 29,008 ft   **(b)** shorter   **21.** 63.7 m
22. 19.87° or 19° 52'   **23.** 25° 00'   **24.** 173 ft   **25.** 41° 40'   **26.** 55.5 m
27. 70.9 m   **28.** 26.5° or 26° 30'   **29.** 54° 20' or 125° 40'   **30.** 148 cm
31. 49° 30'   **32.** 32°   **33.** $B = 17° 10'$, $C = 137° 40'$, $c = 11.0$ yd
34. $B_1 = 74.6°$, $C_1 = 43.7°$, $c_1 = 61.9$ m; $B_2 = 105.4°$, $C_2 = 12.9°$, $c_2 = 20.0$ m
35. 2.7 mi   **36.** 43 ft   **37.** 7 km   **38.** 1.91 mi   **39.** 58.6 ft
40. 11 ft   **41.** 15.8 ft   **42.** 77.1°   **43. (a)** 18°   **(b)** 18°   **(c)** 15°
45. (a) $x = \sin u$, $-\frac{\pi}{2} \le u \le \frac{\pi}{2}$ **(b)** side along x-axis: $\sqrt{1 - x^2}$; side perpendicular to x-axis: x; hypotenuse: 1
(c) $\tan u = \frac{x\sqrt{1-x^2}}{1-x^2}$ **(d)** $u = \tan^{-1}(\frac{x\sqrt{1-x^2}}{1-x^2})$ **46. (f)** $\frac{\sqrt{6}+\sqrt{2}}{4}$ **(g)** $\frac{\sqrt{6}-\sqrt{2}}{4}$ **(h)** $2 - \sqrt{3}$
47. (a) approximately 82.6341 meters   **(b)** approximately .017 meter   **48. (a)** $d = \frac{b}{2}(\cot\frac{\alpha}{2} + \cot\frac{\beta}{2})$
(b) approximately 345.3951 meters   **49. (a)** approximately 23.4 feet   **(b)** approximately 48.3 feet   **(c)** The
faster the speed, the more land that needs to be cleared on the side of the curve.   **50.** Suppose, for example,
the date is October 12, 1949, and it is currently 1996.   **(a)** 1949   **(b)** $1996 - 1980 = 16$, so
$1949 - 16 = 1933$.   **(c)** $\sin 1933° \approx .7313537016$ **(d)** $\sin^{-1}(\sin 1933°) = 47$ (degrees)   **(e)** It is the age of the
person when he or she celebrates his or her birthday that year.
51. 153,600 m^2 **52.** 20.3 ft^2 **53.** .234 km^2 **54.** 680 m^2 **55.** $\frac{15}{32}$ second   **56.** 108 radians
57. $\frac{\pi}{20}$ radian per second   **58.** $\frac{4\pi}{75}$ radian per second   **59.** 285.3 cm   **60.** 1260π m per second

CHAPTER 5

Section 5.1 (page 309)

1. m and p; n and r   **3.** m and p equal 2t, or t is $\frac{1}{2}$m or $\frac{1}{2}$p; also m = 1p and n = 1r
5. 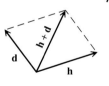 **7.** **9.** 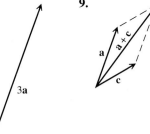 **11.** 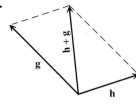 **13.**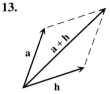

15. **17.** **19.** 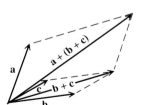 **21.**

23. Yes, vector addition is associative.   **25.** 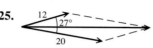   **27.**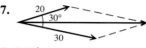

29. 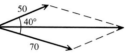   **31.** $\langle 9.5, 7.4 \rangle$ **33.** $\langle 17, 20 \rangle$ **35.** $\langle 13.7, 7.11 \rangle$

37. $\langle -123, 155 \rangle$ **39.** 530 newtons   **41.** 27.2 pounds   **43.** 7   **45.** -3 **47.** 20
49. $135°$ **51.** $90°$ **53.** $36.87°$ **55.** -6 **57.** -24 **59.** orthogonal
61. not orthogonal   **63.** not orthogonal   **65.** 190, 283 pounds, respectively
67. $173.1°$ **69.** 39.2 km   **71.** $358°$; 170 mph   **73.** 230 km per hr; $167°$
75. $P = (2,160,893.0, \quad 963,895.3)$ **77.** magnitude: 9.52082827; direction angle: 119.0646784°
79. $\langle -.520944533, -2.954423259 \rangle$ **81.** magnitude: 9.52082827; direction angle: 119.0646784°

Section 5.2 (page 323)

1.
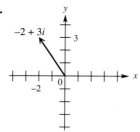

3.
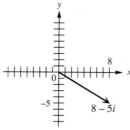

5.
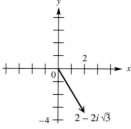

7.
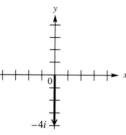

9.
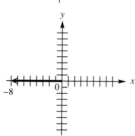

11. The imaginary part must be 0.   **13.** 0

15. $3 - i$ **17.** $-3 + 3i$ **19.** $2 + 4i$ **21.** $\sqrt{2} + i\sqrt{2}$ **23.** $10i$ **25.** $-2 - 2i\sqrt{3}$
27. $\frac{\sqrt{3}}{2} + \frac{1}{2}i$ **29.** $\frac{5}{2} - \frac{5\sqrt{3}}{2}i$ **31.** $-\sqrt{2}$ **33.** $3\sqrt{2}(\cos(-45°) + i \sin(-45°))$
35. $6(\cos(-120°) + i \sin(-120°))$ **37.** $8(\cos 30° + i \sin 30°)$ **39.** $2(\cos 135° + i \sin 135°)$
41. $2\sqrt{2}(\cos 45° + i \sin 45°)$ **43.** $4(\cos 180° + i \sin 180°)$ **45.** $2(\cos (-90°) + i \sin(-90°))$
47. 5.830951895 cis $59.03624347°$ **49.** $-3\sqrt{3} + 3i$ **51.** $-4i$ **53.** $12\sqrt{3} + 12i$
55. $-\frac{15\sqrt{2}}{2} + \frac{15\sqrt{2}}{2}i$ **57.** $-3i$ **59.** $\sqrt{3} - i$ **61.** $-1 - i\sqrt{3}$ **63.** $-\frac{1}{6} - \frac{\sqrt{3}}{6}i$ **65.** 2
67. 2 cis $0°$ **69.** $-i$ **71.** $-i$; It is the same.   **73.** yes   **75.** no   **77.** $110 + 32i$
79. $E = 30 + 60i$ **81.** $Z = \frac{233}{37} + \frac{119}{37}i$ **83.** $\approx 27.43 + 11.5i$ **85.** $1.2 - .14i$

Section 5.3 (page 331)

1. $27i$ **3.** 1   **5.** $\frac{27}{2} - \frac{27\sqrt{3}}{2}i$ **7.** $-16\sqrt{3} + 16i$ **9.** $-128 + 128i\sqrt{3}$
11. $128 + 128i$ **13.** cis $0°$, cis $120°$, cis $240°$ **15.** 2 cis $20°$, 2 cis $140°$, 2 cis $260°$

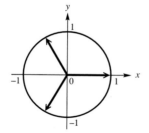   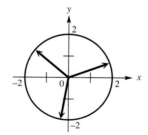

17. 2 cis 90°, 2 cis 210°, 2 cis 330°   **19.** 4 cis 60°, 4 cis 180°, 4 cis 300°

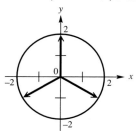   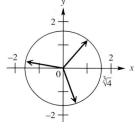

21. $\sqrt[3]{2}$ cis 20°, $\sqrt[3]{2}$ cis 140°, $\sqrt[3]{2}$ cis 260°   **23.** $\sqrt[3]{4}$ cis 50°, $\sqrt[3]{4}$ cis 170°, $\sqrt[3]{4}$ cis 290°

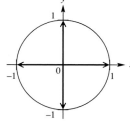   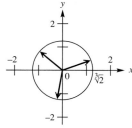

25. (a) $1, i, -1, -i$ **(b)** $1, \frac{1}{2} + \frac{\sqrt{3}}{2}i, -\frac{1}{2} + \frac{\sqrt{3}}{2}i, -1, -\frac{1}{2} - \frac{\sqrt{3}}{2}i, \frac{1}{2} - \frac{\sqrt{3}}{2}i$

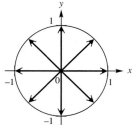

(c) $1, \frac{\sqrt{2}}{2} + \frac{\sqrt{2}}{2}i, i, -\frac{\sqrt{2}}{2} + \frac{\sqrt{2}}{2}i, -1, -\frac{\sqrt{2}}{2} - \frac{\sqrt{2}}{2}i, -i, \frac{\sqrt{2}}{2} - \frac{\sqrt{2}}{2}i$

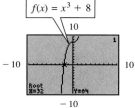

27. The trigonometric form of a positive real number will have $\theta = 0°$. When we divide $\frac{0° + 360° \cdot k}{n}$ for $k = 0$ and $n =$ the root index, we get 0°. Thus, one of the roots will also be a positive real number.

29. $(x + 2)(x^2 - 2x + 4)$ **31.** $x^2 - 2x + 4 = 0$ implies $x = 1 + i\sqrt{3}$ or $x = 1 - i\sqrt{3}$

33. $1 + i\sqrt{3}, -2, 1 - i\sqrt{3}$

35. The x-intercept, -2, is the real cube root of -8.   **37.** {cis 45°, cis 135°, cis 225°, cis 315°}

39. {cis 18°, cis 90°, cis 162°, cis 234°, cis 306°}

41. blue   **43.** yellow

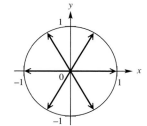

Further Explorations

1. Let $Y_1 = \frac{135 + 360X}{3}$ and show that for $X = 0$, 1, and 2, $Y_1 = 45$, 165, and 285 (degrees).
3. cis 22.5°, cis 112.5°, cis 202.5°, cis 292.5°

Chapter 5 Review Exercises (page 335)

1.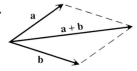
    2.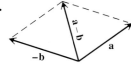
    3.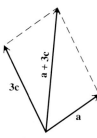

4. $\langle 25\sqrt{2}, 25\sqrt{2} \rangle$ **5.** $\langle 17.9, 66.8 \rangle$ **6.** $\langle 869, 418 \rangle$ **7.** 209 newtons
8. 135 newtons   **9. (a)** 14   **(b)** 52.13°   **10. (a)** $20\sqrt{3}$ **(b)** 30°   **11.** bearing: 306°; speed: 524 mph
12. speed: 21 km per hr; bearing: 118°   **13.** bearing: 7° 20'; speed; 8.3 mph
14.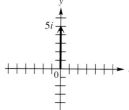
   15.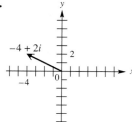
   16.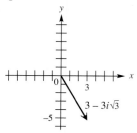

17. $5 + 4i$ **18.** $7 - 3i$ **19.** y

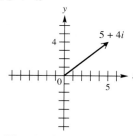   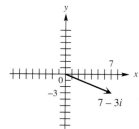

20. The absolute value of a real number represents its distance from 0 on the number line, while the absolute value (or modulus) of a complex number represents its distance from (0, 0) in the complex plane.
21. $-30i$ **22.** $-3 - 3i\sqrt{3}$ **23.** $-\frac{1}{8} + \frac{\sqrt{3}}{8}i$ **24.** -2 **25.** $8i$ **26.** $-128 + 128i$
27. $-\frac{1}{2} - \frac{\sqrt{3}}{2}i$ **28.** $\frac{1}{2} + \frac{\sqrt{3}}{2}i$ **29.** $2\sqrt{2}$ cis 135°   **30.** $3i$ **31.** $-\sqrt{2} - i\sqrt{2}$ **32.** 8 cis 120°
33. $\sqrt{2}$ cis(−45°)   **34.** $-2 - 2i\sqrt{3}$ **35.** 4 cis(−90°)   **36.** -2
37. 3 cis 90°, 3 cis 210°, 3 cis 330°   **38.** 2 cis 22.5°, 2 cis 112.5°, 2 cis 202.5°, 2 cis 292.5°

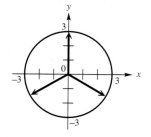   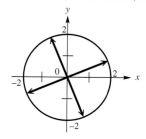

39. 2 cis 0°, 2 cis 72°, 2 cis 144°, 2 cis 216°, 2 cis 288°

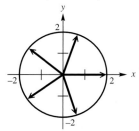

40. {cis 67.5°, cis 157.5°, cis 247.5°, cis 337.5°}   **41.** They are the same because the two factors in each product are the same. They only appear to be different. (Since the argument 45° is coterminal with −315°, and 90° is coterminal with −270°, the products will have coterminal arguments.)   **42.** The imaginary parts b and d must be equal.

CHAPTER 6

Section 6.1 (page 341)

Answers for polar coordinates may vary in Exercises 1–9.

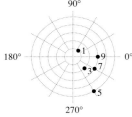

1. $(1, 405°)$, $(−1, 225°)$; $(\frac{\sqrt{2}}{2}, \frac{\sqrt{2}}{2})$ **3.** $(−2, 495°)$, $(2, 315°)$; $(\sqrt{2}, −\sqrt{2})$
5. $(5, 300°)$, $(−5, 120°)$; $(\frac{5}{2}, −\frac{5\sqrt{3}}{2})$ **7.** $(−3, 150°)$, $(3, −30°)$; $(\frac{3\sqrt{3}}{2}, −\frac{3}{2})$ **9.** $(3, 0°)$, $(−3, 180°)$; $(3, 0)$
11. quadrantal

Answers may vary in Exercises 13–21.

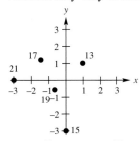

13. $(\sqrt{2}, 45°)$; $(−\sqrt{2}, 225°)$ **15.** $(3, 270°)$; $(−3, 90°)$ **17.** $(2, 135°)$; $(−2, 315°)$
19. $(1, 210°)$; $(−1, 30°)$ **21.** $(3, 180°)$; $(−3, 0°)$

Section 6.2 (page 348)

1.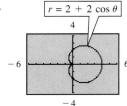

3.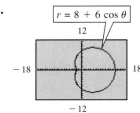

5.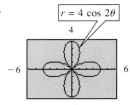

7.

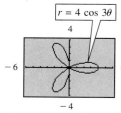

$r = 4 \cos 3\theta$

9.

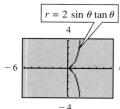

$r^2 = 4 \cos 2\theta$

11.

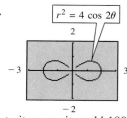

$r = 2 \sin \theta \tan \theta$

13. One way would be to change r to its opposite, add $180°$ to θ, and then move $|r|$ units along the terminal ray of angle $\theta + 180°$, when it is in standard position.

15. $x^2 + (y - 1)^2 = 1$

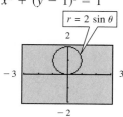

$r = 2 \sin \theta$

17. $y^2 = 4(x + 1)$

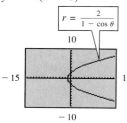

$r = \dfrac{2}{1 - \cos \theta}$

19. $(x + 1)^2 + (y + 1)^2 = 2$

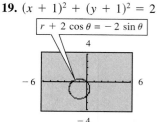

$r + 2 \cos \theta = -2 \sin \theta$

21. $x = 2$

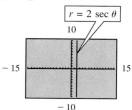

$r = 2 \sec \theta$

23. $x + y = 2$

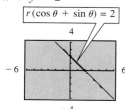

$r(\cos \theta + \sin \theta) = 2$

25. $r = \dfrac{4}{\cos \theta + \sin \theta}$

27. $r = 4$

29. $r = \dfrac{2}{\sin \theta}$ or $r = 2 \csc \theta$

31.

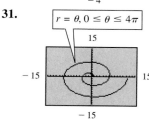

$r = \theta, \ 0 \le \theta \le 4\pi$

33.

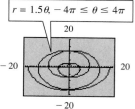

$r = 1.5\theta, \ -4\pi \le \theta \le 4\pi$

35. While a matter of preference, most people would probably prefer the rectangular form.

Further Explorations

1. In each case, r (the *r*adius) is 8, so the graph is a set points 8 units from the pole. This is a circle.
3. (a) twelve; yes **(b)** five   **5. (a)** It looks like a circle. Most people *would* be surprised.   **(b)** This should look like a rose with 96 petals. The TABLE shows that only the points with $r = 8$ have been graphed, due to the choice of θstep and ΔTbl. As we saw in Exercise 1 in these explorations, this describes a circle.

Section 6.3 (page 355)

1. $y = \frac{1}{2}x + 1$, for x in $[-4, 6]$

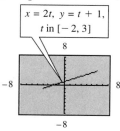

$x = 2t, \ y = t + 1,$ t in $[-2, 3]$

3. $y = 3x^2 - 4$, for x in $[0, 2]$

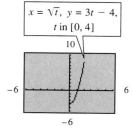

$x = \sqrt{t}, \ y = 3t - 4,$ t in $[0, 4]$

5. $y = x - 2$, for x in $[-26, 28]$

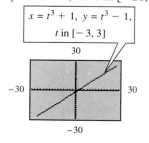

$x = t^3 + 1, \ y = t^3 - 1,$ t in $[-3, 3]$

7. $x = 2^{(y^2 + 1)/3}$, for y in $[0, \sqrt{11}]$ or $y^2 = \frac{3 \ln x}{\ln 2} - 1$, for x in $[\sqrt[3]{2}, 16]$ **9.** $x^2 + y^2 = 4$, for x in $[-2, 2]$

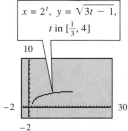

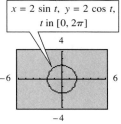

11. $\frac{y^2}{4} - \frac{x^2}{9} = 1$, for x in $(-\infty, \infty)$ **13.** $y = \frac{1}{x}$, for x in $(0, 1]$ **15.** $y = \sqrt{x^2 + 2}$, for x in $(-\infty, \infty)$

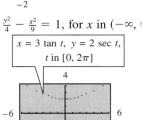

Dot Mode

17. $y = \frac{1}{x}$, for x in $(0, \infty)$ **19.** $x^2 + y^2 - 4x - 2y + 4 = 0$, for x in $[1, 3]$ **21.** $y = \frac{1}{x}$, for $x \neq 0$

23. $y = \ln x$, for x in $(0, \infty)$ **25.** 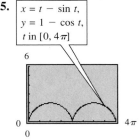   **27. (a)** 17.7 seconds   **(b)** 5000 feet   **(c)** 1250 feet

31. Many answers are possible, two of which are $x = t$, $y - y_1 = m(t - x_1)$ and $t = x - x_1$, $y = mt + y_1$.

33. Many answers are possible, two of which are $x = a \sec t$, $y = b \tan t$ and $x = t$, $y^2 = b^2(\frac{t^2}{a^2} - 1)$.

Section 6.4 (page 365)

1. (a) $x = 24t$ and $y = -16t^2 + 24\sqrt{3}\,t$ **(b)** 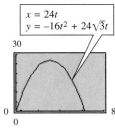   **(c)** approximately 2.6 seconds
(d) 27 feet
(e) $y = -\frac{1}{36}x^2 + \sqrt{3}\,x$

3. (a) 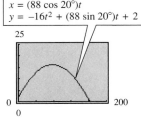   **(b)** approximately 16 feet; approximately 1 second
(c) approximately 160 feet; approximately 2 seconds

5. yes   **7.** yes   **9.** 70°   **11. (a)** approximately 4.6 seconds   **(b)** approximately 4.55 seconds
13. 23,000 feet   **15. (a)** approximately 46,800 feet   **(b)** 25° or 65°   **(c)** no
17. (a)
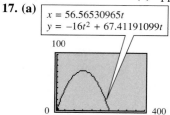
$x = 56.56530965t$
$y = -16t^2 + 67.41191099t$

(b) 50.0°   **(c)** $x = (88 \cos 50.0°)\, t$ and $y = -16t^2 + (88 \sin 50.0°)t$

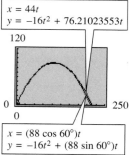
$x = 56.56530965t$
$y = -16t^2 + 67.41191099t$

$x = (88 \cos 50.0°)t$
$y = -16t^2 + (88 \sin 50.0°)t$

19. (a)
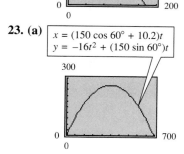
$x = 44t$
$y = -16t^2 + 76.21023553t$

(b) 60°   **(c)** $x = (88 \cos 60°)t$ and $y = -16t^2 + (88 \sin 60°)t$

$x = 44t$
$y = -16t^2 + 76.21023553t$

$x = (88 \cos 60°)t$
$y = -16t^2 + (88 \sin 60°)t$

21. (a)
$x = (88 \cos 20° + 7.2)t$
$y = -16t^2 + (88 \sin 20°)t + 2$

(b) approximately 16 feet; approximately 1 second
(c) approximately 175 feet; approximately 2 seconds

23. (a)
$x = (150 \cos 60° + 10.2)t$
$y = -16t^2 + (150 \sin 60°)t$

(b) approximately 263.7 feet; approximately 4.1 seconds
(c) approximately 692 feet; approximately 8.1 seconds

25. If 90° < θ < 180°, the ball will go behind the batter (that is, a foul ball). If −90° < θ < 0°, the ball will be hit onto the ground (that is, a ground ball).   **27. (a)** $x = 90t$ and $y = 4000 - 30t$ **(b)** 12,000 feet
(c) approximately 133 seconds   **(d)** 18.4°

Chapter 6 Review Exercises (page 368)

1. $(3\sqrt{3}, 3)$ **2.** $(-6\sqrt{2}, -6\sqrt{2})$ **3.** $(-4, 4\sqrt{3})$ **4.** $(6\sqrt{2}, 135°)$ **5.** $(5, -90°)$

6. $(\sqrt{10}, -135°)$ **7.**

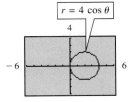
$r = 4 \cos \theta$

8.

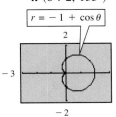
$r = -1 + \cos \theta$

9.

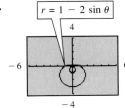
$r = 1 - 2 \sin \theta$

10.

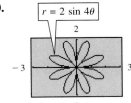
$r = 2 \sin 4\theta$

11.

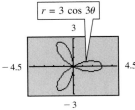
$r = 3 \cos 3\theta$

12. $y^2 + 6x - 9 = 0$ **13.** $-x - 2y = 4$ **14.** $x^2 - x + y^2 - y = 0$ **15.** $x^2 + y^2 = 4$

16. $r = \frac{-3}{\cos \theta}$ or $r = -3 \sec \theta$ **17.** $\tan \theta = 1$ **18.** $r = \frac{\tan \theta}{\cos \theta}$ **19.** $r = \frac{\cos \theta}{\sin^2 \theta}$ **20.**

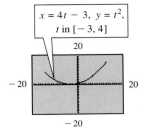
$x = 4t - 3,\ y = t^2,\ t$ in $[-3, 4]$

21.

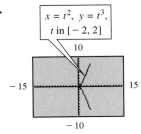
$x = t^2,\ y = t^3,\ t$ in $[-2, 2]$

22.

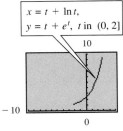
$x = t + \ln t,\ y = t + e^t,\ t$ in $(0, 2]$

23.

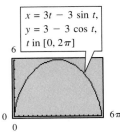
$x = 3t - 3 \sin t,\ y = 3 - 3 \cos t,\ t$ in $[0, 2\pi]$

24. $x - 3y = 5$ for x in $[-13, 17]$ **25.** $y = \sqrt{x^2 + 1}$ for x in $[0, \infty)$

26. $\frac{y^2}{9} - \frac{x^2}{25} = 1$ for x in $(-\infty, \infty)$ **27. (a)** approximately 358 feet; approximately 3.4 seconds
(b) approximately 48 feet   **28.** yes   **29. (a)** $x = 100t, y = 20t$ **(b)** 11.3°   **30.** 37.0°
31. yes   **32.** yes